Sensors for Human Activity Recognition II

Sensors for Human Activity Recognition II

Guest Editors
Hui Liu
Hugo Gamboa
Tanja Schultz

Basel • Beijing • Wuhan • Barcelona • Belgrade • Novi Sad • Cluj • Manchester

Guest Editors

Hui Liu	Hugo Gamboa	Tanja Schultz
Universität Bremen	NOVA University of Lisbon	Universität Bremen
Bremen	Lisbon	Bremen
Germany	Portugal	Germany

Editorial Office
MDPI AG
Grosspeteranlage 5
4052 Basel, Switzerland

This is a reprint of the Special Issue, published open access by the journal *Sensors* (ISSN 1424-8220), freely accessible at: https://www.mdpi.com/journal/sensors/special_issues/671TM05M9J.

For citation purposes, cite each article independently as indicated on the article page online and as indicated below:

Lastname, A.A.; Lastname, B.B. Article Title. *Journal Name* **Year**, *Volume Number*, Page Range.

ISBN 978-3-7258-2803-6 (Hbk)
ISBN 978-3-7258-2804-3 (PDF)
https://doi.org/10.3390/books978-3-7258-2804-3

© 2025 by the authors. Articles in this book are Open Access and distributed under the Creative Commons Attribution (CC BY) license. The book as a whole is distributed by MDPI under the terms and conditions of the Creative Commons Attribution-NonCommercial-NoDerivs (CC BY-NC-ND) license (https://creativecommons.org/licenses/by-nc-nd/4.0/).

Contents

About the Editors vii

Hui Liu, Hugo Gamboa and Tanja Schultz
Human Activity Recognition, Monitoring, and Analysis Facilitated by Novel and Widespread Applications of Sensors
Reprinted from: *Sensors* **2024**, *24*, 5250, https://doi.org/10.3390/s24165250 1

Peng Su and Dejiu Chen
Adopting Graph Neural Networks to Analyze Human–Object Interactions for Inferring Activities of Daily Living
Reprinted from: *Sensors* **2024**, *24*, 2567, https://doi.org/10.3390/s24082567 6

Nouf Abdullah Almujally, Danyal Khan, Naif Al Mudawi, Mohammed Alonazi, Abdulwahab Alazeb, Asaad Algarni, et al.
Biosensor-Driven IoT Wearables for Accurate Body Motion Tracking and Localization
Reprinted from: *Sensors* **2024**, *24*, 3032, https://doi.org/10.3390/s24103032 18

Xiankai Cheng, Benkun Bao, Weidong Cui, Shuai Liu, Jun Zhong, Liming Cai and Hongbo Yang
Classification and Analysis of Human Body Movement Characteristics Associated with Acrophobia Induced by Virtual Reality Scenes of Heights
Reprinted from: *Sensors* **2023**, *23*, 5482, https://doi.org/10.3390/s23125482 44

Georgios Sopidis, Michael Haslgrübler and Alois Ferscha
Counting Activities Using Weakly Labeled Raw Acceleration Data: A Variable-Length Sequence Approach with Deep Learning to Maintain Event Duration Flexibility
Reprinted from: *Sensors* **2023**, *23*, 5057, https://doi.org/10.3390/s23115057 63

Sabrina Gado, Katharina Lingelbach, Maria Wirzberger and Mathias Vukelić
Decoding Mental Effort in a Quasi-Realistic Scenario: A Feasibility Study on Multimodal Data Fusion and Classification
Reprinted from: *Sensors* **2023**, *23*, 6546, https://doi.org/10.3390/s23146546 79

Rok Novak, Johanna Amalia Robinson, Tjaša Kanduč, Dimosthenis Sarigiannis, Sašo Džeroski and David Kocman
Empowering Participatory Research in Urban Health: Wearable Biometric and Environmental Sensors for Activity Recognition
Reprinted from: *Sensors* **2023**, *23*, 9890, https://doi.org/10.3390/s23249890 105

Nuno Bento, Joana Rebelo, André V. Carreiro, François Ravache and Marília Barandas
Exploring Regularization Methods for Domain Generalization in Accelerometer-Based Human Activity Recognition
Reprinted from: *Sensors* **2023**, *23*, 6511, https://doi.org/10.3390/s23146511 127

Weiling Zheng, Yu Zhang, Landu Jiang, Dian Zhang and Tao Gu
MeshID: Few-Shot Finger Gesture Based User Identification Using Orthogonal Signal Interference
Reprinted from: *Sensors* **2024**, *24*, 1978, https://doi.org/10.3390/s24061978 139

Teng-Wen Chang, Hsin-Yi Huang, Cheng-Chun Hong, Sambit Datta and Walaiporn Nakapan
SENS+: A Co-Existing Fabrication System for a Smart DFA Environment Based on Energy Fusion Information
Reprinted from: *Sensors* **2023**, *23*, 2890, https://doi.org/10.3390/s23062890 163

Yifan Bian, Dennis Küster, Hui Liu and Eva G. Krumhuber
Understanding Naturalistic Facial Expressions with Deep Learning and Multimodal Large Language Models
Reprinted from: *Sensors* **2024**, *24*, 126, https://doi.org/10.3390/s24010126 . . . . **184**

About the Editors

Hui Liu

Hui Liu, Dr.-Ing., studied Electrical Engineering at the Technical University of Berlin (Dipl.-Ing., 2009) and Information and Communication Systems at Shanghai Jiao Tong University (M. Sc., 2011). He has been a researcher of biosignal processing, human activity recognition, virtual reality, music information retrieval, and the multidisciplinary application of AI at the Cognitive Systems Lab of the University of Bremen since 2016, where he received his Ph.D. (2021). He is co-responsible for three multipartner projects, developed the first smart knee bandage that recognizes users' activities in real-time (Best Student Paper, BIODEVICES 2019) and proposed Motion Units, an activity-modeling method with interpretability, generalizability, and expandability, rationalizing kinesiology and speech recognition knowledge into human activity research (2021). He collected and released the 19-channel, 22-activity, 20-subject human activity dataset CSL-SHARE (2021). A dataset of ECG acquisition artifacts called sensORder was published under his guidance (2024). He received the following third-party funding and awards: the Erasmus Teaching Mobility Award (2018), the CAMPUSiDEEN Public Choice Award for advanced sensing and recognition technology (2022), and the YERUN Research Mobility Award (2022/2023). He served as a PC for 17th BIOSIGNALS and an AC for 33rd ICANN. He was also given the award of Sensors 2023 Outstanding Reviewer.

Hugo Gamboa

Hugo Gamboa, Prof. Dr., received his Ph.D. degree in Electrical and Computer Engineering from the Instituto Superior Técnico of the University of Lisbon (IST UL) in 2007. He is a Co-Founder and the President of PLUX, a company that develops biosignal monitoring wearable technology. He is currently a Researcher at the Laboratory for Instrumentation, Biomedical Engineering and Radiation Physics (LIBPhys), Faculdade de Ciências e Tecnologia of the NOVA University of Lisbon (FCT NOVA), where he is also a full-time Professor with habilitation at the Physics Department. Since 2014, he has been a Senior Researcher at the Fraunhofer Center for Assistive Information and Communication Solutions (AICOS). His research interests include biosignal processing and instrumentation.

Tanja Schultz

Tanja Schultz, Prof. Dr.-Ing., received her doctoral and diploma degree in Informatics from the University of Karlsruhe, in 2000 and 1995, respectively, and successfully passed the German state examination for teachers of Mathematics, Sports, and Educational Science at Heidelberg University in 1990. In 2000, she joined Carnegie Mellon University in Pittsburgh, where she holds a position as Research Professor at the Language Technologies Institute. From 2007 to 2015, she was a Full Professor at the Department of Informatics of the Karlsruhe Institute of Technology before she accepted an offer from the University of Bremen in April 2015. Since 2007, she has directed the Cognitive Systems Lab, where her research activities focus on human–machine communication, with a particular emphasis on multilingual speech processing and human–machine interfaces. She has received the FZI award for an outstanding Ph.D. thesis (2001), the Allen Newell Medal for Research Excellence from Carnegie Mellon (2002), the ISCA Best Journal Award for her publication on language-independent acoustic modeling (2002) and on silent speech interfaces (2015), the Plux Wireless Award (2011) for the development of Airwriting, the Alcatel–Lucent Research Award for Technical Communication (2012), the Otto Haxel Award (2013), a Google Research Award (2013 and

2020), as well as several best paper awards. She serves as an IEEE Fellow, as Associate Editor of IEEE Transactions (2002–2004), as an Associate Editor for ACM TALLIP (since 2013), as an editorial board member of Speech Communication (since 2001), and has served as a board member and elected president of the International Speech Communication Association (ISCA) from 2007 to 2015.

Editorial

Human Activity Recognition, Monitoring, and Analysis Facilitated by Novel and Widespread Applications of Sensors

Hui Liu [1,2,*], Hugo Gamboa [2] and Tanja Schultz [1]

[1] Cognitive Systems Lab, University of Bremen, 28359 Bremen, Germany; tanja.schultz@uni-bremen.de
[2] Laboratory for Instrumentation, Biomedical Engineering and Radiation Physics (LIBPhys-UNL), Faculty of Sciences and Technology, NOVA University of Lisbon, 2820-001 Caparica, Portugal; h.gamboa@fct.unl.pt
* Correspondence: hui.liu@uni-bremen.com

Citation: Liu, H.; Gamboa, H.; Schultz, T. Human Activity Recognition, Monitoring, and Analysis Facilitated by Novel and Widespread Applications of Sensors. *Sensors* **2024**, *24*, 5250. https://doi.org/10.3390/s24165250

Received: 1 July 2024
Accepted: 8 July 2024
Published: 14 August 2024

Copyright: © 2024 by the authors. Licensee MDPI, Basel, Switzerland. This article is an open access article distributed under the terms and conditions of the Creative Commons Attribution (CC BY) license (https://creativecommons.org/licenses/by/4.0/).

1. Introduction

The Special Issue *Sensors for Human Activity Recognition* has received a total of 30 submissions so far, and from these, this new edition will publish 10 academic articles.

Scientists in the fields of sensor applications and human activity recognition (HAR) have once again collaborated for this shared academic endeavor. After the first edition of this Special Issue [1] received significant attention, at the request of numerous authors and readers, a second series was launched, for which we received submissions and chose many excellent manuscripts based on a rigorous review, revision, and selection process. This edition brings together the hard work of 50 authors from 14 countries across three continents: Oceania (Australia), Europe (France, Greece, Italy, Australia, Germany, Portugal, Slovenia, Sweden, and the UK), and Asia (China, Pakistan, Saudi Arabia, and Thailand). Of the 51 authors who participated in the first edition [2], less than ten have once again presented their valuable research achievements; the remaining authors are first-time participants.

From hardware to software, from pipelines to applications, from handcrafted features to domain generalization, and from shallow learning to deep and even large language models, the collected literature will provide readers in related fields with state-of-the-art approaches to many challenges in HAR.

2. Overview of the Contributions

In the first edition, the literature was introduced in the order specified in the up-to-date HAR research pipeline proposed in [3]. However, such an approach is not adopted in this new edition because, on the one hand, most of the articles selected for this edition involve the vast majority of the links in the pipeline, and some of them even followed the overall HAR research pipeline to formulate their research workflow and plan; on the other hand, the breadth and novelty of the articles' research fields mean that each have a thoroughly different perspective and reference significance. They will be introduced in alphabetical order of the first letters of the titles; therefore, the order in which the articles appear in this section is not based on their topic.

The subsection titles may not be the most comprehensive summary of the articles' contributions, but they attempt to highlight their most prominent features and information that sets them apart from other research.

2.1. Video-Based Human Activity Recognition (HAR) Using Graph Neural Networks

Research on the use of video as an external sensing method for HAR is increasing and facilitates multiple daily application scenarios. Su and Chen, from Sweden, innovatively used graph neural networks (GNNs) as their model, incorporating semantic content within relational data rather than directly relying on high-dimensional data. They effectively improved the recognition accuracy of activities of daily living (ADLs).

2.2. Wearable-Based HAR Plus Localization Using Inertial Plus GPS Data

As carriers of wearable sensors, mobile phones are receiving increasing attention in HAR research [4,5]. As the real-time characteristics of wearable HAR systems face more requirements and challenges [6], smartphones are bound to attract more study. However, the vast majority of this research is based on unimodality, and in particular, accelerometers. Research on superposing localization on HAR is on the rise. First, localization provides not only an auxiliary reference for HAR but also additional motion information for the ending system as well as the HAR results. The authors of this article incorporated GPS sensors into the study of inertial data, expanding mobile phone-based HAR on the above two points: (1) multimodal sensing; (2) superposing localization.

2.3. Human Body Movement Characteristics for Virtual Reality-Based Acrophobia Study

Virtual reality should be a potentially safe means of studying and treating acrophobia, as it effectively avoids the environmental risks of physical scenarios of simulated acrophobia in the real world. The Chinese authors set up a series of motion tasks in a virtual reality high-altitude scene to investigate human body movement characteristics. The experimenters established a high-precision classification model, which confirmed that there are differences in the movement patterns of people with and without acrophobia in a high-altitude environment, facilitating quick screening of patients. In addition, a quantitative analysis was presented to provide targeted training guidance to aviation personnel in the future.

2.4. Human Activity Counting Using Deep Learning (DL) Maintaining Duration Flexibility

Activity counting is not widely studied but is a potentially useful area of research, similar to the task of event-based automatic segmentation in HAR. In general, data mining methods can be used to count activities in a black-box manner, such as feature-based information retrieval using self-similarity matrices to discover change points and time-series subsequence search to mark pattern reproducibility. Considering event duration flexibility, dynamic time warping and its advanced variants can also be used to count activities through subsequent queries. In this study, the author transformed the task of activity counting from pure statistical analysis to deep learning and achieved robust counting performance on weakly calibrated IMU data for hand-performed activities while maintaining the flexibility of event duration.

2.5. Warship Commander Activities for Multisensory Mental Analysis

The authors of this study stated that human performance varies depending on the psychological resources required to successfully complete tasks. In order to monitor users' cognitive resources in natural scenes, it is necessary not only to measure the needs caused by the task itself, but also to consider the contextual and environmental impacts. A multisource perception dataset of 18 participants was collected based on the warship commander task. The sensor modalities involved include functional near-infrared spectroscopy (fNIRS), electroencephalography (EEG), electrocardiography (ECG), temperature, respiration, and eye tracking, which are used to decode mental effort. The experimental investigation utilized multimodal machine learning approaches that include feature engineering, model optimization, and model selection steps, achieving a reliable classification of mental effort states.

2.6. Complex HAR in the Context of Urban Environmental Exposure Research

This work links HAR with the important research field of participatory exposure research, which transforms it into a novel and interesting interdisciplinary task. This type of research usually uses an activity diary, while the authors innovatively used multiple wearable and environmental sensors to perform complex activity recognition in urban stressor exposure studies. At the same time, parameters such as particle concentration, temperature, and humidity were also measured. The recognition experiments were conducted on three shallow learning models. Because the innovation in this research lies in the application scenarios rather than the sensor types, sensor positions, data processing, and machine learning algorithms, the experimental results are not as impressive for less complex activities, and even worse for fuzzy activities such as resting and playing. However, for more complex activities with clear definitions, such as smoking and cooking, strong precision was demonstrated, and these activities are actually essential for the research context.

2.7. Accelerometer-Based HAR Using Domain Generalization with Regularization Methods

The concept of domain generalization has been on the research agenda in many fields of artificial intelligence, including HAR. Scientists are trying to find universal means to smoothly connect excellent HAR models to diverse dataset types, including different sensor types, wearable schemes, and data modalities, among others. For example, Motion Units attempt to provide a generalizable, interpretable, and expandable methodology from a human activity modeling perspective by finely distinguishing the stages of daily and sports activities in order to efficiently model the same and different activities in multiple datasets, respectively.

In the article included in this edition, the authors utilized regularization methods to study the domain generalization of HAR and improved the performance of the model. Interestingly, the article confirms again that in the case of sufficient background information, handcrafted features still demonstrate their powerful advantages in HAR [7,8], which is consistent with the conclusions drawn by the authors in the last edition.

It should be supplemented to readers who have conducted in-depth research in this field that recently, studies have shown that designing high-level features based on activity characteristics is a novel approach between traditional handcrafted feature extraction and deep-learning-based automatic feature learning to further improve model interpretability and recognition efficiency [9].

2.8. Finger Gesture-Based User Identification Using Radio Frequency Technology

Gesture analysis is also one of the hot topics closely related to HAR [10,11], including finger gesture research. This work applied radio frequency (RF) technology to perceive finger motions to perform the user identification task. The drawback of RF signals is not only their low resolution [12] but also the associated user heterogeneity. To address these challenges, the sensing sensitivity against RF signal interference has been significantly improved through orthogonal signal interference, and subtle individual features have been extracted from less distinct finger motions, such as air-writing digits, through velocity distribution profiling. Under the few-shot model retraining framework based on the first component reverse module, the experimenters efficiently and effectively achieved extensive model robustness and performance in complex environments.

2.9. Associating Human Behavior, Manufacture, and Digital Interaction with Fabrication

The integration of coexisting interactive fabrication tools and a dynamic interactive process for fabrication design is proposed in this article. Via virtual–physical integration approaches, designers, manufacturers, and assemblers could be supported in digital fabrication, comprehensively bringing Internet of Things technologies into the co-fabrication space. Human behavior, physical manufacturing, and digital interaction, as the three major components of the system, were focused on, and a seeing–moving–seeing thinking framework was applied to convey the design results.

2.10. Facial Expression Understanding Using DL and Multimodal Large Language Models

Facial expressions are subtle changes that occur on the human face. External sensing [13–16] and physiological signals [17] are increasingly facilitating this area. Similarly to the gesture research introduced in Section 2.8, research on facial expressions is also strongly related to HAR in scientific fields. A notable example is the world-renowned annual FG conference, i.e., the IEEE International Conference on Automatic Face and Gesture Recognition, which clearly prioritizes face, gestures, and body activity computing, analysis, synthesis, and recognition as an equally important direction of research, and which is also implied in their study's correlations and method cross-referencing.

As the only review included in this edition, this article provides a detailed introduction to the current research status of deep learning and multimodal large language models for facial expression recognition, presenting abundant reference materials for subsequent peer researchers.

3. Conclusions and Acknowledgments

Based on the introduction in Section 2, we have extracted ten interesting individual words: graph, localization, acrophobia, counting, mental, exposure, generalization, identification, fabrication, expression. At first glance, none of them seem to be conventionally related to traditional "Sensors for HAR" research, but after careful consideration, they certainly show some subtle connections to HAR which are worth investigating more deeply. The ten articles in this edition present these discoveries in detail to readers without reservation.

In addition to paying tribute to and expressing gratitude for the 50 outstanding contributing scientists, we also want to thank all the diligent and responsible reviewers, as well as MDPI and its editors, who have always supported this Special Issue in the background.

Author Contributions: Writing—original draft preparation, H.L.; writing—review and editing, H.L. and H.G.; Conceptualization, T.S. All authors have read and agreed to the published version of the manuscript.

Conflicts of Interest: The authors declare no conflict of interest.

List of Contributions

1. Su, P.; Chen, D. Adopting Graph Neural Networks to Analyze Human–Object Interactions for Inferring Activities of Daily Living. *Sensors* **2024**, *24*, 2567. https://doi.org/10.3390/s24082567.
2. Almujally, N.A.; Khan, D.; Al Mudawi, N.; Alonazi, M.; Alazeb, A.; Algarni, A.; Jalal, A.; Liu, H. Biosensor-Driven IoT Wearables for Accurate Body Motion Tracking and Localization. *Sensors* **2024**, *24*, 3032. https://doi.org/10.3390/s24103032.
3. Cheng, X.; Bao, B.; Cui, W.; Liu, S.; Zhong, J.; Cai, L.; Yang, H. Classification and Analysis of Human Body Movement Characteristics Associated with Acrophobia Induced by Virtual Reality Scenes of Heights. *Sensors* **2023**, *23*, 5482. https://doi.org/10.3390/s23125482.
4. Sopidis, G.; Haslgrübler, M.; Ferscha, A. Counting Activities Using Weakly Labeled Raw Acceleration Data: A Variable-Length Sequence Approach with Deep Learning to Maintain Event Duration Flexibility. *Sensors* **2023**, *23*, 5057. https://doi.org/10.3390/s23115057.
5. Gado, S.; Lingelbach, K.; Wirzberger, M.; Vukelić, M. Decoding Mental Effort in a Quasi-Realistic Scenario: A Feasibility Study on Multimodal Data Fusion and Classification. *Sensors* **2023**, *23*, 6546. https://doi.org/10.3390/s23146546.
6. Novak, R.; Robinson, J.A.; Kanduč, T.; Sarigiannis, D.; Džeroski, S.; Kocman, D. Empowering Participatory Research in Urban Health: Wearable Biometric and Environmental Sensors for Activity Recognition. *Sensors* **2023**, *23*, 9890. https://doi.org/10.3390/s23249890.
7. Bento, N.; Rebelo, J.; Carreiro, A.V.; Ravache, F.; Barandas, M. Exploring Regularization Methods for Domain Generalization in Accelerometer-Based Human Activity Recognition. *Sensors* **2023**, *23*, 6511. https://doi.org/10.3390/s23146511.
8. Zheng, W.; Zhang, Y.; Jiang, L.; Zhang, D.; Gu, T. MeshID: Few-Shot Finger Gesture Based User Identification Using Orthogonal Signal Interference. *Sensors* **2024**, *24*, 1978. https://doi.org/10.3390/s24061978.
9. Chang, T.W.; Huang, H.Y.; Hong, C.C.; Datta, S.; Nakapan, W. SENS+: A Co-Existing Fabrication System for a Smart DFA Environment Based on Energy Fusion Information. *Sensors* **2023**, *23*, 2890. https://doi.org/10.3390/s23062890.
10. Bian, Y.; Küster, D.; Liu, H.; Krumhuber, E.G. Understanding Naturalistic Facial Expressions with Deep Learning and Multimodal Large Language Models. *Sensors* **2024**, *24*, 126. https://doi.org/10.3390/s24010126.

References

1. Liu, H.; Gamboa, H.; Schultz, T. (Eds.) *Sensors for Human Activity Recognition*; MDPI: Basel, Switzerland, 2023. [CrossRef]
2. Liu, H.; Gamboa, H.; Schultz, T. Sensor-Based Human Activity and Behavior Research: Where Advanced Sensing and Recognition Technologies Meet. *Sensors* **2023**, *23*, 125. [CrossRef] [PubMed]
3. Liu, H.; Hartmann, Y.; Schultz, T. A Practical Wearable Sensor-Based Human Activity Recognition Research Pipeline. In Proceedings of the 15th International Joint Conference on Biomedical Engineering Systems and Technologies—Volume 5: HEALTHINF, Vienna, Austria, 9–11 February 2022; pp. 847–856. [CrossRef]
4. Kwon, Y.; Kang, K.; Bae, C. Analysis and evaluation of smartphone-based human activity recognition using a neural network approach. In Proceedings of the IJCNN 2015—International Joint Conference on Neural Networks, Killarney, Ireland, 12–17 July 2015; pp. 1–5. [CrossRef]
5. Straczkiewicz, M.; James, P.; Onnela, J.P. A systematic review of smartphone-based human activity recognition methods for health research. *npj Digit. Med.* **2021**, *4*, 148. [CrossRef] [PubMed]
6. Hartmann, Y.; Liu, H.; Schultz, T. Interactive and Interpretable Online Human Activity Recognition. In Proceedings of the PERCOM 2022—20th IEEE International Conference on Pervasive Computing and Communications Workshops and other Affiliated Events (PerCom Workshops), Pisa, Italy, 21–25 March 2022; pp. 109–111. [CrossRef]
7. Liu, H.; Schultz, T. ASK: A Framework for Data Acquisition and Activity Recognition. In Proceedings of the 11th International Joint Conference on Biomedical Engineering Systems and Technologies—Volume 3: BIOSIGNALS, Madeira, Portugal, 19–21 February 2018; pp. 262–268. [CrossRef]
8. Hartmann, Y.; Liu, H.; Schultz, T. Feature Space Reduction for Multimodal Human Activity Recognition. In Proceedings of the 13th International Joint Conference on Biomedical Engineering Systems and Technologies—Volume 4: BIOSIGNALS, Valletta, Malta, 24–26 February 2020; pp. 135–140. [CrossRef]
9. Hartmann, Y.; Liu, H.; Schultz, T. High-Level Features for Human Activity Recognition and Modeling. *Biomed. Eng. Syst. Technol.* **2023**, 141–163. [CrossRef]
10. Saini, R.; Maan, V. Human Activity and Gesture Recognition: A Review. In Proceedings of the ICONC3 2020—International Conference on Emerging Trends in Communication, Control and Computing, Lakshmangarh, India, 21–22 February 2020; pp. 1–2. [CrossRef]
11. Mahbub, U.; Ahad, M.A.R. Advances in human action, activity and gesture recognition. *Pattern Recognit. Lett.* **2022**, *155*, 186–190. [CrossRef]
12. Godyak, V. RF discharge diagnostics: Some problems and their resolution. *J. Appl. Phys.* **2021**, *129*, 041101. [CrossRef]
13. Cohen, I.; Sebe, N.; Garg, A.; Lew, M.S.; Huang, T.S. Facial expression recognition from video sequences. In Proceedings of the IEEE International Conference on Multimedia and Expo, Lausanne, Switzerland, 26–29 August 2002; Volume 2, pp. 121–124. [CrossRef]
14. Michael, P.; El Kaliouby, R. Real time facial expression recognition in video using support vector machines. In Proceedings of the 5th International Conference on Multimodal Interfaces, New York, NY, USA, 5–7 November 2003; pp. 258–264. [CrossRef]
15. Chen, J.; Chen, Z.; Chi, Z.; Fu, H. Facial Expression Recognition in Video with Multiple Feature Fusion. *IEEE Trans. Affect. Comput.* **2018**, 38–50. [CrossRef]
16. Cohen, I.; Sebe, N.; Garg, A.; Chen, L.S.; Huang, T.S. Facial expression recognition from video sequences: temporal and static modeling *Comput. Vis. Image Underst.* **2003**, *91*, 160–187. [CrossRef]
17. Veldanda, A.; Liu, H.; Koschke, R.; Schultz, T.; Küster D. Can Electromyography Alone Reveal Facial Action Units? A Pilot EMG-Based Action Unit Recognition Study with Real-Time Validation. In Proceedings of the 17th International Joint Conference on Biomedical Engineering Systems and Technologies—BIODEVICES, Rome, Italy, 21–23 February 2024; pp. 142–151. [CrossRef]

Disclaimer/Publisher's Note: The statements, opinions and data contained in all publications are solely those of the individual author(s) and contributor(s) and not of MDPI and/or the editor(s). MDPI and/or the editor(s) disclaim responsibility for any injury to people or property resulting from any ideas, methods, instructions or products referred to in the content.

Article

Adopting Graph Neural Networks to Analyze Human–Object Interactions for Inferring Activities of Daily Living

Peng Su and Dejiu Chen *

Department of Engineering Design, KTH Royal Institute of Technology, 100 44 Stockholm, Sweden; pensu@kth.se
* Correspondence: chendj@kth.se

Abstract: Human Activity Recognition (HAR) refers to a field that aims to identify human activities by adopting multiple techniques. In this field, different applications, such as smart homes and assistive robots, are introduced to support individuals in their Activities of Daily Living (ADL) by analyzing data collected from various sensors. Apart from wearable sensors, the adoption of camera frames to analyze and classify ADL has emerged as a promising trend for achieving the identification and classification of ADL. To accomplish this, the existing approaches typically rely on object classification with pose estimation using the image frames collected from cameras. Given the existence of inherent correlations between human–object interactions and ADL, further efforts are often needed to leverage these correlations for more effective and well justified decisions. To this end, this work proposes a framework where Graph Neural Networks (GNN) are adopted to explicitly analyze human–object interactions for more effectively recognizing daily activities. By automatically encoding the correlations among various interactions detected through some collected relational data, the framework infers the existence of different activities alongside their corresponding environmental objects. As a case study, we use the Toyota Smart Home dataset to evaluate the proposed framework. Compared with conventional feed-forward neural networks, the results demonstrate significantly superior performance in identifying ADL, allowing for the classification of different daily activities with an accuracy of 0.88. Furthermore, the incorporation of encoded information from relational data enhances object-inference performance compared to the GNN without joint prediction, increasing accuracy from 0.71 to 0.77.

Keywords: graph neural network; scene understanding; activities of daily living analysis

Citation: Su, P.; Chen, D. Adopting Graph Neural Networks to Analyze Human–Object Interactions for Inferring Activities of Daily Living. *Sensors* **2024**, *24*, 2567. https://doi.org/10.3390/s24082567

Received: 29 February 2024
Revised: 28 March 2024
Accepted: 13 April 2024
Published: 17 April 2024

Copyright: © 2024 by the authors. Licensee MDPI, Basel, Switzerland. This article is an open access article distributed under the terms and conditions of the Creative Commons Attribution (CC BY) license (https://creativecommons.org/licenses/by/4.0/).

1. Introduction

Human Activity Recognition (HAR) involves multiple techniques to analyze sensory data [1]. These sensory data constitute a basis for assessing and predicting human activities. In the field of Human Activity Recognition (HAR), the applications of smart homes and assistive robotic systems are paving the way to support individuals in performing their Activities of Daily Living (ADL), therefore facilitating and monitoring their quality of life [2]. Various equipment collect operational conditions and human status by employing wearable sensors like wrist-worn accelerometers [3] and non-wearable sensors like cameras [2,4] to attain the recognition of ADL. Compared to wearable sensors, the adoption of camera frames to analyze and classify Activities of Daily Living (ADL) presents a promising solution due to the inherently multifarious features found in image data [5–7]. Most of the approaches utilize image frames to detect ADL by combining pose estimation with skeleton-based action recognition [8,9]. Methods based on Convolutional Neural Networks (CNN) typically demand significant effort to identify key points and joints of human bodies. As shown by [10,11], complex human motion capture systems can be used to support annotating the key points through extensive data. With such data, a variety of CNN architectures can be trained to estimate pose by formulating body joints and extracting features [12,13]. Many Graph Neural Networks (GNN)-based solutions have been considered to be support for alleviating the need for deep architectures to extract the features from the images, as such solutions capture the key points and joints with graph models [14–17]. Through the analysis of graph models representing skeleton-based human bodies, GNN can be used to estimate the likelihood of human actions. However, the uncertainties stemming from the probabilistic nature of neural networks [18,19] often necessitate extensive training data with high sensory resolution for accurately identifying the human body parts [8,9,13,14,16]. These requirements restrict the applicability of cameras for recognizing daily activities in the context of assisting at-home scenarios.

To address this issue, we propose a framework where GNN are adopted to explicitly analyze human–object interactions for inferring human activities of daily living alongside the corresponding environmental objects. Specifically, the framework first extracts the relational data on the interactions between humans and environmental objects from the collected image frames. Next, GNN automatically encodes the correlations among the interactions indicated by the respective relational data and, therefore, detects the presence of activities and their environmental objects, leading to a more effective analysis of ADL. We present the contribution of this paper as follows:

- Designing a conceptual framework to construct graph-based data by image frames to infer the ADL within assisting at-home applications.
- Proposing a GNN architecture to jointly predict environmental objects and ADL by comprehending the relational data.
- Enhancing the prediction accuracy of ADL and environmental objects by aggregating the encoded information from the semantics of relational data.

The rest of the paper is organized as follows: Section 2 presents prior work related to GNN with environmental scene understanding. Section 3 describes the proposed framework. Section 4 presents a case study by verifying the proposed framework with the Toyota Smart Home dataset. Section 5 presents the conclusion of the proposed framework and discusses the future work.

2. Related Work

This section first provides background information on GNN. Next, we present previous work on GNN applied in the applications related to the topic. In addition, we exhibit current efforts to apply image frames to relational data in scene understanding.

2.1. Background of GNN

GNN are specifically designed for processing non-Euclidean data, supporting the analysis of graph-based data [20]. Such graph-based data structures usually consist of nodes and edges to represent a set of objects and relations. Specifically, graphs can be classified into heterogeneous graphs, which typically connect nodes with different types of edges, and homogeneous graphs, where edges do not convey additional information [20]. A variety of GNN models are used to analyze these two graphs regarding their spatial and temporal properties [21]. Spatial models support the transformation of graph-based data into a spectrum space using Graph Laplacian [22,23] or encoding information from local neighbors of specific nodes through aggregation operations [24] with Graph Convolutional Networks (GCN). Building on the spatial models, the adoption of gate mechanisms from RNN and LSTM is a common solution to enable temporal analysis of graph-based data [21].

2.2. GNN to Cope with HAR and ADL

Most GNN integrate different models to analyze human activities by synthesizing spatial–temporal features. As mentioned earlier, some of them recognize the key points of the human body by analyzing unstructured high-dimensional data such as video clips [9,25]. These high-dimensional data could either contain video clips with depth information as 3D data or solely rely on raw 2D images captured by cameras [26,27]. Depending on the input data formats, these GNN can be roughly categorized into the following trends [9]: (1) Spatio-temporal GCNs encode the key points of human bodies as nodes in graphs, while the evolution of human activities is usually interpreted as attributes of edges among the nodes within the graphs [28]. This method usually requires the analysis of the graphs, including all elements, such as edges and nodes, to identify human activities. However, to accurately identify the key points of human bodies, such a method usually requires high-resolution data or additional depth information. As an example in [29], the input data requires annotating bones and joints within human bodies with depth information, which decreases the generalization of the proposed framework. (2) Temporal-aware GCNs focus on extracting contextual dependencies in sequential data by adopting and optimizing attention mechanisms. This method typically analyzes contextual information across video sequences with similar lengths. However, due to the diversity of activities within video sequences, attention-based methods could become more time-consuming and less efficient [30,31]. (3) Multi-stream GCN refers to an integration with different inputs for identifying human activities. A typical example in [15,17] usually uses video clips and skeleton-based data as two-stream input for GCN to extract features. This method aims to identify human daily activities by aggregating image frames and incomplete skeleton-based data, reducing the reliance on high-resolution and well-annotated datasets. While these methods enhance the efficiency of detecting human activities, further efforts are needed to understand the interaction between humans and environmental objects. Towards this direction, we also investigate previous work on scene understanding through the utilization of GNN.

2.3. Applying Relational Data to Scene Understanding

One common solution is to adopt GNN to analyze and understand scenes in image frames. Such GNN support inferring common-sense relationships among objects within scenes [32–34]. Therefore, a critical step in utilizing GNN for scene understanding is to convert high-dimensional unstructured data (e.g., image frames) into relational context within a graph-based structure. A basic process for constructing such graph-based data is to extract objects within image frames as nodes. The edges between nodes represent pairwise relations between the objects, depicting their spatial and temporal evolution. The semantics of graph-based data are analyzed through the adoption of GCN. However, this implementation could be insufficient for understanding the task-specific scene. For example, when a human detected to be overlapping with a motorcycle is represented in graph-based data and analyzed by the GCN, their relationship is highly likely to be recognized as the human riding the motorcycle in a public area. However, when this human is riding a motorcycle without a helmet, these methods may not capture insights into unsafe behaviors. Hence, combining task-specific scene understanding with certain prior knowledge aids in achieving specific tasks. The presentation of such prior knowledge could be categorized as follows: (1) Explicit rules refer to directly leveraging human knowledge imposed into the graph-based data. In [35–37], objects from Bird's-Eye Views (BEV) within dynamic driving scenarios are converted into graph-based data to facilitate analysis by GCN, incorporating specific traffic rules and common-sense knowledge. A typical human-understandable rule is exemplified in [36], where the weighted edge within the node represents the relative distance. GCN are used to analyze potential node pairs whose relative distance violates specified rules. However, these methods usually require landmarks (e.g., static objects) to annotate the relationships among objects, which limits their generalization for extension in ADL-related applications. (2) Encoded formal knowledge refers to the process of interpreting human knowledge into machine-readable specifications. For example, in [38], common-sense knowledge is converted into propositional logic to be incorporated with GCN in the context of recommendation systems.

Inspired by the aforementioned methods of understanding scenes, we introduce a GNN-based framework designed to comprehend scenarios within ADL-related applications. Unlike the conventional approach of relying solely on pose estimation for daily activities prediction [15,17], our proposed method achieves joint prediction by mapping the interactions, alleviating the need for skeleton-based data as part of the input. Compared with existing methods adopted in [35–37], the proposed method interprets common-sense knowledge into temporal logic specifications without relying on landmarks for further annotating the relationships.

3. Methodology

In this section, we present the framework shown in Figure 1 to infer activities of daily living. We describe the main workflow of the proposed work as follows:

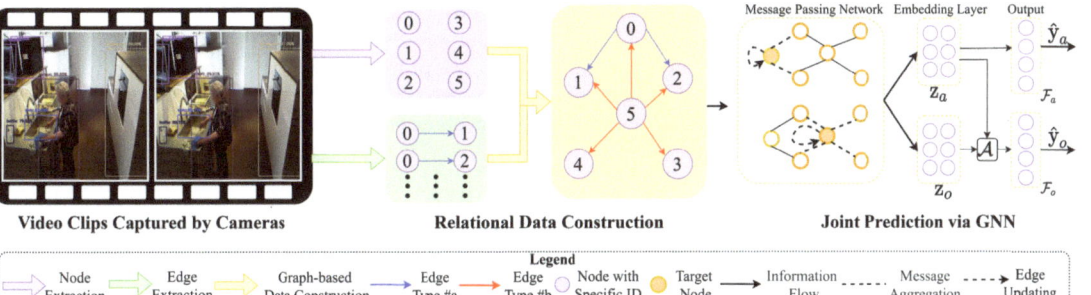

Figure 1. An example to present the overall process of extracting and constructing relational data. The edge types #a and #b refer to interactions with different features extracted from temporal specifications, as defined by Equation (3).

3.1. Relational Data Construction

We construct relational data for GNN analysis by extracting interactions from image frames. Specifically, the relational data in terms of graph-based data consists of nodes and edges. The objects in the video clips are extracted as nodes in the graph models, while the interactions within these objects are represented as edges. Therefore, the following steps outline the process to obtain these graph models:

3.1.1. Node Extraction

At this step, we obtain the node information required for creating graph-based data. We define the nodes based on the information presented in image frames. Specifically, we formulate $\mathcal{D}^{a_i}$ for a video clip collected from a scenario a_i as follows:

$$\mathcal{D}^{a_i} = \{d_1^{a_i}, d_2^{a_i}, \ldots, d_n^{a_i}\} \tag{1}$$

where n refers to the number of frames in the video clip $\mathcal{D}^{a_i}$. a_i refers to a specific daily activity obtaining a label $\mathbf{y}_a \in \mathcal{T}_a$. $\mathcal{T}_a$ represents a set of labels for daily activities collected in the dataset.

An object-detection module $\mathcal{M}_n(\cdot)$ is used to identify the nodes of the graph model by extracting the objects in any frames $d_k^{a_i}$ of $\mathcal{D}^{a_i}$. We formulate the process as follows:

$$\mathcal{O}_k^{a_i} = \mathcal{M}_n(d_k^{a_i}) \tag{2}$$

where $\mathcal{O}_k^{a_i} = \{o_1^k, o_2^k, \ldots, o_j^k\}$ refers to the collection containing the objects extracted from the video clips. Each o_i^k from $\mathcal{O}_k^{a_i}$ is a vector denoting the features of an object, such as its bounding box sizes and object types. Each detected object o_i^k obtains a label $\mathbf{y}_o \in \mathcal{T}_o$ indicating the types of object. $\mathcal{T}_o$ represents a set of labels for environmental objects collected in the dataset.

3.1.2. Edge Extraction

To represent the relationships within the video clip $\mathcal{D}^{a_i}$, it is critical to analyze the spatial and temporal properties of human and environmental objects. We label these relationships via the edges across nodes. As mentioned earlier, existing studies typically employ data-driven approaches, such as LSTM, to extract relationships by encoding input features from extensive graphs [35–37]. However, the duration periods within different daily activities could exhibit extreme variety [17]. For example, drinking water in the kitchen could be captured in a few image frames, while recognizing activities like washing dishes in the same place may require more images. Therefore, using data-driven methods could be inefficient for encoding an entire video clip. In contrast, the knowledge could enhance the efficiency of data-driven methods in task-specific scenarios (e.g., human action reasoning [16,32] and recommendation systems [39]) that involve possible known relationships. Since activities of daily living typically involve well-known interactions between humans and environmental objects, we propose a rule-based method for extracting the relationships of nodes. Similar rule-based methods also can be found in [16,37]. Specifically, we formulate the rule to identify the interactions by temporal logic specifications:

$$\Diamond(\phi \cup (T \wedge (\neg \phi \cup \rho))) \tag{3}$$

where T refers to the time duration, and $\rho ::= (Occ_{m_{ij}} \geq n)$, where $Occ_{m_{ij}}$ refers to the number of appearances in the video clip $\mathcal{D}^{a_i}$, n refers to the threshold of occurrence number. $\phi ::= (m_{ij} \geq \tau)$ denotes the condition when the interaction rate m_{ij} for objects o_i^k, o_j^k in a single frame k exceeds a threshold τ.

We formulate the interaction rate m_{ij} as Equation (4), which is identified by the Intersection over Union (IoU) areas between a pair of objects with non-maximal suppression [32,33].

$$m_{ij} = \frac{I(xy_i^k, xy_j^k)}{U(xy_i^k, xy_j^k)} \tag{4}$$

where xy_i^k, xy_j^k refer to the bounding box sizes of o_i^k, o_j^k. These sizes are obtained by the object-detection module $\mathcal{M}_n(\cdot)$. $I(xy_i^k, xy_j^k)$ refers to the intersection area within the objects, while $U(xy_i^k, xy_j^k)$ refers to the union area within the objects. Once m_{ij} satisfy the rule defined by Equation (3), we denote the interaction as $< o_i^{a_i}, r_{i,j}^{a_i}, o_j^{a_i} >$, where $r_{ij} \in \mathcal{M}^{a_i}$. $\mathcal{M}^{a_i}$ denotes a set of identified interactions within detected objects from the video clip $\mathcal{D}^{a_i}$. Furthermore, we denote all interaction pairs in the context of a graph $\mathcal{G}^{a_i}$ as follows [24]:

$$\mathcal{G}^{a_i} = \{(o_i^{a_i}, r_{i,j}^{a_i}, o_j^{a_i})\} \tag{5}$$

Additionally, each generated graph $\mathcal{G}^{a_i}$ obtains a label $\mathbf{y}_a \in \mathcal{T}_a$ indicating the type of daily activities. $\mathcal{T}_a$ refers to a set of labels for the daily activities.

3.2. Joint Prediction via GNN

After the relational data construction phase, we utilize Message-Passing Neural Networks (MPNN) [40] to integrate GNN models for the joint prediction (see Figure 1).

3.2.1. Message-Passing Phase

This step involves the computation for aggregating and updating information from the neighbors of a specific node along with the edges of shared relationships. Specifically, we model the message aggregating process in the layer l as follows:

$$m_i^{l+1} = \sum_{j \in N(i)} \mathcal{M}_l(h_i^l, h_j^l, r_{i,j}) \tag{6}$$

where i, j are the same as Equation (5), $r_{i,j} \in t_{i,j}^{a_i}$ refers to the edge types connecting from $o_i^{a_i}$ to $o_j^{a_i}$. We denote h_i^l, h_j^l as the encoded information of the node $o_i^{a_i}, o_j^{a_i}$ in layer l. This encoded information is dependent on the configuration of the message-passing network. As an example, h_i^l, h_j^l are equivalent to the features within $o_i^{a_i}$ and $o_j^{a_i}$, respectively, when

$l = 1$. $N(i)$ refers to the set of all neighboring nodes of the node $o_i^{a_i}$ whose example is shown in Figure 1. $\mathcal{M}_l(\cdot)$ refers to message-passing functions, such as concatenation and multiplication operations. Equation (6) shows that by computing all the neighboring nodes $N(i)$ in terms of message passing, m_i^{l+1} merges the information from the features of both the target node and their contextual nodes.

To further encode the aggregated relational data, the network propagates the edge information within the neighbors by creating an edge (vertex) updating function $\mathcal{U}_l$ as follows:

$$h_i^{l+1} = \mathcal{U}_l(h_i^l, m_i^{l+1}) \tag{7}$$

where $\mathcal{U}_l$ refers to a composition of non-linear functions, such as a ReLU function and recurrent units.

3.2.2. Readout Phase

In this step, the readout operation approximates feature vectors $\mathbf{z}$ for the graph-based data $\mathcal{G}^{a_i}$. We use multiple embedding $\mathbf{z} \in \{\mathbf{z}_a^{a_i}, \mathbf{z}_o^{a_i}\}$ to encode the information of activities $\hat{y}_a$ and environmental objects $\hat{y}_o$ within the context of the graph $\mathcal{G}^{a_i}$. The embedding vectors $\mathbf{z}$ are formulated as follows:

$$\mathbf{z} = \mathcal{R}(\{h_i^L | i \in \mathcal{G}^{a_i}\}) \tag{8}$$

where $\mathcal{R} \in \{\mathcal{R}_a, \mathcal{R}_o\}$. $\mathcal{R}_a$ and $\mathcal{R}_o$ refer to readout functions, configurable with various operations, such as a linear layer and sum operation, to generate $\mathbf{z}_a^{a_i}$ and $\mathbf{z}_o^{a_i}$, respectively. L refers to the running steps in the message-passing phase.

Considering daily activities involving interactions between humans and objects, the predicted object classes are often correlated with these activities. For instance, eating in a kitchen is a typical daily activity commonly associated with specific environmental objects such as bowls [17]. However, detecting bowls in the kitchen is insufficient to confirm that humans are eating. Therefore, we propose an aggregation operation $\mathcal{A}(\cdot)$ to enhance the performance of predicting environmental objects by synthesizing embeddings $\mathbf{z}_a^{a_i}$ and $\mathbf{z}_o^{a_i}$ as follows:

$$\mathbf{z}_c^{a_i} = \mathcal{A}(\mathbf{z}_a^{a_i}, \mathbf{z}_o^{a_i}) \tag{9}$$

$\mathbf{z}_c^{a_i}$ refers to an aggregated embedding to predict environmental objects. To this end, we model the output layers as follows:

$$\hat{y} = \mathcal{F}(\mathbf{z}_e) \tag{10}$$

where $\mathbf{z}_e \in \{\mathbf{z}_c^{a_i}, \mathbf{z}_a^{a_i}\}$, $\mathcal{F}(\cdot)$ refers to the configuration of output functions to predict activities and objects, where $\mathcal{F} \in \{\mathcal{F}_a, \mathcal{F}_o\}$, $\hat{y}$ refers to the predicted results, where $\hat{y} \in \{\hat{y}_a, \hat{y}_o\}$. $\hat{y}_a$ denotes the predicted activities of daily living using the output function $\mathcal{F}_a$ with embedding $\mathbf{z}_a^{a_i}$, while $\hat{y}_o$ represents the predicted classes of environmental objects using the output function $\mathcal{F}_o$ with the aggregated embedding $\mathbf{z}_c^{a_i}$.

4. Case Study

In this section, we elaborate on the implementation of the proposed framework. First, we provide a brief introduction along with an explanation for selecting the Toyota Smart Home dataset. Next, we present the configuration of relational data construction and joint prediction based on this dataset. Finally, we present the results in comparison with baseline methods.

4.1. Overview of Toyota Dataset

The Toyota Smart Home dataset [17] is a set of video clips collected from different locations of an apartment whose Bird Eye View (BEV) is shown in Figure 2. The reasons for selecting this dataset to evaluate the proposed methods are as follows: (1) It contains over 10,000 video clips captured from different locations in the apartment, providing diversity to record various daily activities. (2) The resolution of video clips captured by cameras is

640 × 480, challenging the identification of human body parts. In this case, understanding between humans and environmental objects provides a promising solution for detecting daily activities.

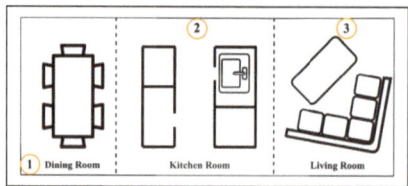 

(a) Bird Eye Views of the Smart Home Configurations.　　(b) Samples of Video Clips from the Cameras in Different Rooms

Figure 2. Bird Eye View (BEV) of the apartment. The numbers in the figure refer to the location of the camera installation. ①, ②, and ③ refer to the camera locations used to capture video clips of activities.

Specifically, we choose three camera views shown in Figure 2 recording from the dining room, living room, and kitchen. To evaluate the proposed framework, we particularly selected 8 daily activities, including eating meals, calling phones, and using laptops. These activities commonly involve the interaction between humans and environmental objects. To reduce the correlation between the daily activities and locations, these activities could occur in multiple locations. Additionally, we select video clips that feature multiple types of daily activities occurring in the same location, as well as the same activity taking place in different locations. For example, a person could use a cell phone in both locations shown in Figure 2 while also engaging in cooking and cleaning in the kitchen.

4.2. Constructing Relational Data

Existing deep neural networks designed for object detection could be employed in the object-detection module $\mathcal{M}_n$. In this paper, we adopt a pre-trained Fast-RCNN model to detect objects in the video clips [41]. Every node consists of the types and bounding box sizes from the detected objects. Moreover, we assign an ID to each detected object to prevent duplicating the same types of objects occurring in the images. To extract relationships from the video clips, we set the IoU threshold to $\tau = 0.4$. If m_{ij} exceeds the threshold until more than $n = 20$ instances or appears continuously for more than $T = 0.2$ length in the image sequences throughout the entire video clips, we annotate that the relationships between objects i and j are engaged in interaction. In particular, we annotate relationships between people and environmental objects when constructing graph-based relational data. Furthermore, we incorporate the location information of the video clips to enrich these data and facilitate the GNN in aggregating node features.

As a result, we extract 33 different types of environmental objects. The following daily activities are extracted from the dataset: cleaning, cooking, watching TV, eating food, reading books, using the telephone, using a laptop, and drinking water. Except for cleaning, cooking, and watching TV, the rest of the activities could occur in multiple locations. As illustrated in Figure 3, we present the graph-based relational data of daily activities extracted from various locations. From Figure 3, we note that even though the person is cleaning and cooking in the same location, the edges in the graph for these two daily activities still depict different connections. Specifically, when the person is cooking, there is more interaction between the person and the bowls and the refrigerator. In contrast, when the person is cleaning, the edges are more connected to the person, bottles, and sink. Moreover, the remaining activities also manifest significant features within the context of relational data. For instance, during eating, interactions typically occur with items such as tables, chairs, and dishes. Similarly, when watching TV, interactions involve remotes and humans.

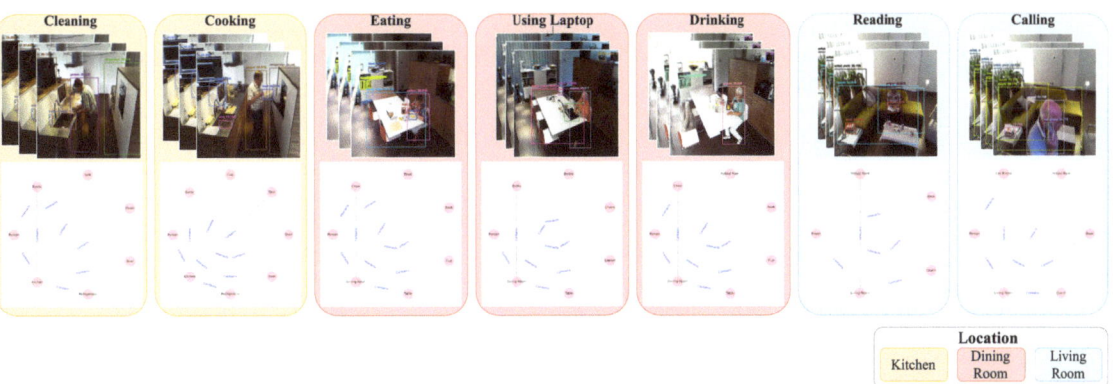

Figure 3. Samples of graph-based relational data generation based on the image frames.

4.3. Implementing Joint Prediction via GNN

We adopt two-layer message-passing networks whose layout is shown in Figure 1 to encode the information from input graphs. We use GraphSAGE in the first layer to attain encoding the features of the edges and nodes [40,42]. Specifically, We use a mean aggregator shown in Equation (11) as the message-passing function M_l.

$$m_i^{l+1} = \bigoplus_{j \in N(i)} (h_i^l, h_j^l, r_{i,j}) \tag{11}$$

where $\bigoplus$ refers to approximate element-wise mean value from the encoded information h_i, h_j with their edge type $r_{i,j}$.

We adopt graph convolutional operators (GCNConv) with Laplacian-based methods based on [23] to attain message-passing functions in the embedding layers. Specifically, we model the message function as follows:

$$m_i^{l+1} = D^{\frac{1}{2}} A D^{-\frac{1}{2}} h^l W^l \tag{12}$$

D refers to the degree matrix. A refers to the adjacency matrix. W^l refers to layer-wise learnable parameters in the l-th layer [23,40].

This layer consists of two parallel GCNConv, which are used to separately generate the embedding $\mathbf{z}_a^{a_i}$ and $\mathbf{z}_o^{a_i}$ from a video clip a_i. We use tanh functions as the edge updating function $\mathcal{U}_l$ in each layer. We propose an element-wise multiplication operation as $\mathcal{A}(\cdot)$ to aggregate the correlated features within $\mathbf{z}_a^{a_i}$, $\mathbf{z}_o^{a_i}$ and generate $\mathbf{z}_c^{a_i}$. To this end, we use SoftMax classifiers as the output layers to generate the likelihood of prediction results $\hat{y}_a, \hat{y}_o$ from $\mathbf{z}_a^{a_i}, \mathbf{z}_c^{a_i}$, respectively. Sequentially, we define the loss function $\mathcal{L}$ as follows:

$$\mathcal{L} = \mathcal{L}_c(\hat{y}_a, y_a) + \mathcal{L}_c(\hat{y}_o, y_o) \tag{13}$$

We train the parameters in the network by optimizing the loss function $\mathcal{L}$, where $\mathcal{L}_c$ refers to the cross-entropy between the predicted results and the ground-truth label. To this end, we develop a GNN-based framework to classify the graph-based content $\hat{y}_a$ under-recognized nodes and edges and to predict nodes $\hat{y}_o$ within a given graph. This framework synthesizes human–object interaction to infer activities of daily living.

4.4. Ablation Study

The training platform is configured with an AMD Ryzen 7 5800 and NVIDIA RTX-3070. During the training of the proposed methods, we collect all these daily activities, with each activity containing 600 graphs. We configure the training ratio to 0.8, and the training epoch is 800. We select multiple baseline methods to evaluate the proposed method. Specifically, we employ an MLP with two hidden layers to infer activities and objects by solely analyzing the features of nodes. This MLP configuration is equivalent to concatenating the intermediate embeddings from Fast-RCNN in Equation (2) to dense layers. In addition, we introduce two GCN designs, GNN with Split Prediction (S-GNN) and Attention-based GNN (Att-GNN), to evaluate their performance using the same dataset as the comparison. S-GNN shares the same network topology in [28] to analyze spatial properties of the graph-based data. This S-GNN adopts graph convolutional and dense layers to concatenate the features within the nodes from graphs. Att-GNN identifies correlations by modeling an energy function and attention distributions within spatial and temporal properties, enabling the analysis of graph and node patterns. In our case, we implement a similar network architecture used in [37], wherein a self-attention layer is connected behind the graph convolutional layers by replacing the multiplication operation $\mathcal{A}(\cdot)$. As an ablation test, we additionally construct a Joint-Prediction Network (JP-GNN) by removing the operation $\mathcal{A}(\cdot)$ and directly predicting the data.

The final results are shown in Table 1, where we conclude that the proposed method demonstrates significantly superior performance compared with MLP. Such results indicate that the relationships within the nodes empower the capability to infer daily activities and objects. Unlike GNN-based approaches, the inference process of MLP does not explicitly incorporate semantic context within graphs, owing to the inherent properties of feed-forward networks. Among GNN-based approaches trained for the same number of epochs, our proposed method achieves higher accuracy compared to the attention-based method, which also analyzes correlations within the embeddings. The possible reason for this situation could be that the attention-based method requires more time to attain convergence in the attention mechanism (e.g., learnable parameters in score functions). Compared with the JP-GNN which does not include the aggregation function, our proposed method shows significant improvement in object inference. These results indicate that the activity classification embedding aids in inferring objects. Additionally, the embeddings of activities and objects share the same layer, therefore affecting the convergence of the network. As a result, the TOP-1 accuracy of activities classification of JP-GNN is lower than that of our methods and the S-GNN which infers objects and activities separately. We also observe that the TOP-1 accuracy of activity inference from the proposed method is slightly higher than those of S-GNN. We believe that the reason could be the implementation of multiple embeddings serving as regularization to optimize networks. Similar situations also could be observed in prior studies, such as [22,39]. To further evaluate the performance of the proposed method, we also utilize the F1-score in Equation (14) by leveraging the Confusion Matrix in multi-classification cases [43,44].

$$Pr@y_a^k = \frac{TP@y_a^k}{TP@y_a^k + FP@y_a^k}$$
$$Re@y_a^k = \frac{TP@y_a^k}{TP@y_a^k + FN@y_a^k} \quad (14)$$
$$F1@y_a^k = \frac{2 \times Re@y_a^k \times Pr@y_a^k}{Re@y_a^k + Pr@y_a^k}$$

where $TP@y_a^k, TN@y_a^k, FP@y_a^k, FN@y_a^k$ refer to True Positive, True Negative, False Positive, and False Negative in the Confusion Matrix. $Pr@y_a^k, Re@y_a^k, F1@y_a^k$ refer to the precision, recall and F1-Score at any activities $\mathbf{y}_a$ with label k. Table 2 presents the overall results with Equation (14). Compared to the other activities, the proposed method shows poorer performance in identifying cooking and cleaning. This situation could be implied by the presence of common interacting objects in these two activities. For instance, both cooking and cleaning involve bowls and dishes in the same location. Additionally, the location of the camera in the kitchen, as shown in Figure 2, may introduce some uncertainty in efficiently detecting interactions between cookstoves and humans during cooking. This situation could be improved by utilizing image frames from multiple camera views with different locations.

Table 1. TOP-1 Accuracy of Different Methods.

	MLP	GNN-Based Methods			
		Our Method	Att-GNN	JP-GNN	S-GNN
Activities Inference	0.49	**0.88**	0.82	0.83	0.86
Objects Inference	0.56	**0.77**	0.65	0.71	0.68

Table 2. Precision, Recall and F1-Score Comparison.

	Reading	Cooking	Cleaning	Eating	Drinking	Using Laptop	Calling	Wathcing TV	Average
Precision	0.94	0.71	0.75	0.89	0.78	0.95	0.90	0.92	0.86
Recall	0.66	0.63	0.67	0.72	0.84	0.91	0.91	0.83	0.77
F1-Score	0.77	0.67	0.71	0.85	0.81	0.93	0.90	0.87	0.81

Additionally, we evaluate the time consumption of training each method. With the same hyper-parameters (e.g., training epoch, batch sizes), S-GNN takes approximately 9 and 33 min to train the network to attain stable performance, respectively. Att-GNN requires more than 25 min to train the joint prediction. The proposed method takes around 21 min. These results indicate that compared with baseline methods, the proposed method spends less time to attain better performance.

5. Discussion and Future Work

This paper presents a framework to jointly infer the daily activities and environmental objects. Specifically, compared to the baseline methods, our framework demonstrates competitive performance in terms of TOP-1 accuracy and training efficiency. The proposed method supports incorporating semantic content within relational data rather than directly relying on high-dimensional data. This approach offers an explicit solution for inferring human daily activities and environmental objects. Compared to prior work on GCN related to the identification of human daily activities, the proposed method avoids the need for skeleton-based data and reduces reliance on complex training data. However, the proposed work relies on the semantics in the context of interaction between humans and the environment to identify the objects and daily activities. Such a mechanism could be inefficient in specific scenarios (e.g., entering and leaving).

Therefore, the following aspects could be future works: (1) Combining knowledge-aware approaches (e.g., knowledge graphs) with embedding to enhance the explainability and performance of the proposed networks. In contrast to the temporal logic constraints imposed in the proposed framework, domain knowledge can be encoded within the GCN-based framework to offer flexible constraints. (2) Utilizing recurrent units (e.g., LSTM) to reduce the labeling data and improve the generalization by encoding the temporal evolution. The proposed method can integrate various embeddings to encode and analyze temporal correlations. This encoded evolution is expected to enhance the granularity of daily activities, enabling the decomposition of activities (e.g., entering can be decomposed into opening doors and walking). (3) Extending the proposed framework to diverse datasets with complicated scenarios such as dynamic driving scenarios. In such scenarios, environmental objects exhibit various correlated behaviors, posing challenges in modeling and analyzing relational data in terms of their relationships and types. An extension of the proposed work targeting heterogeneous graphs with weighted edges could address these scenarios.

Author Contributions: The method design, software implementation and testing by P.S.; The conceptualization, overall framework and method design, research supervision and funding acquisition by D.C.; P.S. contributed to the manuscript writing, with D.C. to the reviewing and refining. All authors have read and agreed to the published version of the manuscript.

Funding: This work is partially supported by the industrial research project ADinSOS (2019065006), KTH Royal Institute of Technology, Sweden.

Informed Consent Statement: Not applicable.

Data Availability Statement: Data available on request from the authors

Conflicts of Interest: The authors declare no conflict of interest.

References

1. Zhang, S.; Li, Y.; Zhang, S.; Shahabi, F.; Xia, S.; Deng, Y.; Alshurafa, N. Deep learning in human activity recognition with wearable sensors: A review on advances. *Sensors* **2022**, *22*, 1476. [CrossRef]
2. Petrich, L.; Jin, J.; Dehghan, M.; Jagersand, M. A quantitative analysis of activities of daily living: Insights into improving functional independence with assistive robotics. In Proceedings of the 2022 International Conference on Robotics and Automation (ICRA), Philadelphia, PA, USA, 23–27 May 2022; IEEE: Piscateville, NJ, USA, 2022; pp. 6999–7006.
3. Masud, M.T.; Mamun, M.A.; Thapa, K.; Lee, D.; Griffiths, M.D.; Yang, S.H. Unobtrusive monitoring of behavior and movement patterns to detect clinical depression severity level via smartphone. *J. Biomed. Inform.* **2020**, *103*, 103371. [CrossRef]
4. Johnson, D.O.; Cuijpers, R.H.; Juola, J.F.; Torta, E.; Simonov, M.; Frisiello, A.; Bazzani, M.; Yan, W.; Weber, C.; Wermter, S.; et al. Socially assistive robots: A comprehensive approach to extending independent living. *Int. J. Soc. Robot.* **2014**, *6*, 195–211. [CrossRef]
5. Chen, K.; Zhang, D.; Yao, L.; Guo, B.; Yu, Z.; Liu, Y. Deep learning for sensor-based human activity recognition: Overview, challenges, and opportunities. *ACM Comput. Surv.* **2021**, *54*, 1–40. [CrossRef]
6. Ferrari, A.; Micucci, D.; Mobilio, M.; Napoletano, P. Deep learning and model personalization in sensor-based human activity recognition. *J. Reliab. Intell. Environ.* **2023**, *9*, 27–39. [CrossRef]
7. Borkar, P.; Wankhede, V.A.; Mane, D.T.; Limkar, S.; Ramesh, J.; Ajani, S.N. Deep learning and image processing-based early detection of Alzheimer disease in cognitively normal individuals. *Soft Comput.* **2023**. [CrossRef]
8. Munea, T.L.; Jembre, Y.Z.; Weldegebriel, H.T.; Chen, L.; Huang, C.; Yang, C. The progress of human pose estimation: A survey and taxonomy of models applied in 2D human pose estimation. *IEEE Access* **2020**, *8*, 133330–133348. [CrossRef]
9. Zheng, C.; Wu, W.; Chen, C.; Yang, T.; Zhu, S.; Shen, J.; Kehtarnavaz, N.; Shah, M. Deep learning-based human pose estimation: A survey. *ACM Comput. Surv.* **2023**, *56*, 1–37. [CrossRef]
10. Ionescu, C.; Papava, D.; Olaru, V.; Sminchisescu, C. Human3. 6m: Large scale datasets and predictive methods for 3d human sensing in natural environments. *IEEE Trans. Pattern Anal. Mach. Intell.* **2013**, *36*, 1325–1339. [CrossRef]
11. Mandery, C.; Terlemez, Ö.; Do, M.; Vahrenkamp, N.; Asfour, T. Unifying representations and large-scale whole-body motion databases for studying human motion. *IEEE Trans. Robot.* **2016**, *32*, 796–809. [CrossRef]
12. Toshev, A.; Szegedy, C. Deeppose: Human pose estimation via deep neural networks. In Proceedings of the IEEE Conference on Computer Vision and Pattern Recognition, Columbus, OH, USA, 23–28 June 2014; pp. 1653–1660.
13. Duan, H.; Zhao, Y.; Chen, K.; Lin, D.; Dai, B. Revisiting skeleton-based action recognition. In Proceedings of the IEEE/CVF Conference on Computer Vision and Pattern Recognition, New Orleans, LA, USA, 18–24 June 2022; pp. 2969–2978.
14. Yan, S.; Xiong, Y.; Lin, D. Spatial temporal graph convolutional networks for skeleton-based action recognition. In Proceedings of the AAAI Conference on Artificial Intelligence, New Orleans, LA, USA, 2–7 February 2018; Volume 32.
15. Shi, L.; Zhang, Y.; Cheng, J.; Lu, H. Two-stream adaptive graph convolutional networks for skeleton-based action recognition. In Proceedings of the IEEE/CVF Conference on Computer Vision and Pattern Recognition, Long Beach, CA, USA, 16–17 June 2019; pp. 12026–12035.
16. Ma, Y.; Wang, Y.; Wu, Y.; Lyu, Z.; Chen, S.; Li, X.; Qiao, Y. Visual knowledge graph for human action reasoning in videos. In Proceedings of the 30th ACM International Conference on Multimedia, Lisboa, Portugal, 10–14 October 2022; pp. 4132–4141.
17. Das, S.; Dai, R.; Koperski, M.; Minciullo, L.; Garattoni, L.; Bremond, F.; Francesca, G. Toyota smarthome: Real-world activities of daily living. In Proceedings of the IEEE/CVF international Conference on Computer Vision, Seoul, Republic of Korea, 27 October–2 November 2019; pp. 833–842.
18. Su, P.; Chen, D. Using fault injection for the training of functions to detect soft errors of dnns in automotive vehicles. In Proceedings of the International Conference on Dependability and Complex Systems, Wrocław, Poland, 27 June–1 July 2022; Springer: Berlin/Heidelberg, Germany, 2022; pp. 308–318.
19. Su, P.; Warg, F.; Chen, D. A simulation-aided approach to safety analysis of learning-enabled components in automated driving systems. In Proceedings of the 2023 IEEE 26th International Conference on Intelligent Transportation Systems (ITSC), Bilbao, Spain, 24–28 September 2023; IEEE: Piscateville, NJ, USA, 2023; pp. 6152–6157.
20. Zhou, J.; Cui, G.; Hu, S.; Zhang, Z.; Yang, C.; Liu, Z.; Wang, L.; Li, C.; Sun, M. Graph neural networks: A review of methods and applications. *AI Open* **2020**, *1*, 57–81. [CrossRef]

21. Liu, Z.; Zhou, J. *Introduction to Graph Neural Networks*; Springer Nature: Berlin/Heidelberg, Germany, 2022.
22. Yang, Z.; Cohen, W.; Salakhudinov, R. Revisiting semi-supervised learning with graph embeddings. In Proceedings of the International Conference on Machine Learning, PMLR, New York, NY, USA, 20–22 June 2016; pp. 40–48.
23. Kipf, T.N.; Welling, M. Semi-supervised classification with graph convolutional networks. *arXiv* **2016**, arXiv:1609.02907.
24. Berg, R.v.d.; Kipf, T.N.; Welling, M. Graph convolutional matrix completion. *arXiv* **2017**, arXiv:1706.02263.
25. Ahmad, T.; Jin, L.; Zhang, X.; Lai, S.; Tang, G.; Lin, L. Graph convolutional neural network for human action recognition: A comprehensive survey. *IEEE Trans. Artif. Intell.* **2021**, *2*, 128–145. [CrossRef]
26. Elias, P.; Sedmidubsky, J.; Zezula, P. Understanding the gap between 2D and 3D skeleton-based action recognition. In Proceedings of the 2019 IEEE International Symposium on Multimedia (ISM), San Diego, CA, USA, 9–11 December 2019; IEEE: Piscateville, NJ, USA, 2019; pp. 192–1923.
27. Liu, Y.; Zhang, H.; Xu, D.; He, K. Graph transformer network with temporal kernel attention for skeleton-based action recognition. *Knowl.-Based Syst.* **2022**, *240*, 108146. [CrossRef]
28. Li, B.; Li, X.; Zhang, Z.; Wu, F. Spatio-temporal graph routing for skeleton-based action recognition. In Proceedings of the AAAI Conference on Artificial Intelligence, Honolulu, HI, USA, 29–31 January 2019; Volume 33, pp. 8561–8568.
29. Tasnim, N.; Baek, J.H. Dynamic edge convolutional neural network for skeleton-based human action recognition. *Sensors* **2023**, *23*, 778. [CrossRef]
30. Liu, Y.; Zhang, H.; Li, Y.; He, K.; Xu, D. Skeleton-based human action recognition via large-kernel attention graph convolutional network. *IEEE Trans. Vis. Comput. Graph.* **2023**, *29*, 2575–2585. [CrossRef] [PubMed]
31. Wu, L.; Zhang, C.; Zou, Y. SpatioTemporal focus for skeleton-based action recognition. *Pattern Recognit.* **2023**, *136*, 109231. [CrossRef]
32. Krishna, R.; Zhu, Y.; Groth, O.; Johnson, J.; Hata, K.; Kravitz, J.; Chen, S.; Kalantidis, Y.; Li, L.J.; Shamma, D.A.; et al. Visual genome: Connecting language and vision using crowdsourced dense image annotations. *Int. J. Comput. Vis.* **2017**, *123*, 32–73. [CrossRef]
33. Yang, J.; Lu, J.; Lee, S.; Batra, D.; Parikh, D. Graph r-cnn for scene graph generation. In Proceedings of the European Conference on Computer Vision (ECCV), Munich, Germany, 8–14 September 2018; pp. 670–685.
34. Tang, K.; Niu, Y.; Huang, J.; Shi, J.; Zhang, H. Unbiased scene graph generation from biased training. In Proceedings of the IEEE/CVF Conference on Computer Vision and Pattern Recognition, Seattle, WA, USA, 13–19 June 2020; pp. 3716–3725.
35. Yu, S.Y.; Malawade, A.V.; Muthirayan, D.; Khargonekar, P.P.; Al Faruque, M.A. Scene-graph augmented data-driven risk assessment of autonomous vehicle decisions. *IEEE Trans. Intell. Transp. Syst.* **2021**, *23*, 7941–7951. [CrossRef]
36. Jin, K.; Wang, H.; Liu, C.; Zhai, Y.; Tang, L. Graph neural network based relation learning for abnormal perception information detection in self-driving scenarios. In Proceedings of the 2022 International Conference on Robotics and Automation (ICRA), Philadelphia, PA, USA, 23–27 May 2022; IEEE: Piscateville, NJ, USA, 2022; pp. 8943–8949.
37. Mylavarapu, S.; Sandhu, M.; Vijayan, P.; Krishna, K.M.; Ravindran, B.; Namboodiri, A. Understanding dynamic scenes using graph convolution networks. In Proceedings of the 2020 IEEE/RSJ International Conference on Intelligent Robots and Systems (IROS), Las Vegas, NV, USA, 25–29 October 2020; IEEE: Piscateville, NJ, USA, 2020; pp. 8279–8286.
38. Chang, Y.; Zhou, W.; Cai, H.; Fan, W.; Hu, L.; Wen, J. Meta-relation assisted knowledge-aware coupled graph neural network for recommendation. *Inf. Process. Manag.* **2023**, *60*, 103353. [CrossRef]
39. Wang, H.; Zhang, F.; Zhang, M.; Leskovec, J.; Zhao, M.; Li, W.; Wang, Z. Knowledge-aware graph neural networks with label smoothness regularization for recommender systems. In Proceedings of the 25th International Conference on Knowledge Discovery & Data Mining, Anchorage, AK, USA, 4–8 August 2019; pp. 968–977.
40. Gilmer, J.; Schoenholz, S.S.; Riley, P.F.; Vinyals, O.; Dahl, G.E. Neural message passing for quantum chemistry. In Proceedings of the International conference on machine learning. PMLR, Sydney, Australia, 6–11 August 2017; pp. 1263–1272.
41. Girshick, R. Fast r-cnn. In Proceedings of the IEEE International Conference on Computer Vision, Santiago, Chile, 7–13 December 2015; pp. 1440–1448.
42. Hamilton, W.; Ying, Z.; Leskovec, J. Inductive representation learning on large graphs. In *Advances in Neural Information Processing Systems*; Curran Associates, Inc.: Red Hook, NY, USA, 2017.
43. Zhu, Z.; Su, P.; Zhong, S.; Huang, J.; Ottikkutti, S.; Tahmasebi, K.N.; Zou, Z.; Zheng, L.; Chen, D. Using a vae-som architecture for anomaly detection of flexible sensors in limb prosthesis. *J. Ind. Inf. Integr.* **2023**, *35*, 100490. [CrossRef]
44. Su, P.; Lu, Z.; Chen, D. Combining Self-Organizing Map with Reinforcement Learning for Multivariate Time Series Anomaly Detection. In Proceedings of the 2023 IEEE International Conference on Systems, Man, and Cybernetics (SMC), Hyatt Regency Maui, HI, USA, 1–4 October 2023; IEEE: Piscateville, NJ, USA, 2023; pp. 1964–1969.

Disclaimer/Publisher's Note: The statements, opinions and data contained in all publications are solely those of the individual author(s) and contributor(s) and not of MDPI and/or the editor(s). MDPI and/or the editor(s) disclaim responsibility for any injury to people or property resulting from any ideas, methods, instructions or products referred to in the content.

Article

Biosensor-Driven IoT Wearables for Accurate Body Motion Tracking and Localization

Nouf Abdullah Almujally [1], Danyal Khan [2], Naif Al Mudawi [3], Mohammed Alonazi [4], Abdulwahab Alazeb [3], Asaad Algarni [5], Ahmad Jalal [2,*] and Hui Liu [6,*]

1. Department of Information Systems, College of Computer and Information Sciences, Princess Nourah bint Abdulrahman University, P.O. Box 84428, Riyadh 11671, Saudi Arabia; naalmujaly@pnu.edu.sa
2. Faculty of Computing ad AI, Air University, E-9, Islamabad 44000, Pakistan; 211651@students.au.edu.pk
3. Department of Computer Science, College of Computer Science and Information System, Najran University, Najran 55461, Saudi Arabia; nalmdwi@nu.edu.sa (N.A.M.); afalzeb@nu.edu.sa (A.A.)
4. Department of Information Systems, College of Computer Engineering and Sciences, Prince Sattam bin Abdulaziz University, Al-Kharj 16273, Saudi Arabia; mn.alonazi@psau.edu.sa
5. Department of Computer Sciences, Faculty of Computing and Information Technology, Northern Border University, Rafha 91911, Saudi Arabia; asad.algani@nbu.edu.sa
6. Cognitive Systems Lab, University of Bremen, 28359 Bremen, Germany
* Correspondence: ahmadjalal@mail.au.edu.pk (A.J.); hui.liu@uni-bremen.de (H.L.)

Abstract: The domain of human locomotion identification through smartphone sensors is witnessing rapid expansion within the realm of research. This domain boasts significant potential across various sectors, including healthcare, sports, security systems, home automation, and real-time location tracking. Despite the considerable volume of existing research, the greater portion of it has primarily concentrated on locomotion activities. Comparatively less emphasis has been placed on the recognition of human localization patterns. In the current study, we introduce a system by facilitating the recognition of both human physical and location-based patterns. This system utilizes the capabilities of smartphone sensors to achieve its objectives. Our goal is to develop a system that can accurately identify different human physical and localization activities, such as walking, running, jumping, indoor, and outdoor activities. To achieve this, we perform preprocessing on the raw sensor data using a Butterworth filter for inertial sensors and a Median Filter for Global Positioning System (GPS) and then applying Hamming windowing techniques to segment the filtered data. We then extract features from the raw inertial and GPS sensors and select relevant features using the variance threshold feature selection method. The extrasensory dataset exhibits an imbalanced number of samples for certain activities. To address this issue, the permutation-based data augmentation technique is employed. The augmented features are optimized using the Yeo–Johnson power transformation algorithm before being sent to a multi-layer perceptron for classification. We evaluate our system using the K-fold cross-validation technique. The datasets used in this study are the Extrasensory and Sussex Huawei Locomotion (SHL), which contain both physical and localization activities. Our experiments demonstrate that our system achieves high accuracy with 96% and 94% over Extrasensory and SHL in physical activities and 94% and 91% over Extrasensory and SHL in the location-based activities, outperforming previous state-of-the-art methods in recognizing both types of activities.

Keywords: machine learning; segmentation; feature fusion; multi-layer perceptron; Yeo–Johnson

Citation: Almujally, N.A.; Khan, D.; Al Mudawi, N.; Alonazi, M.; Alazeb, A.; Algarni, A.; Jalal, A.; Liu, H. Biosensor-Driven IoT Wearables for Accurate Body Motion Tracking and Localization. *Sensors* **2024**, *24*, 3032. https://doi.org/10.3390/s24103032

Received: 20 March 2024
Revised: 26 April 2024
Accepted: 29 April 2024
Published: 10 May 2024

Copyright: © 2024 by the authors. Licensee MDPI, Basel, Switzerland. This article is an open access article distributed under the terms and conditions of the Creative Commons Attribution (CC BY) license (https://creativecommons.org/licenses/by/4.0/).

1. Introduction

Human locomotion activity recognition is a rapidly emerging field that analyzes and classifies many types of physical activity and locomotion activity using smartphone sensors [1]. Smartphones are an ideal platform for this research because they are widely available, equipped with a variety of sensors [2], and are commonly carried by individuals. In modern smartphones, a rich array of sensors including accelerometers, gyroscopes, magnetometers, GPS, light sensors, barometers, and microphones, are utilized for comprehensive human locomotion activity recognition. These sensors allow for the accurate detection and analysis of user movements and environmental interactions, enhancing the performance of activity recognition systems. Human activity recognition using smartphone sensors has a wide range of applications in various fields. Some of the most common applications include fitness and health monitoring, and elderly care (to monitor the movements of elderly individuals and detect any signs of falls or other accidents [3]. This can provide peace of mind for caregivers and family members and can also be used to trigger an alert if assistance is needed), transportation, environmental monitoring, sports, and marketing (to track and analyze the movements of individuals in different retail environments. This information can be used to better understand consumer behavior and make decisions about product placement and advertising), safety, and security [4].

Researchers have used different machine-learning approaches to recognize human activity through smartphone sensors. These approaches have several advantages, including the ability to recognize complex patterns and classify multiple activities simultaneously. Additionally, machine learning algorithms can improve over time as they are fed more data, making them more accurate and reliable [5,6]. However, there are also some disadvantages to using machine learning for activity recognition. One of the main disadvantages is the requirement for large amounts of labeled data to train the algorithms effectively. This can be time-consuming and expensive, especially when dealing with numerous activities or users. Another potential issue is the sensitivity of the algorithms to changes in sensor placement or sensor data quality. If the sensor is not positioned correctly or the data are noisy, the accuracy of the machine-learning algorithm can be significantly reduced [7].

The field of HAR is impeded by a variety of challenges, including sensor heterogeneity across devices [8], which complicates the development of universal application. Different smartphone models are equipped with various types of sensors that have differing specifications and capabilities. This variation complicates the development of universal applications that perform consistently across all devices. Another big challenge researchers face is that of noise in the raw sensor data [9]. Smartphone sensors frequently capture data contaminated with noise, which can significantly affect accuracy. The precision of sensors varies widely between devices, often depending on the hardware quality and sensor calibration. Additionally, variations in sensor sampling frequencies [10], mean that the rate at which sensor sample data can fluctuate is influenced by other processes running on the device. This inconsistency can lead to challenges in capturing the real-time, high-resolution data necessary for precise motion recognition. Similarly, sensor characteristics can change over time due to aging hardware or software updates, leading to a data drift. This phenomenon can degrade the performance of motion recognition algorithms that were trained on data from newer or different sensors. Continuous sensor data collection is resource-intensive, consuming significant battery life and processing power. Managing these resources efficiently while maintaining accurate motion detection is a major challenge. Moreover, different users may carry their smartphones in various positions (e.g., pocket, hand, or bag) [11], leading to vastly different data profiles. Algorithms must be robust enough to handle these variations to ensure accurate motion recognition. Moving forward, another challenge is the privacy of the user [12], collecting and analyzing sensor data increases privacy concerns, as such data can inadvertently reveal sensitive information about a user's location and activities. Ensuring data privacy and security while collecting and processing sensor data is critical. Lastly, the complexity of human activities and the limited spatial coverage of sensors add to the difficulty of capturing a comprehensive range of motions, highlighting the multifaceted nature of these technological hurdles. In this study, we developed a system that recognizes human movements along with location, and ultimately, can provide valuable insights into an individual's physical and localization activity levels. We processed the raw sensor data for physical and location-based activity separately. In the first stage, we denoised the data using the Butterworth filter [13] and Median filter [14]. In the second stage, we segmented the long sequence signal data into small pieces using the Hamming windowing technique [15]. After that, features were extracted. The Variance threshold feature selection method [16,17] is used for feature selection, the selected feature vectors are balanced using the data augmentation technique, and after that, the augmented data are well optimized before classification using the Yeo–Johnson power transformation technique. Finally, physical and localization activity classification is performed by a multi-layer perceptron (MLP). The contribution of this research is described below:

- Implemented separate denoising filters for inertial and GPS sensors, significantly enhancing data cleanliness and accuracy.
- Developed a robust methodology for concurrent feature extraction from human locomotion and localization data, improving processing efficiency and reliability.
- Established dedicated processing streams for localization and locomotion activities, allowing for more precise activity recognition by reducing computational interference.
- Applied a novel data augmentation technique to substantially increase the dataset size of activity samples, enhancing the robustness and generalizability of the recognition algorithms.
- Utilized an advanced feature optimization algorithm to adjust the feature vector distribution towards normality, significantly improving the accuracy of activity recognition.

The research has been divided into the following sections: Section 2 discusses some literature review in the field of human activity recognition, and then material and methods, including noise removal, signal windowing and segmentation, feature extraction, feature selection, and optimization presented in Section 3. Section 5 analyzes the computational complexity of the proposed system, Section 6 presents a discussion and limitations. Finally, in Section 7, the research study is concluded.

2. Related Work

Scientists have explored different approaches to studying how to analyze human motion, both inside and outside [18]. These approaches can be divided into two main groups: ones that depend on motion-based analysis and ones that depend on vision-based analysis [19,20], whilst some are based on sensors-based methods that make use of sensors such as accelerometers, gyroscopes, GPS, light, mic, magnetometers, mechanomyography, ECG (electrocardiogram), EMG (electromyogram), and geomagnetic sensors. Vision-based methods, on the other hand, use cameras such as Microsoft Kinect [21], Intel Real sense, video cameras, and dynamic vision sensors [22].

The related work by Hsu et al. [23] involved a method of human activity recognition that utilized a pair of wearable inertial sensors. On the subject's wrist, one sensor was mounted, while the other was mounted on the ankle. The collected sensor data, including accelerations and angular velocities, was wirelessly sent to a central computer for processing. Using a nonparametric weighted feature extraction algorithm and principal component analysis, the system could differentiate between various activities. While the method boasted the high points of portability and wireless data transmission, it was limited by the use of only two sensors, potentially missing out on capturing the full spectrum of human movement and requiring a dependable wireless connection to function effectively. To improve upon this, the proposed solution in the research includes the deployment of additional sensors on various body parts such as the torso, backpack, hand, and pocket to provide a more comprehensive capture of human motion. Additionally, integrating sensors embedded within smartphones eliminates the need for a continuous wireless connection, facilitating the recognition of human activities and locations with enhanced reliability and context awareness. In the research by A-Basset et al. [24], a novel approach to human activity recognition is introduced where sensor data is treated as visual information. Their method considers human activity recognition as analogous to image classification, converting sensor data into an RGB image format for processing. This enables the use of multiscale hierarchical feature extraction and a channel-wise attention mechanism to classify activities. The strength of this system lies in its innovative interpretation of sensor data, which allows for the application of image classification techniques. However, its reliance on small datasets for training raises questions about how well it can be generalized to real-world situations. The uncertainty regarding the computational and space complexity also poses concerns about the system's scalability. The proposed enhancement of this system involves training on larger and more diverse datasets to enhance the system's capacity to generalize across various scenarios. By ensuring that the system is robust when handling larger datasets, the solution seeks to maintain computational efficiency while scaling up to more complex applications. Konak et al.'s [25] method for evaluating human activity recognition performance employs accelerometer data, which is categorized into three distinct classes based on motion, orientation, and rotation. The system utilizes these categories either individually or in combination to assess activity recognition performance and employs a variety of classification techniques, such as decision trees, naive Bayes, and random forests. The primary limitation of this method is its training on a dataset derived from only 10 subjects, raising concerns about its generalizability to a broader population. Additionally, the study relies on common machine learning classifiers, which may not be as effective as more advanced models. In contrast, the proposed model in the research under discussion utilizes the Extrasensory and Huawei dataset, which includes data from more subjects, thus providing a more robust and generalizable system that achieves state-of-the-art performance. The research by Chetty et al. [26] presents an innovative data analytic method for human activity recognition using smartphone inertial sensors, utilizing machine learning classifiers like random forests, ensemble learning, and lazy learning. The system distinguishes itself through its feature ranking process informed by information theory, which optimizes for the most relevant features in activity recognition. Despite the innovative approach, the system's reliance on a single dataset for training is its primary limitation. This constraint could hinder the model's ability to generalize to unobserved scenarios and potentially lead to degraded performance in real-world applications. The proposed solution to these limitations involves a system trained on two benchmark datasets that encompass a wider variety of activities. It includes the Extrasensory dataset, which is notable for being collected in uncontrolled, real-world environments without restrictions on participant behavior. This approach is intended to enhance the system's reliability and applicability to a broader range of real-life situations, thereby making it a more robust solution for activity recognition.

The study by Ehtisham-ul-Haq et al. [27] introduced an innovative context recognition framework that interprets human activity by leveraging physical activity recognition (PAR) and learning patterns from various behavioral scenarios. Their system correlated fourteen different behaviors, including phone positions with five daily living activities, using random forest and other machine learning classifiers for performance evaluation. The high points of this method are its use of human activity recognition to infer context and its integration of additional information such as the subject's location and secondary activities. Nonetheless, the system's primary reliance on accelerometer data makes it less adept at complex activities, and it lacks more comprehensive data sources like GPS and microphone inputs for enhanced location estimation. The proposed enhancement to this framework includes a more integrated sensor approach, utilizing not only the smartphone's accelerometer, magnetometer, and gyroscope, but also the smartwatch's accelerometer and compass, along with smartphone GPS and microphone data. This integration promises increased robustness and accuracy in activity recognition and localization.

2.1. Activity Recognition Using Inertial Sensors

Smartphone sensors are more popular for activity recognition. By using these sensors, human activity can be easily detected. The most significant feature of the smartphone is its portability, which means it can be carried easily in any place. In [28], different supervised machine learning algorithms were used to classify human activity. The classification precision was tested using 5-fold cross-validation. They achieved a good accuracy rate for all classifiers. Ref. [29] presented trends in human activity recognition. The survey discussed different solutions proposed for each phase in human activity classification, i.e., preprocessing, windowing and segmentation, feature extraction, and classification. All the solutions are analyzed, and the weaknesses and strengths are described. The paper also presented how to evaluate the quality of a classifier. In [30], a new method was proposed for recognizing human activity with multi-class SVM using integer parameters. The method used in the research consumes less memory, processor time, and power consumption. The authors in [31] analyzed the performance of the two classifiers, that is, KNN and clustered KNN. The classifiers were evaluated with an online activity recognition system using the Android operating system. The system supports online training and classification by collecting data from one sensor called an accelerometer. They started with KNN and then clustered it. The main rationale for utilizing clustered was to reduce the computational complexity of KNN using clusters. The major goal of the article was to examine the performance of algorithms on the phone with limited training data and memory.

2.2. Activity Recognition Using Computer Vision and Image Processing Techniques

As previously noted, identifying human activity via smartphone is a convenient method because the smartphone is a portable device that can be readily carried anywhere. The use of an RGB camera for activity recognition [32,33] has some limits and constraints. To monitor a person's activity with a camera, for example, the individual must be within range of the camera's eye. Nighttime (changing lighting conditions) is the second most prevalent challenge that researchers face while tracking human activities through a webcam. However, advancements in multimedia tools have mitigated these issues to some extent. To recognize human movement from 2D/3D films and photos, many computer-vision and image-processing algorithms have been utilized [34,35]. Researchers can recognize human activities more easily by employing techniques such as segmentation [36], filtering [37], saliency map detection [38], skeleton extraction [39], and so on. The work described in [40] investigated human activity recognition using a depth camera. The camera first acquired the skeleton data, and then several spatial and temporal features were retrieved. The CNN (Convolutional Neural Network) algorithm was employed to classify the activities. The issue with highlighting an in-depth camera is that noise can occur, leading to misclassification.

3. Proposed System Methodology

Data were collected from various raw sensors. The data were denoised in the first step using the Butterworth filter [40]. The Hamming windowing and segmentation approach [41] is then applied. During the third step, we worked with the data from the inertial and GPS sensors. We picked out various features for each of them. To determine the significance of features, we employed the Variance Threshold for feature selection. We noted that certain activities in the Extrasensory dataset had a limited number of samples. To address this issue, we applied data augmentation and subsequently optimized using the Yeo–Johnson power transformation technique before conducting activity recognition. Finally, the activity recognition was performed by the MLP. The flow diagram of the suggested human physical and localization activity model is shown in Figure 1.

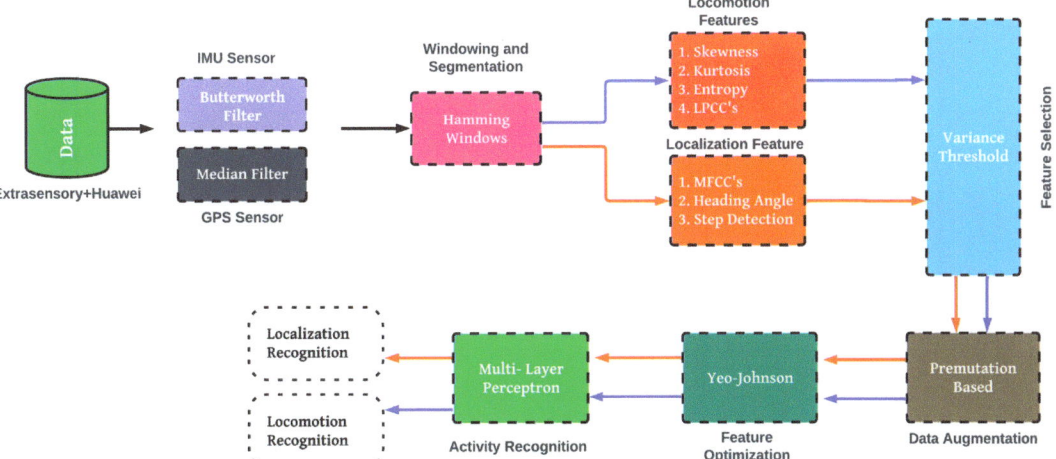

Figure 1. The proposed system architecture.

3.1. Signal Denoising

There is a risk of noise during data collection. Noise is the undesirable portion of data that we do not need to process. Unwanted data processing lengthens and complicates model training. It also reduces the learning model's performance. Therefore, noise removal is crucial in data preprocessing. For this reason, we employed a noise-removal filter.

To get rid of the unwanted disruptions that can happen when collecting data, we applied a low-pass Butterworth filter [42–45] to the inertial sensors. This filter is used in signal processing and aims to make the frequency response as even as possible in the part where it passes signals through. That is why it is called the maximally flat magnitude filter. Equation (1) depicts the general frequency response of the Butterworth filter.

$$H(jw) = \frac{1}{\sqrt{1 + \left(\frac{\omega}{\omega_c}\right)^{2n}}} \quad (1)$$

where n is the order of the filter, ω is the passband frequency (also known as the operational frequency), ω_c is the filter's cut-off frequency, and j is the imaginary unit, used to denote the complex frequency. In Figure 2, the original vs. filtered signal for the inertial sensor is shown. Similarly, for processing our GPS data and to enhance its clarity, we used the median filter [46–49] a robust nonlinear digital filtering technique. The median filter operates by moving a sliding window across each data point in our GPS sequence. Within this window, the data values are arranged in ascending order. The central value, or median, of this sorted list, is then used to replace the current data point. Mathematically, for a given signal, S, and a window of size n, at each point xi in the signal, we consider:

$$W = \{(xi - (n-1))/2 - (xi - (n-1))/(2+1), \ldots (xi + (n-1))/(2-1) + (xi + (n-1))/2\} \quad (2)$$

The median of this set W becomes the new value at xi in the filtered signal. In our experiment on the GPS data, we selected a window size of 3. The selected window size ensures that the filter assesses each data point while considering itself and one neighboring point on either side. This particular size strikes a balance by being large enough to effectively suppress noise, and yet, be compact enough to preserve important details and transitions in the GPS data. But it is important to note that there was not enough noise in the GPS signal as inertial sensors.

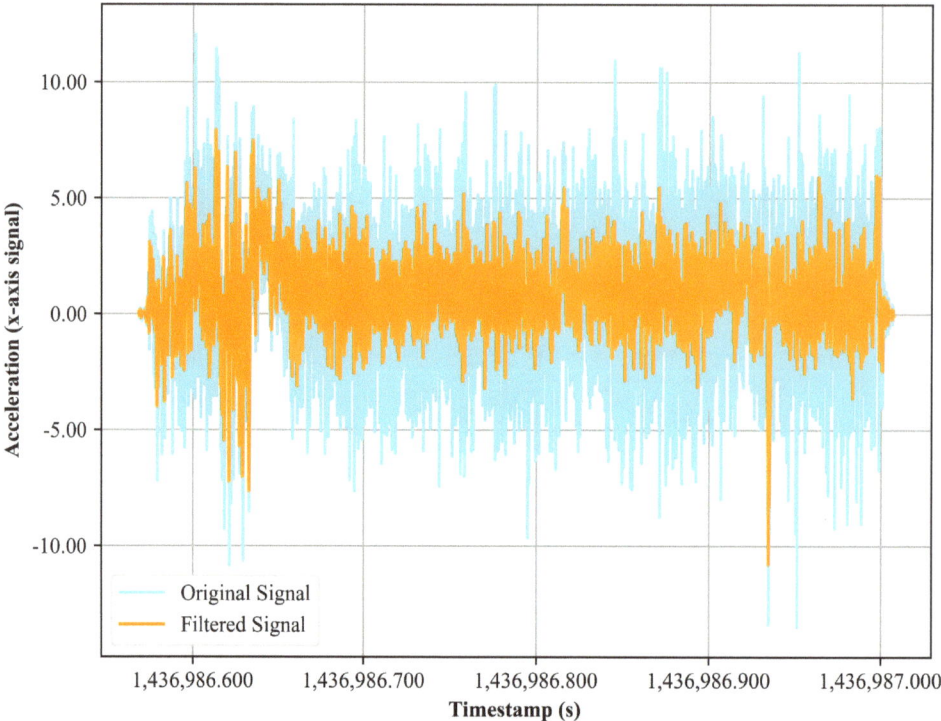

Figure 2. The accelerometer x axis noisy vs. filtered signal.

3.2. Signal Windowing and Segmentation

Segmentation is an important concept used in signal processing. The concept of windowing and segmentation [50–54] involves dividing signals into smaller windows instead of processing complete or long sequences. The advantage of windowing is that it allows for easier data processing, reducing complexity and processing time. This makes it more manageable for machine learning or deep learning models to process. We turned to the Hamming windows technique to modulate the signal. Hamming windows, known for their capacity to reduce spectral leakage [55] during frequency domain operations like the Fourier Transform, effectively tackle the side effects that often arise during such analyses. The principle behind the Hamming window is a simple point-wise multiplication of the signal with the window function, which curtails the signal values at both the start and end of a segment. This modulation ensures a minimized side lobe in the frequency response, which is crucial for accurate spectral analyses.

Mathematically, the Hamming window is represented in Equation (3).

$$W(n) = 0.54 - 0.46 cos\left(\frac{2\pi n}{N-1}\right) \quad (3)$$

where $w(n)$ represents the window function, N signifies the total points in the window, and n spans from 0 to $N-1$. We utilized a window size of 5 s [56,57]. After generating the Hamming window values based on the aforementioned formula, we multiplied each point in our data segments with its corresponding Hamming value. In Figure 3, we visualized the results through distinct line plots, with each of the five windows represented in a unique color.

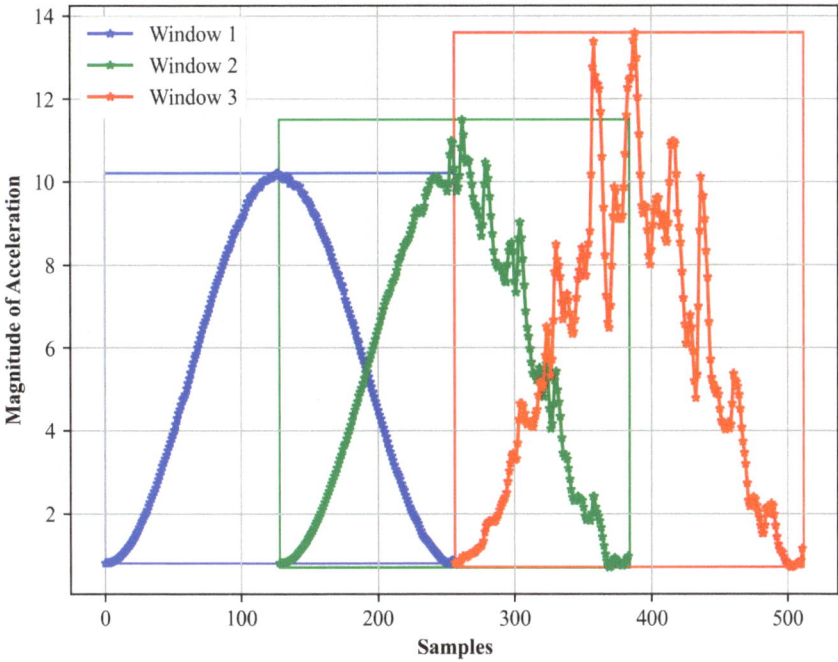

Figure 3. Hamming windows first 3 windows for accelerometer data.

3.3. Feature Extraction

In this section, we listed all the feature lists used in the study specifically aligned with each type of sensor data. We extract separate features for the physical and localization activities. The subsequent section presents each section comprehensively.

3.3.1. Feature Extraction for Physical Activity

For physical recognition, we processed data from three sensors: magnetometer, gyroscope, and accelerometer [58–61]. Various statistical features were extracted.

Shannon Entropy

Shannon entropy is first extracted, as seen in Figure 4. Shannon entropy [62,63] measures the unpredictability [64–66] or randomness of a signal. Mathematically, it can be calculated as:

$$H(P) = -\sum_i p_i log_2(p_i) \quad (4)$$

where p_i represents the probability of occurrence of the different outcomes.

Linear Prediction Cepstral Coefficients (LPCCs)

The extraction of LPCCs from accelerometer signals [67], the primary step involves the application of linear predictive analysis (LPA). Given $s(n)$ as the accelerometer signal, it can be modeled by the relation

$$s(n) = \sum_{k=1}^{p} a_k s(n-k) + e(n) \quad (5)$$

where p represents the order of the linear prediction, a_k are the linear prediction coefficients, and $e(n)$ denotes the prediction error. The linear prediction coefficients, a_k, derived by minimizing the prediction's mean square error, were commonly achieved using the Levinson–Durbin algorithm. After obtaining these coefficients, the transition to cepstral coefficients begins. This conversion entails taking the inverse Fourier transform of the logarithm of the signal's power spectrum. Specifically, the cepstral coefficients are determined through a recurrence relation, where the initial coefficient is the logarithm of the zeroth linear prediction coefficient, and subsequent coefficients are derived using the linear prediction coefficients and previous cepstral coefficients. The LPCCs calculated for different activities can be seen in Figure 4.

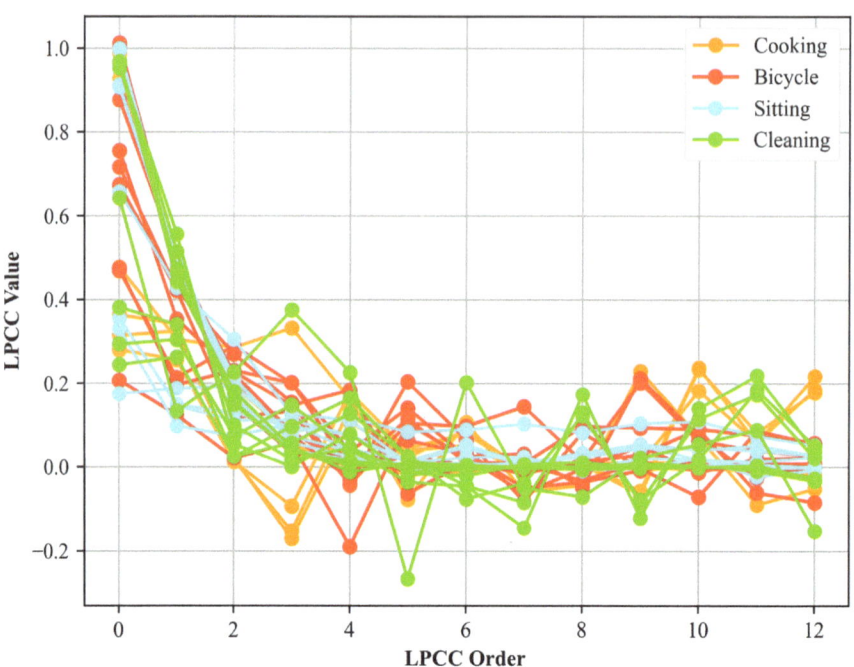

Figure 4. LPCCs are calculated for different activities.

Skewness

In the context of signal processing for accelerometer data, skewness is a crucial statistical measure that captures the asymmetry [68–70] of the signal distribution. To compute the skewness of an accelerometer signal $s(n)$, where n represents the discrete time index, we first calculate the mean (μ) and standard deviation (σ) of the signal. Following this, the skewness (S) is obtained using the formula.

$$S = \frac{1}{X}\sum_{x=1}^{X}\left(\frac{s(x)-\mu}{\sigma}\right)^3 \quad (6)$$

Here, X is the total number of data points in the signal. The formula essentially quantifies the degree to which the signal's distribution deviates from a normal distribution. A skewness value of zero signifies a symmetric distribution. Positive skewness indicates a distribution with an asymmetric tail extending towards more positive values, while negative skewness indicates a tail extending towards more negative values. Computing the skewness of an accelerometer signal can provide insights into the distribution characteristics of the signal. Figure 5 show the skewness for different locomotion activities.

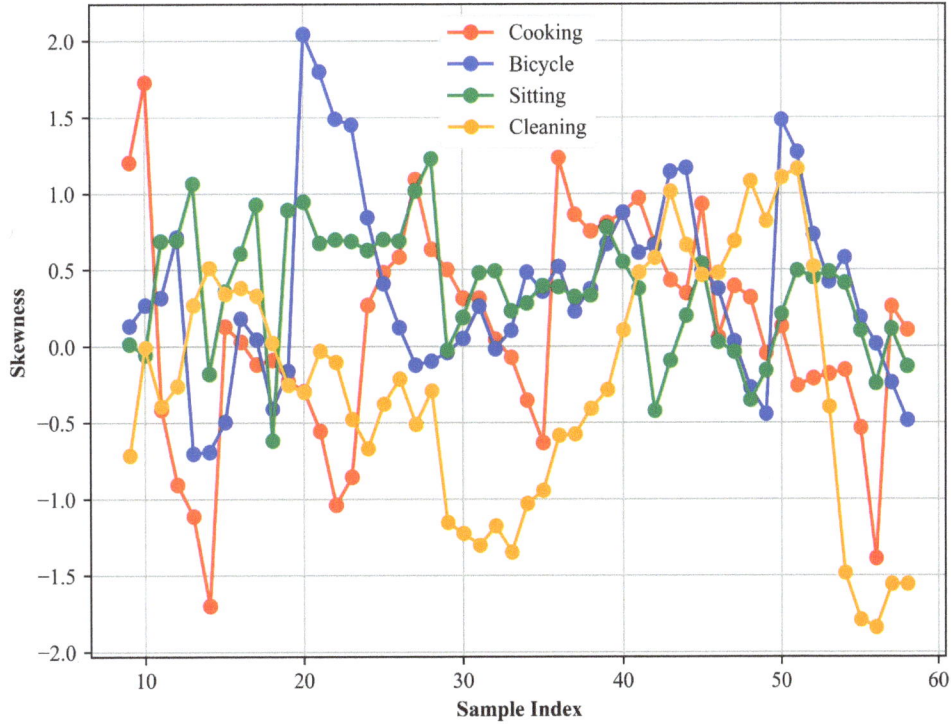

Figure 5. Skewness is calculated for different activities.

Kurtosis

Kurtosis is a statistical measure used to describe the distribution of observed data around the mean. Specifically, it quantifies the probability distribution of a real-valued random variable. In the context of signals, kurtosis [71] can be particularly informative as it can capture the sharpness of the distribution's peak and the heaviness of its tails. This, in turn, can indicate the occurrence of abrupt or high-magnitude changes in the acceleration data, which may be characteristic of specific activities or movements. The formula for kurtosis is given by:

$$\text{Kurtosis}(x) = F\left[\left(\frac{x-\mu}{\sigma}\right)^4\right] - 3 \quad (7)$$

where F denotes the expected value, μ is the mean, and σ is the standard deviation. A kurtosis value greater than zero indicates that the distribution has heavier tails and a sharper peak compared to a normal distribution. Conversely, a value that is less than zero suggests that the distribution has lighter tails. In our analysis, we extracted the kurtosis from the accelerometer signals corresponding to different activities. This enabled us to discern and distinguish the nature of signal distributions for activities such as cooking, sitting, or cleaning. For instance, a sudden or vigorous activity might exhibit a distribution with a higher kurtosis value, indicating rapid changes in acceleration, whereas more steady or uniform activities might have a lower kurtosis value. In Figure 6, the kurtosis plot is presented.

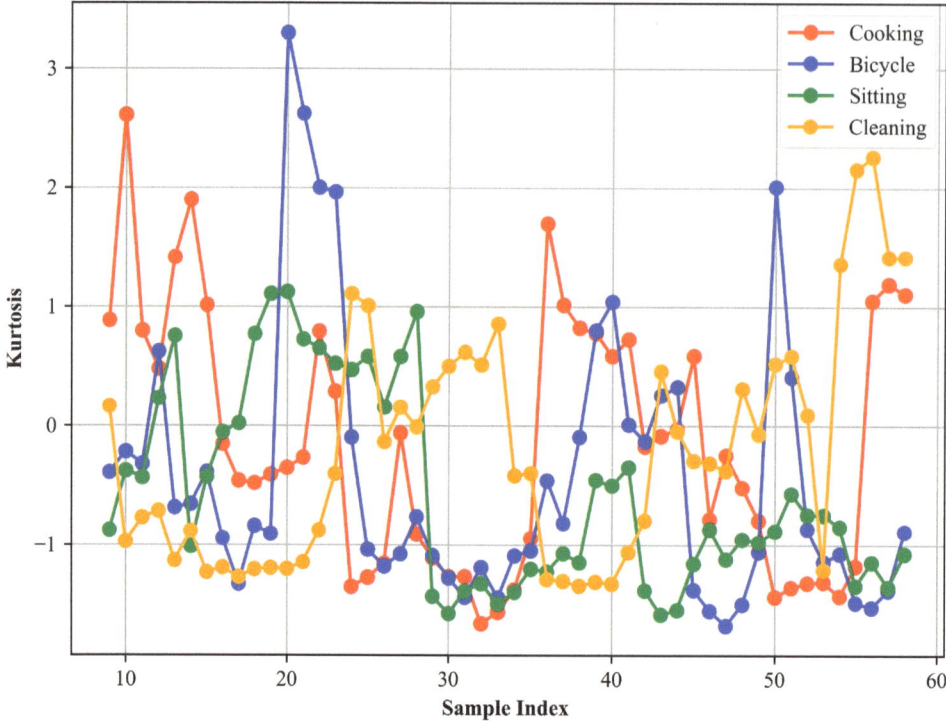

Figure 6. Kurtosis is calculated for different activities.

3.3.2. Feature Extraction for Localization Activity

For localization activities, we try to capture the complicated movement patterns by extracting a set of distinct features. We extracted the Total Distance, Average Speed, Maximum Displacement, Direction Change features, heading angles, skewness, kurtosis, step detection, and MFCCs [72].

Mel-Frequency Cepstral Coefficients (MFCCs)

In human localization using audio signals, MFCCs [73] play a pivotal role in determining the direction, proximity, and potential movement patterns. We begin with the pre-emphasis of the signal $s(n)$, accentuate its high frequencies, a step mathematically represented as:

$$\acute{s} = s(n) - \alpha \times s(n-1) \qquad (8)$$

where α is commonly set to 0.97. This amplification aids in emphasizing the subtle changes in audio signals that may result from human movement or orientation changes. The signal is then split into overlapping frames to analyze temporal variations, and each frame is windowed, often using the Hamming window, to mitigate spectral leakage. The short-time Fourier transform (STFT) offers a frequency domain representation of each frame, and its squared magnitude delivers the power spectrum. As human auditory perception is nonlinear, this spectrum is translated to the Mel scale using triangular filters. This transformation is governed by:

$$m = 2595 \times log_{10}\left(1 + \frac{f}{700}\right) \qquad (9)$$

The mathematics above ensures that the extracted features align with human auditory perception. The logarithm of this Mel spectrum undergoes the Discrete Cosine Transform

(DCT), producing the MFCCs. By retaining only the initial coefficients, one captures the essential spectral shape, pivotal for discerning sound characteristics that aid in human localization. The MFCCs calculated for localization activities can be seen in Figure 7.

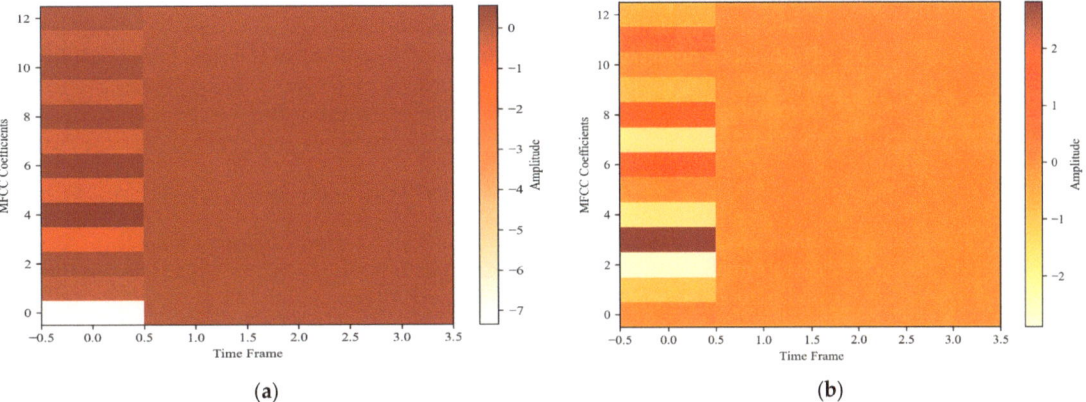

Figure 7. MFCCs are calculated for (**a**) indoor and (**b**) outdoor activity.

Step Detection

To understand the steps [74–78] from accelerometer data, we harness the magnitude of the acceleration vector. This magnitude is essentially a scalar representation of the combined accelerations in the x, y, and z axes. Mathematically, given the acceleration values a_x, a_y, a_z in the respective axes, the magnitude M is calculated using the formula:

$$M = \sqrt{a_x^2 + a_y^2 + a_z^2} \tag{10}$$

Once we have the magnitude of acceleration, the periodic nature of indoor or outdoor environments produces recognizable peaks in this signal. Each peak can correspond to a step, and by detecting these peaks, we can estimate the number of steps taken. The peak detection is anchored on identifying local maxima in the magnitude signal that stand out from their surroundings. The step detected [79] for indoor and outdoor activities can be seen in Figure 8.

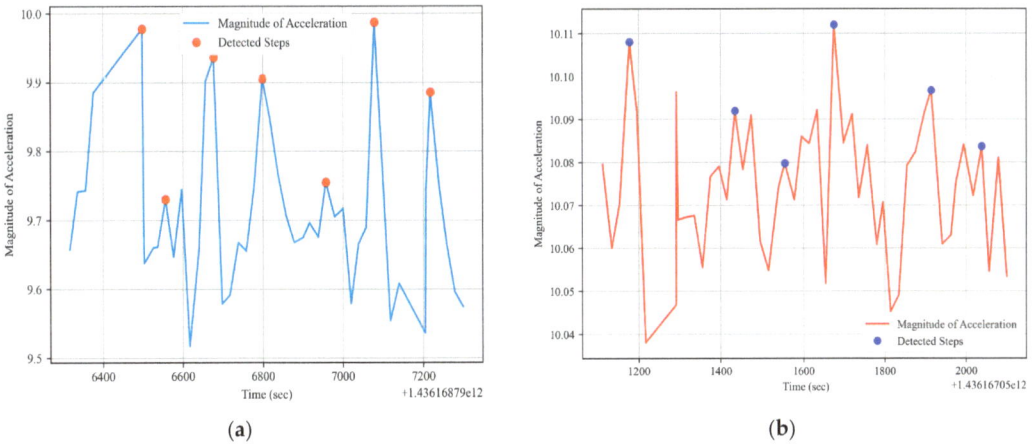

Figure 8. Steps detected for (**a**) indoor and (**b**) outdoor activity.

Heading Angle

The heading angle [80,81] plays a pivotal role in determining the orientation or direction a person is facing. As humans navigate through environments, whether they are indoor spaces like shopping malls or outdoor terrains like city streets, understanding their heading is crucial for applications ranging from pedestrian navigation systems to augmented reality. The heading angle, often termed the azimuth, denotes the angle between the North direction (assuming a geomagnetic North) and the projection of the magnetometer's reading onto the ground plane. Mathematically, the heading angle θ can be calculated using the magnetic field components A and B as:

$$\theta = arctan2(B, A) \tag{11}$$

where *arctan2* is the two-argument arctangent function, ensuring the angle [82] lies in the correct quadrant and providing a result in the range [−180, 180]. The heading for indoor and outdoor activity can be seen in Figure 9.

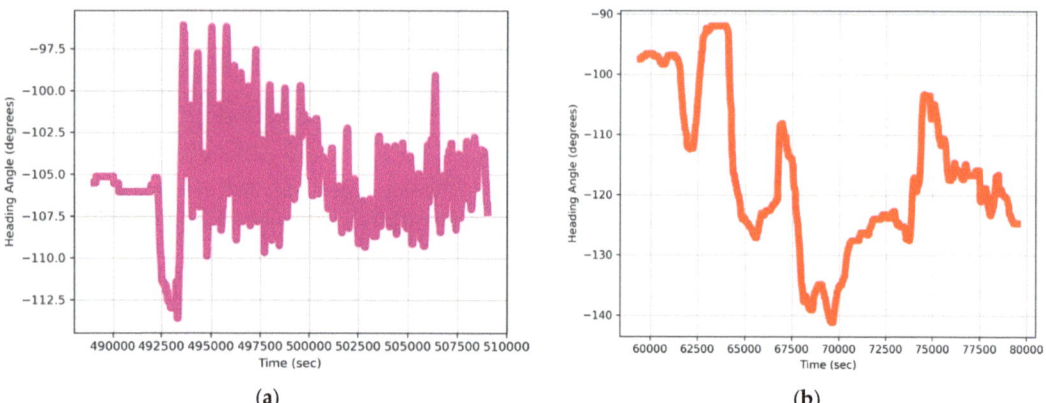

Figure 9. Heading angle calculated for (**a**) indoor and (**b**) outdoor activity.

3.4. Feature Selection Using Variance Threshold

In this experiment, we applied the variance threshold method [83–86] to the feature vector. The goal was to identify and retain only those features that showed significant variation across all features, ensuring that our dataset was as informative as possible. The feature mean, standard deviation, and total distance, exhibited a relatively low variance and were therefore removed, while all other features were retained. Variance Threshold is a simple filter-based feature selection method. It removes all features whose variance across all samples does not meet a specific threshold. The rationale behind this approach is straightforward: features that do not vary much within themselves are less likely to be informative. The variance of a feature X is given by:

$$Var(x) = \frac{1}{n}\Sigma_{i=1}^{n}(xi - \bar{x})^2 \tag{12}$$

where: n is the number of samples; xi is the value of the feature X for the ith sample; and $\bar{x}$ is the mean value of the feature X across all samples. In the context of the variance threshold method, we compare the variance of each feature against a pre-defined threshold. Features with variances below this threshold are considered non-informative and are removed. The algorithm working of the variance threshold is shown in Algorithm 1.

Algorithm 1: Variance Threshold Feature Selection
***
1: Input: Dataset D with m features: $f1, f2, \ldots, fm$.
Variance threshold value τ.
k: Desired number of features to select.
2: Output:
A subset of features whose variance is above
3: Initialization:
Create an empty list R to store the retained features
4: Feature Selection:
    For each feature f_i in (D). Compute the variance v_i of f_i
        Add f_i to the list R
    end for
5: Return:
Return the list R as the subset of features with variance above τ.
6: End
***

3.5. Feature Optimization via Yeo–Johnson Power Transformation

We perform feature optimization before moving on to classification. In simple terms, feature optimization makes the feature clearer to the model. We opted to optimize the specified feature vector after selecting relevant features for the model using the Variance Threshold. For this purpose, we utilized the Yeo–Johnson power transformation method. The Yeo–Johnson power transformation [87] is a statistical method used to transform non-normally distributed data into a normal or Gaussian-like distribution. This method is highly valuable in machine learning, as many algorithms assume that the data follow a normal distribution. By transforming the data using the Yeo–Johnson method, we can enhance the performance of these algorithms and make the results more reliable. The method uses a power transformation to map the original data into a new distribution, with the power being a parameter that is estimated from the data. Mathematically, the Yeo–Johnson optimization is given in Equation (13).

$$\psi(x) = \begin{cases} \frac{(x+1)^\lambda - 1}{\lambda} & \lambda \neq 0 \text{ and } x \geq 0, \\ \ln(x+1), & \lambda = 0 \text{ and } x \geq 0, \\ \frac{-(-x+1)^{2-\lambda} - 1)}{2-\lambda}, & \lambda \neq 2 \text{ and } x < 0, \\ \ln(-x+1), & \lambda = 2 \text{ and } x < 0 \end{cases} \tag{13}$$

3.6. Data Augmentation

In addressing the challenge of class imbalance in datasets, the permutation technique [88–90] emerges as a novel data augmentation method, which is particularly effective for sequential or time-series data. At its core, the permutation technique involves dividing a signal into multiple non-overlapping segments and then rearranging these segments in various orders to generate new samples. For example, given a time-series signal divided into three segments, A, B, and C, permutations can produce sequences such as B–A–C, C–B–A, or even B–C–A. This method capitalizes on the inherent structure and patterns within the data, creating diverse samples that maintain the original signal's fundamental characteristics. When applied to the minority class in an imbalanced dataset, the permutation technique can artificially expand the number of samples, thus bridging the gap between the majority and minority classes. This ensures that the learning algorithm is exposed to a broader spectrum of data variations from the minority class, potentially enhancing its ability to generalize and reducing the bias towards the majority class.

3.7. Proposed Multi-Layer Perceptron Architecture

Our proposed MLP architecture [91–94] was designed to handle the complexity and variability inherent in the sensor data collected. With the manual feature extraction and subsequent optimization processes we employed, our MLP [95–98] was strategically tasked with classifying a refined feature vector that encapsulates essential information for robust activity recognition.

3.7.1. Architecture Overview

- Input Layer: The size of the input layer directly corresponds to the number of features extracted and optimized from the sensor data. In our study, the dimensionality of the input layer was adjusted based on the dataset being processed, aligning it with the feature vector size derived after optimization.
- Hidden Layers: We include three hidden layers. The first and second hidden layers are each composed of 64 neurons, while the third hidden layer contains 32 neurons. We utilized the ReLU (rectified linear unit) activation function across these layers to introduce necessary nonlinearity into the model, which is crucial for learning the complex patterns present in the activity data.
- Output Layer: The size of the output layer varies with the dataset; it comprises nine neurons for the Extrasensory dataset and 10 neurons for the Huawei dataset, each representing the number of activity classes within these datasets. The softmax activation function is employed in the output layer to provide a probability distribution over the predicted activity classes, facilitating accurate activity classification.

3.7.2. Training Process

We trained the MLP using a backpropagation algorithm with a stochastic gradient descent optimizer [99,100]. A categorical cross-entropy [101–103] loss function was employed, suitable for the multi-class classification challenges presented by our datasets. The key elements of our training process included:

- Batch Size: We processed 32 samples per batch, optimizing the computational efficiency without sacrificing the ability to learn complex patterns.
- Epochs: The network was trained for up to 100 epochs. To combat overfitting, we implemented early stopping, which halted training if the validation loss did not improve for 10 consecutive epochs.
- Validation Split: To ensure robust model evaluation and tuning, 20% of our training data were set aside as a validation set. This allowed us to monitor the model's performance and make necessary adjustments to the hyperparameters in real-time.

3.7.3. Model Application and Evaluation

Following the rigorous training phase, we applied the trained MLP model to the test sets from both the Extrasensory and Huawei datasets to critically assess their effectiveness in real-world scenarios. Our evaluation strategy was comprehensive, focusing on a range of metrics that provide the accuracy and robustness of the models.

- Performance Metrics: We evaluated the model based on accuracy, precision, recall, and the F1-score [104,105]. These metrics were calculated to assess the overall effectiveness of the models in correctly classifying the activities.
- Confusion matrix: For each dataset, a confusion matrix was generated to visually represent the performance of the model across all activity classes. The confusion matrix [106,107] helps in identifying not only the instances of correct predictions but also the types of errors made by the model, such as false positives and false negatives. This detailed view allows us to specific activities where the model may require further tuning.
- ROC Curves: We also plotted receiver operating characteristic (ROC) curves for each class within the datasets. The ROC curves provide a graphical representation of the trade-off between the true positive rate and the false positive rate at various threshold settings. The area under the ROC curve (AUC) was calculated to quantify the model's ability to discriminate between the classes under study.

4. Experimental Setup

Evaluation of the proposed system was performed on a benchmark dataset: the Extrasensory dataset and Sussex Huawei locomotion (SHL) datasets. The experiment was performed on a Mac 2017 core i5 with 16 GB of RAM, a 3.2 GHz processor, and 512 GB of SSD.

4.1. Datasets Descriptions

In this section, we delve into the specifics of each dataset, highlighting their diversity and how they reflect real-world scenarios.

4.1.1. The Extrasensory Dataset

The Extrasensory dataset was compiled through the utilization of a variety of sensors, including inertial, GPS, compass, and audio sensors. The data collection process was facilitated by an extra-sensory smartphone app, which aimed to monitor human physical and locomotion activities. The dataset comprises information derived from 36 individual users, with each user contributing a substantial number of instances. Data were collected through both Android and iPhone smartphones, and the dataset includes a comprehensive set of 116 labels for user-reported activities. The details of the dataset are also given in Table 1.

Table 1. Description of the extrasensory dataset.

Sensors	Signal Type	Sampling Rate (Hz)	Duration (sec)	Number of Recordings
Accelerometer	Acceleration	32	2	308,306
Gyroscope	Angular Velocity	32	2	291,883
Magnetometer	Magnetic Field	32	2	282,527
Location	Latitude, Longitude	1	2	273,737

4.1.2. The Sussex Huawei Dataset (SHL)

The Sussex Huawei Locomotion (SHL) dataset [108] is a comprehensive collection of data designed to support research in mobile sensing, particularly for the recognition of human activities and modes of transportation. It was created through a collaboration between the University of Sussex and Huawei Technologies Co., Ltd. The dataset consists of recordings from smartphone sensors, such as accelerometers, gyroscopes, magnetometers, and barometers. These sensors capture movements and environmental characteristics as people go about various activities, including walking, running, cycling, and traveling by car, bus, or train. Participants carried smartphones equipped with these sensors went through a series of movements in real-world settings, ensuring that the data was as realistic and varied as possible.

4.2. First Experiment: Confusion Matrix

We perform the activity classification using MLP. To evaluate the performance, we plotted the confusion matrix. In simple words, a confusion matrix is a table used for classification problems. It is used to see where the model made an error. The confusion matrices calculated for physical and localization activity for both datasets are shown in Tables 2–5.

Table 2. Confusion matrix over the Extrasensory dataset for physical activity.

Obj. Classes	Sitting	Eating	Cooking	Bicycle
sitting	0.95	0.01	0.03	0.00
eating	0.00	1.00	0.00	0.00
cooking	0.00	0.00	1.00	0.00
bicycle	0.03	0.00	0.00	0.97
	Mean Accuracy =		96.61%	

Table 3. Confusion matrix over the Extrasensory dataset for localization activity.

Obj. Classes	Indoors	Outdoors	Home	School	Car
Indoors	**1.00**	0.00	0.00	0.00	0.00
Outdoors	0.00	**1.00**	0.00	0.00	0.00
Home	0.05	0.06	**0.80**	0.02	0.07
School	0.02	0.02	0.03	**0.90**	0.03
Car	0.00	0.00	0.00	0.00	**1.00**
		Mean Accuracy =	94.28%		

Table 4. Confusion matrix over the SHL dataset for physical activity.

Obj. Classes	Sit	Walk	Stand	Run
Sit	**0.96**	0.00	0.04	0.00
Walk	0.03	**0.97**	0.00	0.00
Stand	0.03	0.03	**0.92**	0.02
Run	0.02	0.01	0.03	**0.94**
	Mean Accuracy =	94.75%		

Table 5. Confusion matrix over the SHL dataset for localization activity.

Obj. Classes	Indoor	Outdoor	In Train	In Car	In Bus	In Subway
Indoor	**0.93**	0.00	0.05	0.02	0.00	0.00
Outdoor	0.00	**0.95**	0.04	0.00	0.00	0.01
In train	0.01	0.03	**0.89**	0.02	0.05	0.00
In car	0.00	0.01	0.01	**0.94**	0.00	0.04
In bus	0.03	0.02	0.07	0.00	**0.88**	0.00
In subway	0.03	0.00	0.03	0.00	0.02	**0.92**
	Mean Accuracy	= 91.83%				

4.3. Second Experiment: Precision, Recall, and F1-Score

In this experiment, we evaluated our system by plotting precision, recall, and f1-score for individual activity. In Tables 6 and 7, the evaluation for physical and localization activity can be seen.

Table 6. Precision, recall, and F1-score over physical activity.

Classes	Extrasensory			SHL		
Activities	Precision	Recall	F1-Score	Precision	Recall	F1-Score
Sitting	0.95	1.00	0.92	-	-	-
Eating	1.00	0.80	0.90	-	-	-
Cooking	1.00	0.89	0.95	-	-	-
Bicycle	0.97	0.95	0.96	-	-	-
Sit	-	-	-	0.92	0.96	0.94
Stand	-	-	-	0.94	0.92	0.93
Walking	-	-	-	0.96	0.97	0.97
Run	-	-	-	0.95	0.94	0.92

Table 7. Precision, recall, and F1-score over localization activity.

Classes	Extrasensory			SHL		
Activities	Precision	Recall	F1-Score	Precision	Recall	F1-Score
Indoors	1.00	0.94	0.91	-	-	-
Outdoors	1.00	1.00	0.95	-	-	-
School	0.84	1.00	0.92	-	-	-
Home	0.90	0.85	0.88	-	-	-
Car	1.00	1.00	1.00	-	-	-
Indoor	-	-	-	0.93	0.93	0.93
Outdoor	-	-	-	0.94	0.95	0.94
In train	-	-	-	0.82	0.89	0.85
In car	-	-	-	0.96	0.94	0.95
In subway	-	-	-	0.95	0.92	0.93
In bus	-	-	-	0.93	0.88	0.90

4.4. Third Experiment: Receiver Operating Characteristics (ROC Curve)

To further assess the performance and robustness of our system, we employed the ROC curve, a well-established graphical tool that illustrates the diagnostic ability of a classification system. The ROC curve visualizes the trade-offs between the true positive rate (sensitivity) and false positive rate (1-specificity) across various threshold settings. The area under the ROC curve (AUC) serves as a single scalar value summarizing the overall performance of the classifier. A model with perfect discriminatory power has an AUC of 1, while a model with no discriminatory power (akin to random guessing) has an AUC of 0.5. In Figures 10 and 11 the Roc curve is plotted.

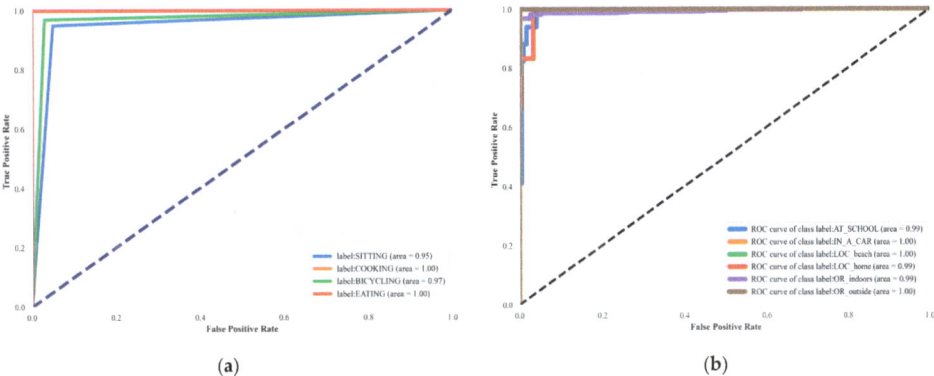

Figure 10. ROC curves: (**a**) physical and (**b**) localization activity over extrasensory dataset.

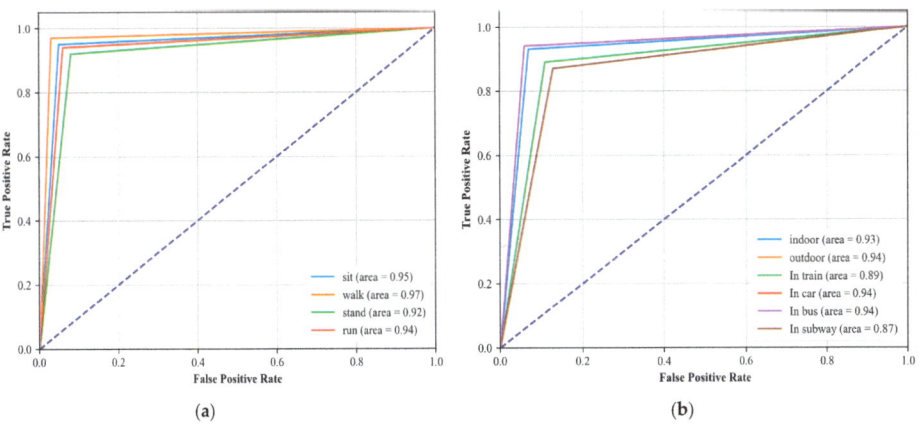

Figure 11. ROC curves: (**a**) physical and (**b**) localization activity over the SHL dataset.

4.5. Fourth Experiment: Comparison with Other Techniques

In the last experiment, the proposed system is compared with the state-of-the-art techniques. Table 8 shows the comparison of the proposed model with other state of the art techniques.

Table 8. Comparison of proposed MLP with other methods.

Method	Accuracy %	
	Extrasensory	SHL
Vaizman et al. [109]	0.83	-
Vaizman et al. [98]	0.83	-
Asim et al. [108]	0.87	-
Sharma et al. [110]	-	0.92
Akbari et al. [111]	-	0.92
Brimacombe et al. [112]	-	0.79
Proposed	**0.94**	**0.91**

5. Computational Analysis

The comparative analysis of time consumption and memory usage between the Extrasensory and Huawei datasets reveals significant differences in efficiency and resource demands. These disparities suggest diverse applicability in real-world scenarios. Specifically, the extrasensory dataset, with its higher time and memory requirements, is best suited for environments where detailed and complex activity recognition is crucial, and computational resources are less constrained, such as in clinical or controlled research settings. On the other hand, the Huawei dataset, with its lower resource demands, demonstrates suitability for consumer electronics and real-time applications, such as smartphones and wearable devices that require efficient processing capabilities. The findings show that, while the system exhibits robust performance, its deployment in resource-limited environments such as low-end smartphones or IoT devices might be challenging. Thus, our system is ideal for scenarios where precision and detailed activity recognition outweigh the need for low resource consumption, and less so for applications requiring minimal power usage and rapid processing. Figure 12 shows the analysis visually.

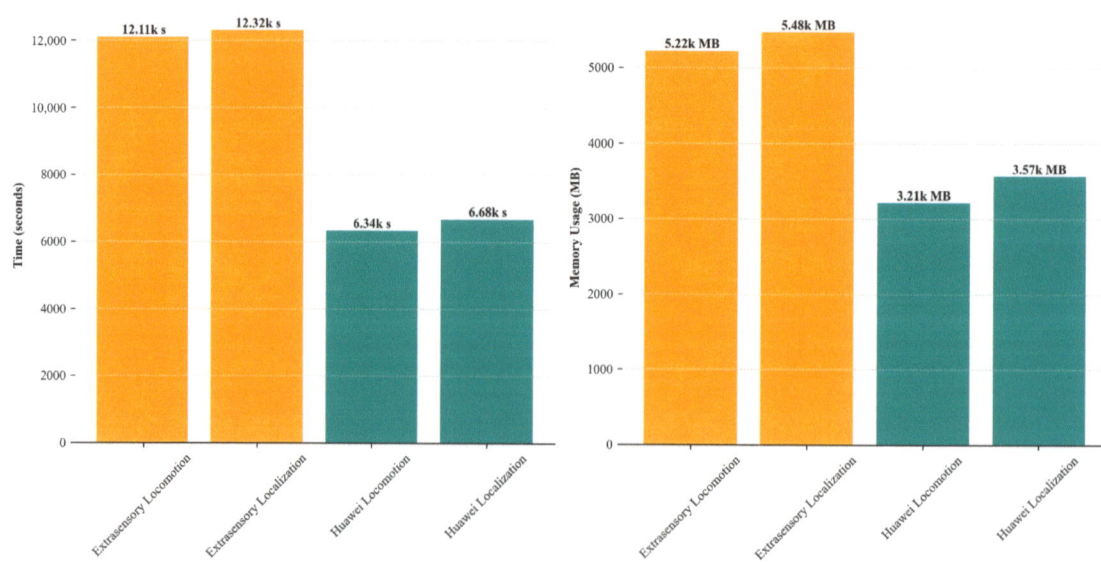

Figure 12. Time and memory usage analysis of the proposed system.

6. Discussion and Limitations

Our research has successfully demonstrated the utilization of smartphone and smartwatch sensors to accurately identify human movements and locations. By methodically cleaning, segmenting, and extracting features from raw sensor data, and employing a multi-layer perceptron for classification, our system achieved high accuracy rates. Specifically, we observed success rates of 96% and 94% for identifying physical activities over the extrasensory and SHL datasets, respectively, and 94% (Extrasensory) and 91% (SHL) for localization activities. These results represent a significant improvement over many existing methods and underscore the potential of our approach in applications where precise activity recognition is crucial.

- Detailed Analysis of Key Findings

The high accuracy rates in physical activity recognition demonstrate the efficacy of the proposed system's feature extraction and machine learning workflow. For localization activities, although slightly lower, the success rates are still competitive, emphasizing our system's capability in varied contexts. These findings suggest that our approach could be particularly beneficial in health monitoring, urban navigation, and other IoT applications that demand reliable human activity and location data.

While our proposed system offers a promising approach for biosensor-driven IoT wearables in human motion tracking and localization, we recognize several inherent challenges that could impact its broader application and effectiveness. These limitations, if not addressed, may curtail the system's reliability and versatility in diverse environments:

- GPS limitations: The GPS technology we utilize, while generally effective, can suffer from significant inaccuracies in environments such as urban canyons or indoors due to signal blockage and multipath interference. These environmental constraints can affect the system's ability to precisely track and localize activities, particularly in complex urban settings.
- Data diversity and completeness: The dataset employed for training our system, though extensive, does not encompass the entire spectrum of human activities, particularly those that are irregular or occur less frequently. This limitation could reduce the model's ability to generalize to activities not represented in the training phase, potentially impacting its applicability in varied real-world scenarios.
- Performance across different hardware: Our system was primarily tested and optimized on a specific computational setup. When considering deployment across diverse real-world devices such as smartphones, smartwatches, or other IoT wearables, variations in processing power, storage capacity, and sensor accuracy must be addressed. The heterogeneity of these devices could result in inconsistent performance, with higher-end devices potentially delivering more accurate results than lower-end counterparts.
- Scalability and real-time processing: Scaling our system to handle real-time data processing across multiple devices simultaneously presents another significant challenge. The computational demands of processing large volumes of sensor data in real time necessitate not only robust algorithms but also hardware capable of efficiently supporting these operations.
- Privacy and security concerns: As with any system handling sensitive personal data, ensuring privacy and security is paramount. Our current model must incorporate more advanced encryption methods and privacy-preserving techniques to safeguard user data against potential breaches or unauthorized access.

7. Conclusions and Future Work

In this study, we successfully developed a comprehensive system capable of effectively recognizing human physical activities and localization through a combination of inertial and GPS sensor data. Our system initiates with denoising the raw signals using Butterworth and median filters to reduce noise while preserving essential signal characteristics. This is followed by the Hamming windowing technique and segmentation processes that structure the data for more effective analysis. Subsequently, we extract and optimize statistical features using the variance threshold selection method and Yeo–Johnson power transformation, respectively, significantly enhancing the relevance and performance of these features in the activity classification process. The final classification of activities is executed through a multilayer perceptron (MLP), which provides a robust model capable of predicting various types of human movements and positions. The findings from our research offer significant implications for the development of smarter, more responsive wearable and mobile technology. By showcasing high accuracy in activity recognition, our system lays a foundation for improved user interaction and monitoring across various applications, spanning from personal fitness tracking to patient health monitoring in medical settings. The successful integration of sensor data for precise activity and location recognition paves the way for more intuitive and context-aware devices.

Moving forward, several enhancements and extensions are proposed to further enrich the capabilities of our system and its applicability to a broader range of real-world scenarios. First, integrating additional types of sensor data, such as environmental and biometric sensors, could provide a more complex understanding of the context and improve the accuracy and reliability of activity recognition. Second, developing adaptive algorithms that can dynamically adjust to changes in the environment or user behavior would make the system more responsive and versatile. Additionally, scalability improvements are crucial, and future work will focus on optimizing the system to more efficiently handle larger, more diverse datasets. This will involve refining our algorithms to manage increased computational demands while enhancing performance. Another important direction for future research involves enhancing the real-time processing capabilities of our system, which is essential for applications requiring immediate responses, such as emergency services or live health monitoring. Furthermore, given the sensitive nature of the data involved in our system, advancing data privacy and security measures will be a priority. We plan to explore sophisticated encryption methods and privacy-preserving data analytics to ensure the security and privacy of user data.

Author Contributions: Conceptualization: D.K., N.A.M. and A.A. (Asaad Algarni); methodology: D.K. and M.A.; software: D.K. and N.A.A.; validation: N.A.M., M.A. and A.A. (Abdulwahab Alazeb); formal analysis: and N.A.M.; resources: N.A.M., A.A. (Asaad Algarni), N.A.M. and A.A. (Abdulwahab Alazeb); writing—review and editing: N.A.M., A.J. and H.L.; funding acquisition: N.A.M., N.A.M., A.A. (Asaad Algarni), H.L., A.A. (Abdulwahab Alazeb) and A.J. All authors have read and agreed to the published version of the manuscript.

Funding: The APC was funded by the Open Access Initiative of the University of Bremen and the DFG via SuUB Bremen. This research is supported and funded by Princess Nourah bint Abdulrahman University Researchers Supporting Project number (PNURSP2024R410), Princess Nourah bint Abdulrahman University, Riyadh, Saudi Arabia. The authors are thankful to the Deanship of Scientific Research at Najran University for funding this work under the Research Group Funding program grant code (NU/GP/SERC/13/30). This study is supported via funding from Prince Sattam bin Abdulaziz University project number (PSAU/2024/R/1445).

Acknowledgments: Princess Nourah bint Abdulrahman University Researchers Supporting Project number (PNURSP2024R410), Princess Nourah bint Abdulrahman University, Riyadh, Saudi Arabia.

Conflicts of Interest: The authors declare no conflicts of interest.

References

1. Qi, M.; Cui, S.; Chang, X.; Xu, Y.; Meng, H.; Wang, Y.; Yin, T. Multi-region Nonuniform Brightness Correction Algorithm Based on L-Channel Gamma Transform. *Secur. Commun. Netw.* **2022**, *2022*, 2675950. [CrossRef]
2. Li, R.; Peng, B. Implementing Monocular Visual-Tactile Sensors for Robust Manipulation. *Think. Ski. Creat.* **2022**, *2022*, 9797562. [CrossRef] [PubMed]
3. Babaei, N.; Hannani, N.; Dabanloo, N.J.; Bahadori, S. A Systematic Review of the Use of Commercial Wearable Activity Trackers for Monitoring Recovery in Individuals Undergoing Total Hip Replacement Surgery. *Think. Ski. Creat.* **2022**, *2022*, 9794641. [CrossRef] [PubMed]

4. Zhao, Q.; Yan, S.; Zhang, B.; Fan, K.; Zhang, J.; Li, W. An On-Chip Viscoelasticity Sensor for Biological Fluids. *Think. Ski. Creat.* **2023**, *4*, 6. [CrossRef] [PubMed]
5. Qu, J.; Mao, B.; Li, Z.; Xu, Y.; Zhou, K.; Cao, X.; Fan, Q.; Xu, M.; Liang, B.; Liu, H.; et al. Recent Progress in Advanced Tactile Sensing Technologies for Soft Grippers. *Adv. Funct. Mater.* **2023**, *33*, 2306249. [CrossRef]
6. Khan, D.; Alonazi, M.; Abdelhaq, M.; Al Mudawi, N.; Algarni, A.; Jalal, A.; Liu, H. Robust human locomotion and localization activity recognition over multisensory. *Front. Physiol.* **2024**, *15*, 1344887. [CrossRef] [PubMed]
7. Jalal, A.; Nadeem, A.; Bobasu, S. Human Body Parts Estimation and Detection for Physical Sports Movements. In Proceedings of the 2019 2nd International Conference on Communication, Computing and Digital systems (C-CODE), Islamabad, Pakistan, 6–7 March 2019; pp. 104–109.
8. Arshad, M.H.; Bilal, M.; Gani, A. Human Activity Recognition: Review, Taxonomy and Open Challenges. *Sensors* **2022**, *22*, 6463. [CrossRef] [PubMed]
9. Elbayoudi, A.; Lotfi, A.; Langensiepen, C.; Appiah, K. Modelling and Simulation of Activities of Daily Living Representing an Older Adult's Behaviour. In Proceedings of the 8th ACM International Conference on Pervasive Technologies Related to Assistive Environments (PETRA '15), Corfu, Greece, 1–3 July 2015; Article 67. Association for Computing Machinery: New York, NY, USA, 2015; pp. 1–8.
10. Azmat, U.; Jalal, A. Smartphone Inertial Sensors for Human Locomotion Activity Recognition based on Template Matching and Codebook Generation. In Proceedings of the 2021 International Conference on Communication Technologies (ComTech), Rawalpindi, Pakistan, 21 September 2021; pp. 109–114.
11. Lara, O.D.; Labrador, M.A. A Survey on Human Activity Recognition using Wearable Sensors. *IEEE Commun. Surv. Tutor.* **2013**, *15*, 1192–1209. [CrossRef]
12. Serpush, F.; Menhaj, M.B.; Masoumi, B.; Karasfi, B. Wearable Sensor-Based Human Activity Recognition in the Smart Healthcare System. *Comput. Intell. Neurosci.* **2022**, *2022*, 1–31. [CrossRef]
13. Yan, L.; Shi, Y.; Wei, M.; Wu, Y. Multi-feature fusing local directional ternary pattern for facial expressions signal recognition based on video communication system. *Alex. Eng. J.* **2023**, *63*, 307–320. [CrossRef]
14. Cai, L.; Yan, S.; Ouyang, C.; Zhang, T.; Zhu, J.; Chen, L.; Ma, X.; Liu, H. Muscle synergies in joystick manipulation. *Front. Physiol.* **2023**, *14*, 1282295. [CrossRef] [PubMed]
15. Li, J.; Li, J.; Wang, C.; Verbeek, F.J.; Schultz, T.; Liu, H. Outlier detection using iterative adaptive mini-minimum spanning tree generation with applications on medical data. *Front. Physiol.* **2023**, *14*, 1233341. [CrossRef] [PubMed]
16. Wang, F.; Ma, M.; Zhang, X. Study on a Portable Electrode Used to Detect the Fatigue of Tower Crane Drivers in Real Construction Environment. *IEEE Trans. Instrum. Meas.* **2024**, *73*, 1–14. [CrossRef]
17. Yu, J.; Dong, X.; Li, Q.; Lu, J.; Ren, Z. Adaptive Practical Optimal Time-Varying Formation Tracking Control for Disturbed High-Order Multi-Agent Systems. *IEEE Trans. Circuits Syst. I Regul. Pap.* **2022**, *69*, 2567–2578. [CrossRef]
18. He, H.; Chen, Z.; Liu, H.; Liu, X.; Guo, Y.; Li, J. Practical Tracking Method based on Best Buddies Similarity. *Think. Ski. Creat.* **2023**, *4*, 50. [CrossRef] [PubMed]
19. Hou, X.; Zhang, L.; Su, Y.; Gao, G.; Liu, Y.; Na, Z.; Xu, Q.; Ding, T.; Xiao, L.; Li, L.; et al. A space crawling robotic bio-paw (SCRBP) enabled by triboelectric sensors for surface identification. *Nano Energy* **2023**, *105*, 108013. [CrossRef]
20. Hou, X.; Xin, L.; Fu, Y.; Na, Z.; Gao, G.; Liu, Y.; Xu, Q.; Zhao, P.; Yan, G.; Su, Y.; et al. A self-powered biomimetic mouse whisker sensor (BMWS) aiming at terrestrial and space objects perception. *Nano Energy* **2023**, *118*, 109034. [CrossRef]
21. Ma, S.; Chen, Y.; Yang, S.; Liu, S.; Tang, L.; Li, B.; Li, Y. The Autonomous Pipeline Navigation of a Cockroach Bio-robot with Enhanced Walking Stimuli. *Think. Ski. Creat.* **2023**, *4*, 0067. [CrossRef] [PubMed]
22. Bahadori, S.; Williams, J.M.; Collard, S.; Swain, I. Can a Purposeful Walk Intervention with a Distance Goal Using an Activity Monitor Improve Individuals' Daily Activity and Function Post Total Hip Replacement Surgery. A Randomized Pilot Trial. *Think. Ski. Creat.* **2023**, *4*, 0069. [CrossRef]
23. Hsu, Y.-L.; Yang, S.-C.; Chang, H.-C.; Lai, H.-C. Human Daily and Sport Activity Recognition Using a Wearable Inertial SensorNetwork. *IEEE Access* **2018**, *6*, 31715–31728. [CrossRef]
24. Abdel-Basset, M.; Hawash, H.; Chang, V.; Chakrabortty, R.K.; Ryan, M. Deep Learning for Heterogeneous Human Activity Recognition in Complex IoT Applications. *IEEE Internet Things J.* **2022**, *9*, 5653–5665. [CrossRef]
25. Konak, S.; Turan, F.; Shoaib, M.; Incel, Ö.D. Feature Engineering for Activity Recognition from Wrist-worn Motion Sensors. In Proceedings of the International Conference on Pervasive and Embedded Computing and Communication Systems, Lisbon, Portugal, 25–27 July 2016.
26. Chetty, G.; White, M.; Akther, F. Smart Phone Based Data Mining for Human Activity Recognition. *Procedia Comput. Sci.* **2016**, *46*, 1181–1187. [CrossRef]
27. Ehatisham-ul-Haq, M.; Azam, M.A. Opportunistic sensing for inferring in-the-wild human contexts based on activity pattern-recognition using smart computing. *Future Gener. Comput. Syst.* **2020**, *106*, 374–392. [CrossRef]
28. Zhang, X.; Huang, D.; Li, H.; Zhang, Y.; Xia, Y.; Liu, J. Self-training maximum classifier discrepancy for EEG emotion recognition. *CAAI Trans. Intell. Technol.* **2023**, *8*, 1480–1491. [CrossRef]
29. Wen, C.; Huang, Y.; Zheng, L.; Liu, W.; Davidson, T.N. Transmit Waveform Design for Dual-Function Radar-Communication Systems via Hybrid Linear-Nonlinear Precoding. *IEEE Trans. Signal Process.* **2023**, *71*, 2130–2145. [CrossRef]

30. Wen, C.; Huang, Y.; Davidson, T.N. Efficient Transceiver Design for MIMO Dual-Function Radar-Communication Systems. *IEEE Trans. Signal Process.* **2023**, *71*, 1786–1801. [CrossRef]
31. Yao, Y.; Shu, F.; Li, Z.; Cheng, X.; Wu, L. Secure Transmission Scheme Based on Joint Radar and Communication in Mobile Vehicular Networks. *IEEE Trans. Intell. Transp. Syst.* **2023**, *24*, 10027–10037. [CrossRef]
32. Jalal, A.; Quaid, M.A.K.; Kim, K. A Wrist Worn Acceleration Based Human Motion Analysis and Classification for Ambient Smart Home System. *J. Electr. Eng. Technol.* **2019**, *14*, 1733–1739. [CrossRef]
33. Hu, Z.; Ren, L.; Wei, G.; Qian, Z.; Liang, W.; Chen, W.; Lu, X.; Ren, L.; Wang, K. Energy Flow and Functional Behavior of Individual Muscles at Different Speeds During Human Walking. *IEEE Trans. Neural Syst. Rehabil. Eng.* **2023**, *31*, 294–303. [CrossRef]
34. Wang, K.; Boonpratatong, A.; Chen, W.; Ren, L.; Wei, G.; Qian, Z.; Lu, X.; Zhao, D. The Fundamental Property of Human Leg During Walking: Linearity and Nonlinearity. *IEEE Trans. Neural Syst. Rehabil. Eng.* **2023**, *31*, 4871–4881. [CrossRef]
35. Jalal, A.; Quaid, M.A.K.; Hasan, A.S. Wearable Sensor-Based Human Behavior Understanding and Recognition in Daily Life for Smart Environments. In Proceedings of the 2018 International Conference on Frontiers of Information Technology (FIT), Islamabad, Pakistan, 17–19 December 2018; pp. 105–110.
36. Zhao, Z.; Xu, G.; Zhang, N.; Zhang, Q. Performance analysis of the hybrid satellite-terrestrial relay network with opportunistic scheduling over generalized fad-ing channels. *IEEE Trans. Veh. Technol.* **2022**, *71*, 2914–2924. [CrossRef]
37. Zhu, T.; Ding, H.; Wang, C.; Liu, Y.; Xiao, S.; Yang, G.; Yang, B. Parameters Calibration of the GISSMO Failure Model for SUS301L-MT. *Chin. J. Mech. Eng.* **2023**, *36*, 1–12. [CrossRef]
38. Qu, J.; Yuan, Q.; Li, Z.; Wang, Z.; Xu, F.; Fan, Q.; Zhang, M.; Qian, X.; Wang, X.; Wang, X.; et al. All-in-one strain-triboelectric sensors based on environment-friendly ionic hydrogel for wearable sensing and underwater soft robotic grasping. *Nano Energy* **2023**, *111*, 108387. [CrossRef]
39. Zhao, S.; Liang, W.; Wang, K.; Ren, L.; Qian, Z.; Chen, G.; Lu, X.; Zhao, D.; Wang, X.; Ren, L. A Multiaxial Bionic Ankle Based on Series Elastic Actuation with a Parallel Spring. *IEEE Trans. Ind. Electron.* **2023**, *71*, 7498–7510. [CrossRef]
40. Liang, X.; Huang, Z.; Yang, S.; Qiu, L. Device-Free Motion & Trajectory Detection via RFID. *ACM Trans. Embed. Comput. Syst.* **2018**, *17*, 1–27. [CrossRef]
41. Liu, C.; Wu, T.; Li, Z.; Ma, T.; Huang, J. Robust Online Tensor Completion for IoT Streaming Data Recovery. *IEEE Trans. Neural Netw. Learn. Syst.* **2022**, *34*, 10178–10192. [CrossRef]
42. Nadeem, A.; Jalal, A.; Kim, K. Automatic human posture estimation for sport activity recognition with robust body parts detection and entropy markov model. *Multimed. Tools Appl.* **2021**, *80*, 21465–21498. [CrossRef]
43. Yu, J.; Lu, L.; Chen, Y.; Zhu, Y.; Kong, L. An Indirect Eavesdropping Attack of Keystrokes on Touch Screen through Acoustic Sensing. *IEEE Trans. Mob. Comput.* **2021**, *20*, 337–351. [CrossRef]
44. Bashar, S.K.; Al Fahim, A.; Chon, K.H. Smartphone-Based Human Activity Recognition with Feature Selection and Dense Neural Network. In Proceedings of the 2020 42nd Annual International Conference of the IEEE Engineering in Medicine & Biology Society (EMBC), Montreal, QC, Canada, 20–24 July 2020; pp. 5888–5891.
45. Xie, L.; Tian, J.; Ding, G.; Zhao, Q. Hu-man activity recognition method based on inertial sensor and barometer. In Proceedings of the 2018 IEEE International Symposium on Inertial Sensors and Systems (INERTIAL), Lake Como, Italy, 26–29 March 2018; pp. 1–4.
46. Lee, S.-M.; Yoon, S.M.; Cho, H. Human activity recognition from accelerometer data using Convolutional Neural Network. In Proceedings of the 2017 IEEE International Conference on Big Data and Smart Computing (BigComp), Jeju, Republic of Korea, 13–16 February 2017; pp. 131–134.
47. Mekruksavanich, S.; Jitpattanakul, A. Recognition of Real-life Activities with Smartphone Sensors using Deep Learning Approaches. In Proceedings of the 2021 IEEE 12th International Conference on Software Engineering and Service Science (ICSESS), Beijing, China, 20–22 August 2021; pp. 243–246.
48. Cong, R.; Sheng, H.; Yang, D.; Cui, Z.; Chen, R. Exploiting Spatial and Angular Correlations with Deep Efficient Transformers for Light Field Image Super-Resolution. *IEEE Trans. Multimed.* **2024**, *26*, 1421–1435. [CrossRef]
49. Liu, H.; Yuan, H.; Liu, Q.; Hou, J.; Zeng, H.; Kwong, S. A Hybrid Compression Framework for Color Attributes of Static 3D Point Clouds. *IEEE Trans. Circuits Syst. Video Technol.* **2022**, *32*, 1564–1577. [CrossRef]
50. Liu, Q.; Yuan, H.; Hamzaoui, R.; Su, H.; Hou, J.; Yang, H. Reduced Reference Perceptual Quality Model with Application to Rate Control for Video-Based Point Cloud Compression. *IEEE Trans. Image Process.* **2021**, *30*, 6623–6636. [CrossRef]
51. Mutegeki, R.; Han, D.S. A CNN-LSTM Approach to Human Activity Recognition. In Proceedings of the International Conference on Artificial Intelligence and Information Communications (ICAIIC), Fukuoka, Japan, 19–21 February 2020; pp. 362–366.
52. Liu, A.-A.; Zhai, Y.; Xu, N.; Nie, W.; Li, W.; Zhang, Y. Region-Aware Image Captioning via Interaction Learning. *IEEE Trans. Circuits Syst. Video Technol.* **2022**, *32*, 3685–3696. [CrossRef]
53. Jaramillo, I.E.; Jeong, J.G.; Lopez, P.R.; Lee, C.-H.; Kang, D.-Y.; Ha, T.-J.; Oh, J.-H.; Jung, H.; Lee, J.H.; Lee, W.H.; et al. Real-Time Human Activity Recognition with IMU and Encoder Sensors in Wearable Exoskeleton Robot via Deep Learning Networks. *Sensors* **2022**, *22*, 9690. [CrossRef]
54. Hussain, I.; Jany, R.; Boyer, R.; Azad, A.; Alyami, S.A.; Park, S.J.; Hasan, M.; Hossain, A. An Explainable EEG-Based Human Activity Recognition Model Using Machine-Learning Approach and LIME. *Sensors* **2023**, *23*, 7452. [CrossRef]
55. Garcia-Gonzalez, D.; Rivero, D.; Fernandez-Blanco, E.; Luaces, M.R. New machine learning approaches for real-life human activity recognition using smartphone sensor-based data. *Knowl. Based Syst.* **2023**, *262*, 110260. [CrossRef]

56. Zhang, J.; Zhu, C.; Zheng, L.; Xu, K. ROSEFusion: Random optimization for online dense reconstruction under fast camera motion. *ACM Trans. Graph.* **2021**, *40*, 1–17. [CrossRef]
57. Zhang, J.; Tang, Y.; Wang, H.; Xu, K. ASRO-DIO: Active Subspace Random Optimization Based Depth Inertial Odometry. *IEEE Trans. Robot.* **2022**, *39*, 1496–1508. [CrossRef]
58. She, Q.; Hu, R.; Xu, J.; Liu, M.; Xu, K.; Huang, H. Learning High-DOF Reaching-and-Grasping via Dynamic Representation of Gripper-Object Interaction. *ACM Trans. Graph.* **2022**, *41*, 1–14. [CrossRef]
59. Xu, J.; Zhang, X.; Park, S.H.; Guo, K. The Alleviation of Perceptual Blindness During Driving in Urban Areas Guided by Saccades Recommendation. *IEEE Trans. Intell. Transp. Syst.* **2022**, *23*, 16386–16396. [CrossRef]
60. Xu, J.; Park, S.H.; Zhang, X.; Hu, J. The Improvement of Road Driving Safety Guided by Visual Inattentional Blindness. *IEEE Trans. Intell. Transp. Syst.* **2022**, *23*, 4972–4981. [CrossRef]
61. Mao, Y.; Sun, R.; Wang, J.; Cheng, Q.; Kiong, L.C.; Ochieng, W.Y. New time-differenced carrier phase approach to GNSS/INS integration. *GPS Solutions* **2022**, *26*, 122. [CrossRef]
62. Jalal, A.; Kim, Y. Dense depth maps-based human pose tracking and recognition in dynamic scenes using ridge data. In Proceedings of the 2014 11th IEEE International Conference on Advanced Video and Signal Based Surveillance (AVSS), Seoul, Republic of Korea, 26–29 August 2014; pp. 119–124.
63. Mahmood, M.; Jalal, A.; Kim, K. WHITE STAG model: Wise human interaction tracking and estimation (WHITE) using spatio-temporal and angular-geometric (STAG) descriptors. *Multimed. Tools Appl.* **2020**, *79*, 6919–6950. [CrossRef]
64. Chen, Z.; Cai, C.; Zheng, T.; Luo, J.; Xiong, J.; Wang, X. RF-Based Human Activity Recognition Using Signal Adapted Convolutional Neural Network. *IEEE Trans. Mob. Comput.* **2023**, *22*, 487–499. [CrossRef]
65. Batool, M.; Alotaibi, S.S.; Alatiyyah, M.H.; Alnowaiser, K.; Aljuaid, H.; Jalal, A.; Park, J. Depth Sensors-Based Action Recognition using a Modified K-Ary Entropy Classifier. *IEEE Access* **2023**, *11*, 58578–58595. [CrossRef]
66. Xu, J.; Pan, S.; Sun, P.Z.H.; Park, S.H.; Guo, K. Human-Factors-in-Driving-Loop: Driver Identification and Verification via a Deep Learning Approach using Psychological Behavioral Data. *IEEE Trans. Intell. Transp. Syst.* **2022**, *24*, 3383–3394. [CrossRef]
67. Xu, J.; Guo, K.; Sun, P.Z. Driving Performance under Violations of Traffic Rules: Novice vs. Experienced Drivers. *IEEE Trans. Intell. Veh.* **2022**, *7*, 908–917. [CrossRef]
68. Liu, H.; Xu, Y.; Chen, F. Sketch2Photo: Synthesizing photo-realistic images from sketches via global contexts. *Eng. Appl. Artif. Intell.* **2023**, *117*, 105608. [CrossRef]
69. Pazhanirajan, S.; Dhanalakshmi, P. EEG Signal Classification using Linear Predictive Cepstral Coefficient Features. *Int. J. Comput. Appl.* **2013**, *73*, 28–31. [CrossRef]
70. Fausto, F.; Cuevas, E.; Gonzales, A. A New Descriptor for Image Matching Based on Bionic Principles. *Pattern Anal. Appl.* **2017**, *20*, 1245–1259. [CrossRef]
71. Alonazi, M.; Ansar, H.; Al Mudawi, N.; Alotaibi, S.S.; Almujally, N.A.; Alazeb, A.; Jalal, A.; Kim, J.; Min, M. Smart healthcare hand gesture recognition using CNN-based detector and deep belief network. *IEEE Access* **2023**, *11*, 84922–84933. [CrossRef]
72. Jalal, A.; Mahmood, M. Students' behavior mining in e-learning environment using cognitive processes with information technologies. *Educ. Inf. Technol.* **2019**, *24*, 2797–2821. [CrossRef]
73. Quaid, M.A.K.; Jalal, A. Wearable sensors based human behavioral pattern recognition using statistical features and reweighted genetic algorithm. *Multimed. Tools Appl.* **2020**, *79*, 6061–6083. [CrossRef]
74. Pervaiz, M.; Jalal, A. Artificial Neural Network for Human Object Interaction System Over Aerial Images. In Proceedings of the 2023 4th International Conference on Advancements in Computational Sciences (ICACS), Lahore, Pakistan, 20–22 February 2023; pp. 1–6.
75. Jalal, A.; Kim, J.T.; Kim, T.-S. Development of a life logging system via depth imaging-based human activity recognition for smart homes. In Proceedings of the International Symposium on Sustainable Healthy Buildings, Seoul, Republic of Korea, 19 September 2012; pp. 91–95.
76. Jalal, A.; Rasheed, Y. Collaboration achievement along with performance maintenance in video streaming. In Proceedings of the IEEE Conference on Interactive Computer Aided Learning, Villach, Austria, 23 December 2007; pp. 1–8.
77. Muneeb, M.; Rustam, H.; Jalal, A. Automate Appliances via Gestures Recognition for Elderly Living Assistance. In Proceedings of the 2023 4th International Conference on Advancements in Computational Sciences (ICACS), Lahore, Pakistan, 20–22 February 2023; pp. 1–6.
78. Szegedy, C.; Ioffe, S.; Vanhoucke, V.; Alemi, A. Inception-v4, inception–ResNet and the impact of residual connections on learning. In Proceedings of the Thirty-First AAAI Conference on Artificial Intelligence, San Francisco, CA, USA, 4–9 February 2017; Volume 3.
79. Azmat, U.; Ghadi, Y.Y.; al Shloul, T.; Alsuhibany, S.A.; Jalal, A.; Park, J. Smartphone Sensor-Based Human Locomotion Surveillance System Using Multilayer Perceptron. *Appl. Sci.* **2022**, *12*, 2550. [CrossRef]
80. Jalal, A.; Batool, M.; Kim, K. Stochastic Recognition of Physical Activity and Healthcare Using Tri-Axial Inertial Wearable Sensors. *Appl. Sci.* **2020**, *10*, 7122. [CrossRef]
81. Tan, T.-H.; Wu, J.-Y.; Liu, S.-H.; Gochoo, M. Human Activity Recognition Using an Ensemble Learning Algorithm with Smartphone Sensor Data. *Electronics* **2022**, *11*, 322. [CrossRef]
82. Hartmann, Y.; Liu, H.; Schultz, T. High-Level Features for Human Activity Recognition and Modeling. In *Biomedical Engineering Systems and Technologies, Proceedings of the BIOSTEC 2022, Virtual Event, 9–11 February 2022*; Roque, A.C.A., Gracanin, D., Lorenz, R., Tsanas, A., Bier, N., Fred, A., Gamboa, H., Eds.; Communications in Computer and In-formation Science; Springer: Cham, Switzerland, 2023; Volume 1814. [CrossRef]

83. Khalid, N.; Gochoo, M.; Jalal, A.; Kim, K. Modeling Two-Person Segmentation and Locomotion for Stereoscopic Action Identification: A Sustainable Video Surveillance System. *Sustainability* **2021**, *13*, 970. [CrossRef]
84. Liu, H.; Yuan, H.; Hou, J.; Hamzaoui, R.; Gao, W. PUFA-GAN: A Frequency-Aware Generative Adversarial Network for 3D Point Cloud Upsampling. *IEEE Trans. Image Process.* **2022**, *31*, 7389–7402. [CrossRef] [PubMed]
85. Jalal, A.; Sharif, N.; Kim, J.T.; Kim, T.-S. Human activity recognition via recognized body parts of human depth silhouettes for residents monitoring services at smart homes. *Indoor Built Environ.* **2013**, *22*, 271–279. [CrossRef]
86. Manos, A.; Klein, I.; Hazan, T. Gravity-based methods for heading computation in pedestrian dead reckoning. *Sensors* **2019**, *19*, 1170. [CrossRef]
87. Jalal, A.; Batool, M.; Kim, K. Sustainable Wearable System: Human Behavior Modeling for Life-logging Activities Using K-AryTree Hashing Classifier. *Sustainability* **2020**, *12*, 10324. [CrossRef]
88. Cruciani, F.; Vafeiadis, A.; Nugent, C.; Cleland, I.; McCullagh, P.; Votis, K.; Giakoumis, D.; Tzovaras, D.; Chen, L.; Hamzaoui, R. Feature learning for human activity recognition using convolutional neural networks: A case study for inertial measurement unit and audio data. *CCF Trans. Pervasive Comput. Interact.* **2020**, *2*, 18–32. [CrossRef]
89. Jalal, A.; Ahmed, A.; Rafique, A.A.; Kim, K. Scene Semantic Recognition Based on Modified Fuzzy C-Mean and Maximum En-tropy Using Object-to-Object Relations. *IEEE Access* **2021**, *9*, 27758–27772. [CrossRef]
90. Won, Y.-S.; Jap, D.; Bhasin, S. Push for More: On Comparison of Data Augmentation and SMOTE with Optimised Deep Learning Architecture for Side-Channel Information Security Applications. In Proceedings of the Information Security Applications: 21st International Conference, WISA 2020, Jeju Island, Republic of Korea, 26–28 August 2020; Volume 12583, ISBN 978-3-030-65298-2
91. Hartmann, Y.; Liu, H.; Schultz, T. Interactive and Interpretable Online Human Activity Recognition. In Proceedings of the 2022 IEEE International Conference on Pervasive Computing and Communications Workshops and other Affiliated Events (PerCom Workshops), Pisa, Italy, 20–25 March 2022; pp. 109–111.
92. Jalal, A.; Khalid, N.; Kim, K. Automatic Recognition of Human Interaction via Hybrid Descriptors and Maximum Entropy Markov Model Using Depth Sensors. *Entropy* **2020**, *22*, 817. [CrossRef] [PubMed]
93. Vaizman, Y.; Ellis, K.; Lanckriet, G. Recognizing Detailed Human Context in the Wild from Smartphones and Smartwatches. *IEEE Pervasive Comput.* **2017**, *16*, 62–74. [CrossRef]
94. Sztyler, T.; Stuckenschmidt, H. Online personalization of cross sub-jects based activity recognition models on wearable devices. In Proceedings of the 2017 IEEE International Conference on Pervasive Computing and Communications (PerCom), Kona, HI, USA, 13–17 March 2017; pp. 180–189.
95. Garcia-Gonzalez, D.; Rivero, D.; Fernandez-Blanco, E.; Luaces, M.R. A public domain dataset for real-life human activi-ty recognition using smartphone sensors. *Sensors* **2020**, *20*, 2200. [CrossRef]
96. Jalal, A.; Kim, Y.-H.; Kim, Y.-J.; Kamal, S.; Kim, D. Robust human activity recognition from depth video using spatiotemporal multi-fused features. *Pattern Recognit.* **2017**, *61*, 295–308. [CrossRef]
97. Sheng, H.; Wang, S.; Yang, D.; Cong, R.; Cui, Z.; Chen, R. Cross-View Recurrence-Based Self-Supervised Super-Resolution of Light Field. *IEEE Trans. Circuits Syst. Video Technol.* **2023**, *33*, 7252–7266. [CrossRef]
98. Wang, L.; Ciliberto, M.; Gjoreski, H.; Lago, P.; Murao, K.; Okita, T.; Roggen, D. Locomotion and Transportation Mode Recognition from GPS and Radio Signals: Summary of SHL Challenge 2021. In *Adjunct Proceedings of the 2021 ACM International Joint Conference on Pervasive and Ubiquitous Computing and Proceedings of the 2021 ACM International Symposium on Wearable Computers (UbiComp/ISWC '21 Adjunct)*, Virtual, 21–26 September 2021; Association for Computing Machinery: New York, NY, USA, 2021.
99. Fu, C.; Yuan, H.; Xu, H.; Zhang, H.; Shen, L. TMSO-Net: Texture adaptive multi-scale observation for light field image depth estimation. *J. Vis. Commun. Image Represent.* **2023**, *90*, 103731. [CrossRef]
100. Luo, G.; Xie, J.; Liu, J.; Luo, Y.; Li, M.; Li, Z.; Yang, P.; Zhao, L.; Wang, K.; Maeda, R.; et al. Highly Stretchable, Knittable, Wearable Fiberform Hydrovoltaic Generators Driven by Water Transpiration for Portable Self-Power Supply and Self-Powered Strain Sensor. *Small* **2023**, *20*, 2306318. [CrossRef]
101. Feng, Y.; Pan, R.; Zhou, T.; Dong, Z.; Yan, Z.; Wang, Y.; Chen, P.; Chen, S. Direct joining of quartz glass and copper by nanosecond laser. *Ceram. Int.* **2023**, *49*, 36056–36070. [CrossRef]
102. Miao, Y.; Wang, X.; Wang, S.; Li, R. Adaptive Switching Control Based on Dynamic Zero-Moment Point for Versatile Hip Exoskeleton Under Hybrid Locomotion. *IEEE Trans. Ind. Electron.* **2022**, *70*, 11443–11452. [CrossRef]
103. Xu, C.; Jiang, Z.; Wang, B.; Chen, J.; Sun, T.; Fu, F.; Wang, C.; Wang, H. Biospinning of hierarchical fibers for a self-sensing actuator. *Chem. Eng. J.* **2024**, *485*, 150014. [CrossRef]
104. Liu, Y.; Fang, Z.; Cheung, M.H.; Cai, W.; Huang, J. Mechanism Design for Blockchain Storage Sustainability. *IEEE Commun. Mag.* **2023**, *61*, 102–107. [CrossRef]
105. Fu, X.; Pace, P.; Aloi, G.; Guerrieri, A.; Li, W.; Fortino, G. Tolerance Analysis of Cyber-Manufacturing Systems to Cascading Failures. *ACM Trans. Internet Technol.* **2023**, *23*, 1–23. [CrossRef]
106. Wang, S.; Sheng, H.; Yang, D.; Zhang, Y.; Wu, Y.; Wang, S. Extendable Multiple Nodes Recurrent Tracking Framework with RTU++. *IEEE Trans. Image Process.* **2022**, *31*, 5257–5271. [CrossRef]
107. Yang, D.; Zhu, T.; Wang, S.; Wang, S.; Xiong, Z. LFRSNet: A Robust Light Field Semantic Segmentation Network Combining Contextual and Geometric Features. *Front. Environ. Sci.* **2022**, *10*, 1443. [CrossRef]
108. Asim, Y.; Azam, M.A.; Ehatisham-Ul-Haq, M.; Naeem, U.; Khalid, A. Context-Aware Human Activity Recognition (CAHAR) in-the-Wild Using Smartphone Accelerometer. *IEEE Sens. J.* **2020**, *20*, 4361–4371. [CrossRef]

109. Vaizman, Y.; Weibel, N.; Lanckriet, G. Context Recognition In-the-Wild: Unified Model for Multi-Modal Sensors and Multi-Label Classification. *Proc. ACM Interact. Mob. Wearable Ubiquitous Technol.* **2017**, *1*, 168. [CrossRef]
110. Sharma, A.; Singh, S.K.; Udmale, S.S.; Singh, A.K.; Singh, R. Early Transportation Mode Detection Using Smartphone Sensing Data. *IEEE Sens. J.* **2021**, *21*, 15651–15659. [CrossRef]
111. Akbari, A.; Jafari, R. Transition-Aware Detection of Modes of Locomotion and Transportation through Hierarchical Segmentation. *IEEE Sens. J.* **2020**, *21*, 3301–3313. [CrossRef]
112. Brimacombe, O.; Gonzalez, L.C.; Wahlstrom, J. Smartphone-Based CO2e Emission Estimation Using Transportation Mode Clas-sification. *IEEE Access* **2023**, *11*, 54782–54794. [CrossRef]

Disclaimer/Publisher's Note: The statements, opinions and data contained in all publications are solely those of the individual author(s) and contributor(s) and not of MDPI and/or the editor(s). MDPI and/or the editor(s) disclaim responsibility for any injury to people or property resulting from any ideas, methods, instructions or products referred to in the content.

Article

Classification and Analysis of Human Body Movement Characteristics Associated with Acrophobia Induced by Virtual Reality Scenes of Heights

Xiankai Cheng [1,2], Benkun Bao [1,2], Weidong Cui [2], Shuai Liu [1,2], Jun Zhong [1,2], Liming Cai [2] and Hongbo Yang [1,2,*]

[1] School of Biomedical Engineering (Suzhou), Division of Life Sciences and Medicine, University of Science and Technology of China, Hefei 230026, China; chengxk@sibet.ac.cn (X.C.); bbk21468@mail.ustc.edu.cn (B.B.); shliu2000@mail.ustc.edu.cn (S.L.); zhongj@sibet.ac.cn (J.Z.)
[2] Suzhou Institute of Biomedical Engineering and Technology, Chinese Academy of Sciences, Suzhou 215163, China; cuiwd@sibet.ac.cn (W.C.); cailm@sibet.ac.cn (L.C.)
* Correspondence: yanghb@sibet.ac.cn; Tel.: +86-159-9588-1535

Abstract: Acrophobia (fear of heights), a prevalent psychological disorder, elicits profound fear and evokes a range of adverse physiological responses in individuals when exposed to heights, which will lead to a very dangerous state for people in actual heights. In this paper, we explore the behavioral influences in terms of movements in people confronted with virtual reality scenes of extreme heights and develop an acrophobia classification model based on human movement characteristics. To this end, we used wireless miniaturized inertial navigation sensors (WMINS) network to obtain the information of limb movements in the virtual environment. Based on these data, we constructed a series of data feature processing processes, proposed a system model for the classification of acrophobia and non-acrophobia based on human motion feature analysis, and realized the classification recognition of acrophobia and non-acrophobia through the designed integrated learning model. The final accuracy of acrophobia dichotomous classification based on limb motion information reached 94.64%, which has higher accuracy and efficiency compared with other existing research models. Overall, our study demonstrates a strong correlation between people's mental state during fear of heights and their limb movements at that time.

Keywords: acrophobia; virtual reality; body movement; machine learning; sensor network

1. Introduction

Human movement data are a valuable information resource. A number of studies have now revealed that human movement information can be used to assess the quality of human health and to classify and identify people's daily activities. For example, inertial sensors were attached to the legs of infants to obtain motion acceleration data and angular velocity data to explore the relationship between motion complexity and developmental outcomes in infants at high familial risk for autism spectrum disorder (ASD) (HR infants) [1], and to develop a classification model between HR infants and normal infants. Statistical analysis of motor acceleration signals at the wrist and ankle in children with normal development and children with attention deficit hyperactivity disorder (ADHD) [2] suggests that the two have different statistical properties of behavior. With the help of motion data from inertial guidance sensors at the human wrist and ankle, or from the accelerometer data that comes with the mobile phone, direct analysis is performed and a classification model [3] is built to enable activity recognition [4,5] for everyday activities such as walking, running, and walking up and down stairs [6], as well as more in-depth gait analysis [7], fall detection [8], and biomedical information detection [9,10]. Further, six-axis inertial sensors [11] are attached to the head, left forearm and right forearm, as well as the left lower tibiae and right lower tibiae, to obtain motion information data and reproduce the human posture through a data network.

At the same time, human movement data in a particular environment can also be used to stimulate and reflect certain current psychological states, such as fear of heights. Acrophobia (fear of heights) is a psychological disorder that refers to an extreme fear and discomfort of heights, even to the point of pathology. People with acrophobia usually feel the fear of losing their balance or falling from high places—even when they are in the safety of a fence or protective barrier—and experience a strong sense of unease and fear. Research indicates that individuals with acrophobia exhibit significant differences in their motor patterns compared to those without acrophobia. Specifically, acrophobia patients tend to exhibit more cautious, defensive postures such as hunching and neck retraction in high places [12,13], while individuals without acrophobia stand or walk more relaxed, confident, and naturally. With the development of society, more and more skyscrapers have emerged [14], creating more opportunities for people to experience high altitude. However, for individuals suffering from acrophobia, this can be dangerous.

Therefore, it is important to assess and analyze individuals with acrophobia to help them understand the extent of their fear and receive personalized treatment when considered necessary. Currently, the mainstream assessment and treatment methods for acrophobia are cognitive-behavioral therapy and virtual reality exposure therapy. Cognitive-behavioral therapy involves face-to-face communication between a psychotherapist and the patient for assessment and treatment [15,16]. While this approach is effective, the one-to-one communication sessions will not only lead to overall inefficiency, but will mean that a large number of psychotherapists will be needed at a social level to focus on the problem; therefore, this approach will be too dependent on the professionalism of the therapist. The second method is virtual reality exposure therapy (VRET) [17], a behavioral therapy used for anxiety disorders, including phobias. Patients immerse themselves in a computer-generated virtual environment that can provide various scenarios that are difficult to create in the real world [18]. The environment allows for controlled safety measures that can be manipulated to meet the individual's needs. As early as 1990, virtual reality technology was used in research related to phobias [19]. In recent years, more research on acrophobia has been based on virtual reality technology. Due to the expanded visual range, high-altitude environments tend to cause vertigo and affect people's control of their posture [20–22]. It has been found that, from a power spectrum perspective, an increase in height in the virtual environment typically reduces low frequency (<0.5 Hz) body sway and increases oscillations in high frequency (>1 Hz) body sway in the frequency range 0–3 Hz, and this change is more pronounced in people who are truly afraid of heights [23]. In addition, the amplitude of body sway decreases with increasing virtual height [24]. This means that people who are afraid of heights and those who are not afraid of heights will have a different sense of motion pattern in an aerial environment.

In conclusion, we hope to investigate automated models for classifying and analyzing acrophobia by examining the movement patterns of acrophobic and non-acrophobic people at height, so that people can quickly and correctly recognize their acrophobic state. To achieve this, we designed common daily movement tasks (walking, ball retrieval, ball release, jumping) in a virtual reality altitude environment and used a set of easy-to-use, miniaturized wireless sparse sensor devices designed to acquire movement information at joints to characterize the movement of acrophobic and non-acrophobic humans under high-altitude conditions, and proposed a set of human movement-based acrophobic and non-acrophobic classification algorithm models, which can achieve efficient and fast classification.

2. Materials and Methods

This section centers on the wireless miniaturized inertial navigation sensor (WMINS) and the development of immersive virtual reality scenario, alongside the associated data collection and analysis classification process.

2.1. Wireless Miniaturized Inertial Navigation Sensor (WMINS)

As shown in Figure 1a, we have developed the WMINS sensor with a housing size of only 23.5 × 18.6 × 9.6 mm and a bare board size of 16.75 × 11.22 mm, the size of a common dollar coin in the market. The weight is amazingly light at 3.8 g, equivalent to a common dollar coin in the market. Due to its lightweight and small size, the WMINS sensor can be attached to any part of the human body to collect motion information without being restricted by joints. The sensor comprises two major functional modules—a motion collection module for collecting the motion information of the attached object and a data transmission module that uses low-power Bluetooth technology for data transmission and reception [25]. Furthermore, by designing a Bluetooth one-to-four wireless sensor network, four WMINS sensors can independently collect and transmit data simultaneously, thus constructing the motion collection system for key body parts. Figure 1b illustrates the specific usage mode of the WMINS sensor in human motion detection and collection. Four WMINS sensors are attached to the upper part of the left and right wrists and lower tibiae of the human body, which are key parts of the limbs, and data acquisition and transmission occur at 30 Hz. Therefore, motion information is obtained through these four joint parts. As the WMINS sensor is remarkably small, it does not require an additional touch switch, as the touch switch is already integrated on the back to save space. With just a light touch of two fingers, WMINS can be effortlessly turned on. Once the sensor is activated, the top left corner of the back will flash blue light at 1 Hz. When the sensor is connected to the receiving device, WMINS will display a steady blue light. Figure 1c showcases the charging mode of WMINS. We have designed a charging box that can accommodate multiple sensors for charging. To minimize the size of WMINS, we have adopted a spring probe charging mode, which incorporates a retractable spring into each charging port that matches the WMINS in Figure 1a. In order to reduce the weight of WMINS, a small circular magnet is placed externally on the portable charging box to fix the sensor in the charging state. The left image in Figure 1c shows the sensor in an uncharged state, and the right image shows the sensor in charging state, with WMINS displaying a yellow light.

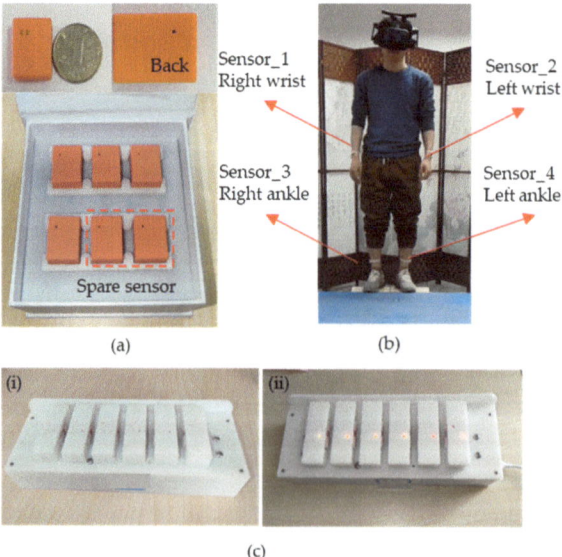

Figure 1. Introduction to wireless miniaturized inertial navigation sensor (WMINS). (**a**) Outline dimensions of the sensor and the introduction of the whole set of equipment. (**b**) Schematic diagram of the attachment position of the sensor to the human body. (**c**) External magnetic charging of the sensor. (**i**) No charging state, (**ii**), charging state.

To collect motion data, we attached four WMINS sensors to specific joints on the human body, as illustrated in Figure 1b. The motion data were collected at a sampling frequency of 30 Hz for a duration of 100 s. The three-axis acceleration and three-axis angular velocity data collected from the left wrist and left lower tibia are displayed in Figure 2.

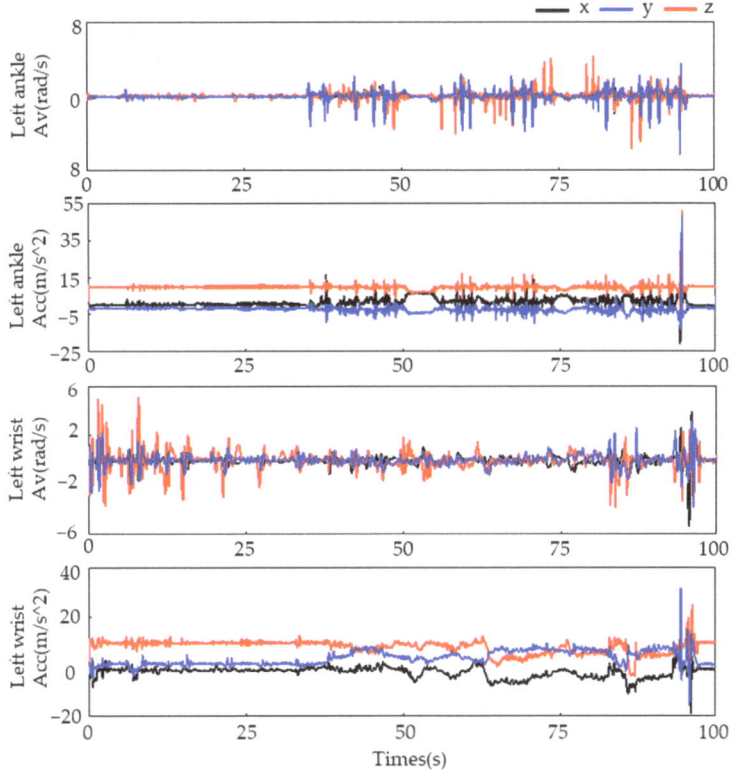

Figure 2. The collected motion data of the left wrist and left lower tibia.

2.2. VR Scene and Task Introduction

We designed a set of virtual scenarios to induce controllable states of acrophobia and explore two different movement states, acrophobia and non-acrophobia, in response to acrophobia. The scenarios were created using VR scenes and WMINS, and were centered around the psychological stress response of individuals to acrophobia environments or events in the real world. Our goal was to create a repeatable and easily operated stress induction paradigm in the laboratory environment, which would elicit subjects' stress response under natural conditions while recording their movement state. The virtual scene mode was comprised of three main parts: the VR scenario, physical tactile stimulation, and data acquisition. The VR scenario consisted of a 60-story high-rise building designed in VR. Subjects were transported to the high-altitude scenario via an elevator and encountered a virtual wooden board extending outward from the high-altitude elevator. At the end of the wooden board, a high-altitude diving platform was designed, with four basketball hoops placed in front, back, left, and right of the wooden board, respectively. Additionally, two virtual tennis balls were designed in the rectangular basket at the end of the wooden board.

In addition to building a high-altitude scene in the virtual environment, we also constructed a real scene that was 25 cm above the ground, based on the virtual coordinates. The size of the plank and basketball hoops in the real scene were designed to be in a 1:1 ratio with the virtual scene, ensuring that the subjects could fully experience the VR scene in the helmet while maintaining safety during the experiment. To increase the realism of the VR scene, we incorporated a vibration module in the elevator area in the physical touch stimulation part. When the subject stood in the elevator area, the vibration module was activated when the elevator in the VR scene ascended. This allowed the subject to feel the vibration of the elevator ascending from the soles of their feet, simulating the ascent and descent of the high-altitude elevator in the virtual environment. Figure 3b shows the left and right swaying of the plank in the experimental scene. In the VR scene, the plank swayed left and right, so we designed a plank rotation mode with the same amplitude and a rotation frequency of 0.5 Hz. The maximum inclination angle of the plank was 16 degrees. When subjects walked on the plank, they could feel the realism of the VR scene. Two handles were placed in the end basket of the plank to simulate the tennis ball models in the VR scene. Throughout the experiment, the subject's motion data were collected in real-time by the WMINS sensors and obtained and saved in real-time by connecting to the VR scene. Overall, our experimental design allowed us to create highly realistic and immersive experience for the subjects while collecting valuable data on their psychological and physiological responses in high-altitude environments and related anxiety disorders.

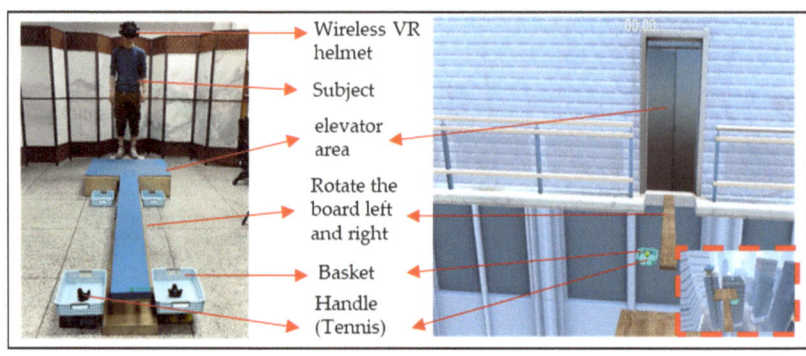

(a)

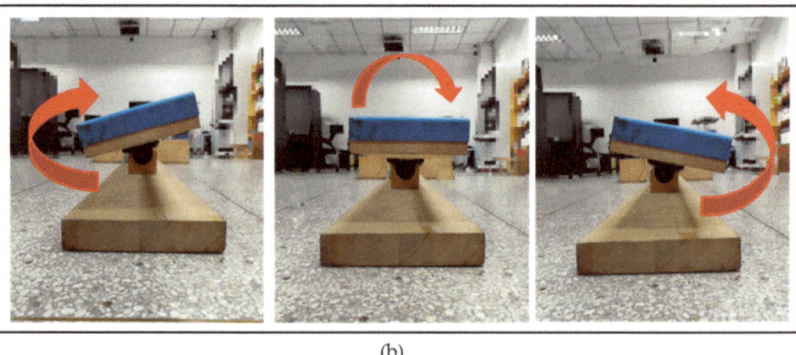

(b)

Figure 3. Introduction to the VR experiment scene. (**a**) The connection between the VR environment and the actual scene construction. (**b**) The plank turns left and right.

2.3. Acrophobia Scale

We utilized two assessment scales, the virtual reality scene quality assessment (VRSQ) [26] and the height interpretation questionnaire (HIQ) [27], to evaluate the subjects' response to specific VR environments and their level of acrophobia, respectively. The VRSQ is a questionnaire that evaluates the level of dizziness and discomfort experienced by users in VR environments. It was developed in 2002 by Kenneth J. Miller et al. We also comprised 16 questions that describe the onset time, severity, and various possible symptoms of dizziness and discomfort. These questions aid evaluators in determining the degree of discomfort experienced by users in VR environments. The HIQ, developed jointly by American psychologists Gary R.

Parker and Robert G. Stumpf, is a questionnaire consisting of 32 questions designed to assess acrophobia. The questionnaire is primarily divided into three parts: high-altitude experiences, high-altitude imagination, and behavioral avoidance. Subjects answer these questions based on their actual experiences, and the answer is selected on a five-level scale. By statistically analyzing the responses of subjects, an overall HIQ score can be obtained, with a higher score indicating a more severe level of acrophobia. The full HIQ scale has a Cronbach alpha of 0.87, meaning that the HIQ is a reliable instrument for measuring height interpretation. In summary, the HIQ is a valid psychometric tool with high reliability and validity that can help clinicians and psychologists understand the level of individual acrophobia and provide a valuable reference for the treatment and prevention of acrophobia.

2.4. Experimental Design

The procedure was as follows: prior to the test, the subjects complete the VRSQ and HIQ scale to assess their acceptance of the VR environment and to make a preliminary assessment of their acrophobia state. The subjects then enter the three-dimensional virtual scene using a wireless VR helmet and WMINS. They stand in the elevator area, where the vibration module is automatically activated to simulate vibration as the elevator rises and falls. When the virtual elevator reaches the 60th floor, the elevator door opens, and subjects exit the elevator area, as detected by the wireless VR helmet's movement tracking. The movement of leaving the lift triggers a rotational movement of the board from side to side, while the participant has to balance on the rotating board at all times and walk to the end of the board, pick up the tennis ball in the rectangular basket at the end of the board, turn around, and walk back to the starting position of the board and place the tennis ball into the rectangular basket at the starting position. After placing the tennis ball in the rectangular basket at the start, they then continue to the end of the board and pick up an-other tennis ball and walk back to the start of the board and place the tennis ball in it. When the last tennis ball has been placed in the rectangular basket at the start position, the left/right switch will automatically turn off and the board will automatically re-turn to a horizontal position. At this point, the participant walks to the end of the board and jumps 40 cm onto a circular diving platform in front of them.

For the purpose of data analysis, we divided the entire VR acrophobia experiment into three main stages: the rest adaptation stage, the board walking stage, and the jumping stage. These three stages were further divided into 11 tasks, as shown in Table 1. In the rest adaptation stage, subjects were divided into four task stages in the VR elevator environment, including resting and waiting in the elevator interior (task0), elevator ascent (task1), elevator door opening (task2), and elevator closing after subjects walked out of the elevator (task3). In the board walking stage, subjects were required to walk back and forth twice on a fixed frequency shaking board, pick up the tennis balls located at the end of the board and place them in the net located in the same direction at the beginning of the board. This stage included tasks 4 to 8. The jumping stage included the jump preparation stage (task9) and the final jump stage (task10). In each stage of the experiment, the movement status of the left and right wrist and the left and right lower tibia of the subjects were recorded and transmitted to the computer receiver in real time by WMINS. Figure 4 details the specific movements of subjects in the 11 tasks.

Table 1. Specific task division of VR acrophobia experiment.

Task Name	Task Content	Adjusted Task Name
Task 0	Rest and wait stage	New task 0 (T0)
Task 1	Elevator ascending stage	New task 1 (T1)
Task 2	Elevator opening stage	
Task 3	Elevator closing stage	New task 2 (T2)

Table 1. Cont.

Task Name	Task Content	Adjusted Task Name
Task 4	Pick up the ball I	
Task 5	Release the ball I	
Task 6	Pick up the ball II	New task 3 (T3)
Task 7	Release the ball II	
Task 8	Go to the end of the single wooden bridge	
Task 9	Pre-jump preparation stage	New task 4 (T4)
Task 10	Jump to the specified position stage	

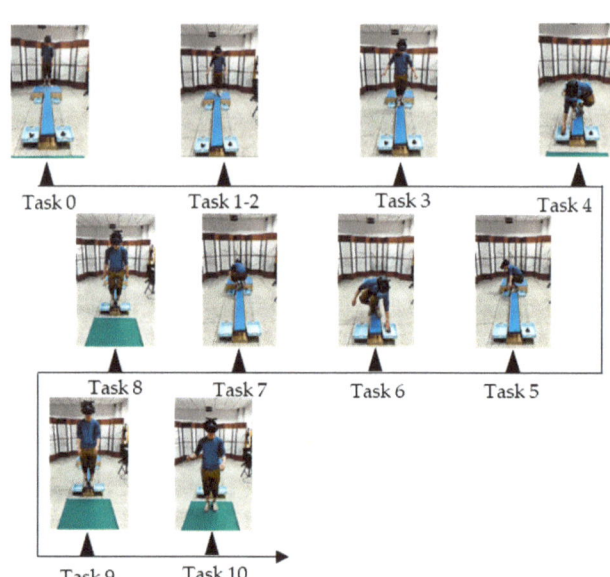

Figure 4. Eleven kinds of task action display under the actual VR acrophobia experiment.

2.5. Acquisition of Experimental Data

A total of 31 subjects were recruited for the present study, and each participant completed 2 experiments. All subjects had no history of neurological disorders or photosensitive epilepsy, and furthermore, were in a normal physical state with basic limb movement abilities during the experiment. Prior to the experiment, subjects were correctly informed of the specific details of the experimental procedure, including the purpose, process, etc., and signed an informed consent form. By triggering each task end node in order among the 11 tasks, a complete VR acrophobia experiment was performed. Finally, 1 experiment can obtain 11 segments of movement data and generate movement data files for each participant for each experiment at key points of the limbs. Due to unexpected situations such as task failure or task timeout during the experiment, 43 usable sample data were obtained after data cleaning, including 24 male subjects and 19 female subjects, with all subjects aged between 23 and 30 years old. According to the HIQ questionnaire administered before and after the experiment to label the subjects, those with HIQ scores greater than 29 were classified as the acrophobia group and those with scores less than 29 were classified as the non-acrophobia group [28]. Finally, there were 16 HIQ scores greater than 29 and 27 scores less than or equal to 29 in the 43-point sample, which included 1 sample with a score of exactly 29.

Considering that the speed at which the subjects perform the tasks in the experiment may vary greatly, hesitant behavior may occur in some tasks for subjects with acrophobia, while non-acrophobia subjects may quickly pass through the experiment, resulting in insufficient movement data for some tasks, making it difficult to obtain accurate quantitative features. Therefore, the 11 tasks were adjusted based on their nature and divided into 5 major tasks, as shown in Table 1. The adjusted Task 3 includes the entire process of picking up and putting down the ball, while Task 4 includes the entire motion data during the jumping period. As shown in Figure 5, the sample data for each subject contains movement data of four joints (right and left wrist and the left and right lower tibia) in the human body, with each joint movement data including five task types, with each data point in each task type including the adjusted task name, acceleration and angular velocity along three axes, and calculated three-axis combined acceleration and angular velocity.

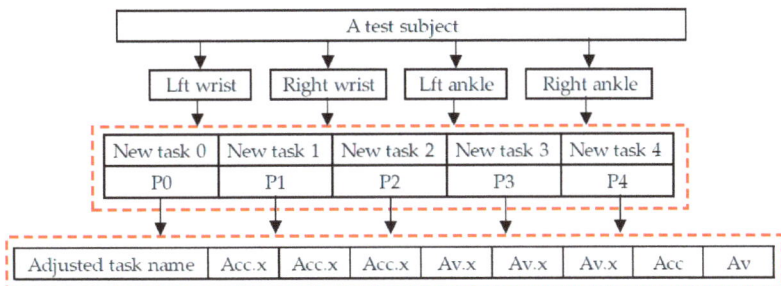

Figure 5. The adjusted experimental data composition display.

3. Results

We divided the labeled motion data of the subjects into four types of joint data based on the attachment position of the sensor, namely left wrist data, right wrist data, left lower tibia data, and right lower tibia data. As illustrated in Figure 6, each joint data type corresponds to a 6-dimensional matrix under 11 task types. Since the motion under the 11 task types is continuous, we preprocessed the individual joint data by interpolating and low-pass filtering according to the data dimension. Then, we adjusted the 11 task types into 5 major task types according to Table 1, and extracted the features from the data to obtain a task-based feature matrix. Finally, we transformed the data to obtain four joint feature matrices. Based on this, sample data were transformed from raw motion data to four feature matrices, and the three-dimensional feature matrices were transposed and concatenated to obtain an overall feature matrix. The feature matrix was subjected to feature selection and over-sampling, and the processed data were fed into machine learning model for training and classification, ultimately achieving a stable and highly accurate acrophobia classification model.

3.1. Feature Analysis

We conducted a detailed feature analysis of the preprocessed data. During the experiment, subjects with acrophobia label generally took longer to perform the tasks compared to non-fearful subjects, particularly in T4, where fearful subjects had more hesitation time during the final jumping moments, and the jumping movements appeared discontinuous. In T0, the subjects were in a closed space elevator where the floor number increased to represent the elevator going up. For fearful subjects, some of them may have had fewer limb movements due to their inner fear. In T2, when the elevator door opened, the sudden appearance of a high-altitude environment could give subjects a visual shock, increasing the sense of reality and stimulating the subjects' belief in the high-altitude environment. Fearful subjects would spend more time building their confidence after the elevator door opened, and most of them would be in a state of external calmness, rather than a natural movement state as observed in non-fearful subjects who could adapt quickly to the environment. In T3, fearful subjects walked more carefully on the wooden board and their walking steps became smaller and slower, resulting in poorer continuity of their walking movements. In addition, their balance will become worse due to the movement of the plank.

During the actual experiment, they keep the center of mass and the plank on one side at all times by a slight outward extension of the arms and an up-and-down movement, while the subject will bend slightly to lower the center of mass and increase stability as they move. In contrast, non-fearful subjects had better balance and control over their body movements.

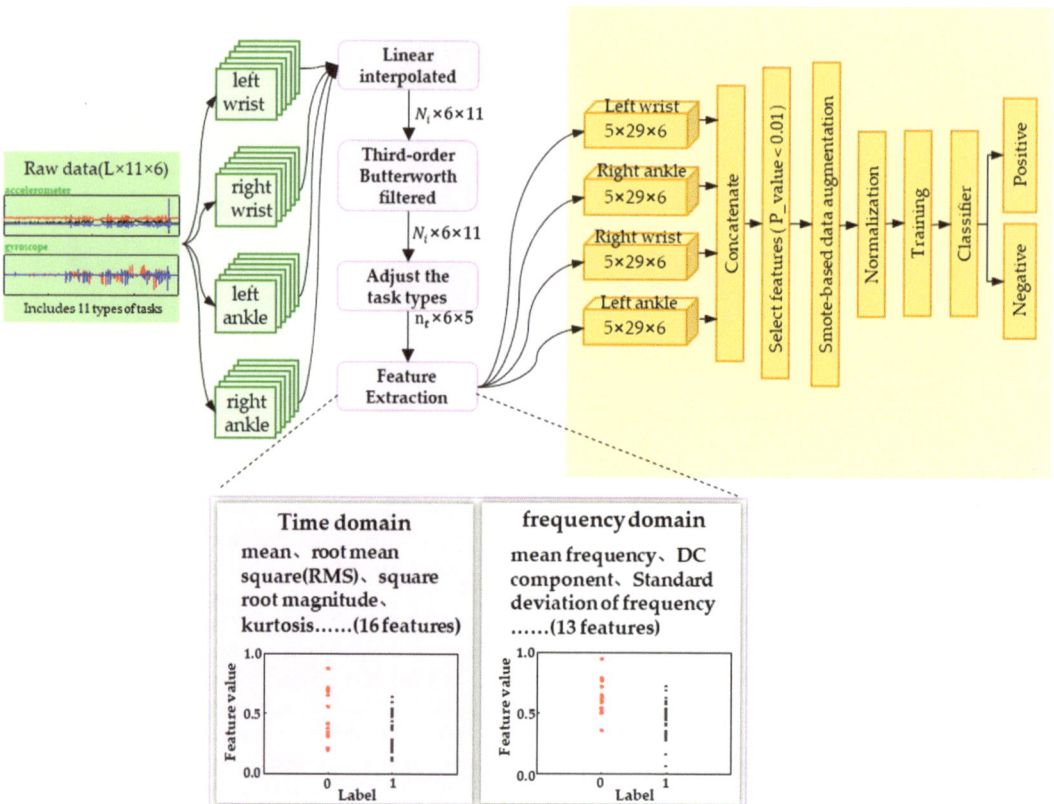

Figure 6. The general diagram of the motion data processing model system for scenes of acrophobia. N_i indicates the data length of each of the 11 tasks, n_t indicates the data length of each task after task adjustment, where i and t both indicate the number of tasks.

Based on the observed behavioral characteristics, we conducted feature analysis on preprocessed data in terms of three-axis acceleration and three-axis angular velocity. Feature extraction was performed in both time and frequency domains to obtain a comprehensive set of features. In the time domain [29], we extracted typical features such as mean, variance, maximum, and minimum values for each data dimension under different task types. In addition, we also obtained specific features such as impulse factor, skewness factor, and peak-to-peak. In total, we extracted 16 features in the time domain. In the frequency domain [30], we extracted 13 different types of features, including frequency center, RMS frequency, and DC component. Ultimately, each data dimension for different task types had a total of 29 features in both time and frequency domains. Table 2 provides detailed information on these 29 features. Finally, we split and combined the feature data according to the joint positions specified by WMINS, resulting in 4 joint positions with $5 \times 29 \times 6$ feature matrices.

Table 2. Selected time-domain and frequency-domain feature names and corresponding notations.

Symbol	Feature
fea1	mean
fea2	root mean square (RMS)
fea3	square root magnitude
fea4	absolute average
fea5	skewness
fea6	Kurtosis
fea7	variance
fea8	max
fea9	min
fea10	peak-to-peak
fea11	form factor
fea12	crest factor
fea13	impulse factor
fea14	margin factor
fea15	skewness factor
fea16	kurtosis factor
fea17	Mean frequency
fea18	Standard deviation of frequency
fea19	the degree of dispersion or concentration of the spectrum
fea20	DC component
fea21	frequency center
fea22	the degree of dispersion or concentration of the spectrum
fea23	RMS frequency
fea24	Indicates a change in the position of the main frequency band
fea25	Indicates a change in the position of the main frequency band
fea26	the degree of dispersion or concentration of the spectrum I
fea27	the degree of dispersion or concentration of the spectrum II
fea28	the degree of dispersion or concentration of the spectrum III
fea29	the degree of dispersion or concentration of the spectrum IV

After feature data segmentation and solving, we obtained 3480 different feature types. However, the number of samples was only 43, much less than the number of feature types. To prevent overfitting in the subsequent classification model, we performed feature selection using p-values, a statistical method [31]. p-value is a probability value obtained from statistical analysis, which represents the probability of observing the results or more extreme results under the null hypothesis. When the p-value is less than the set significance level (e.g., 0.05), the null hypothesis is rejected, and the observed results are considered significant, indicating that the differences observed in the sample may not be due to chance. When the p-value is greater than the set significance level, the null hypothesis cannot be rejected. Here, we have labeled the acrophobia group "0" and the non-acrophobia group "1". Similarly, the features solved for the acrophobia group will be labelled "0" and the features solved for the non-acrophobia group will be labelled "1", and finally, the p-values for each class of features will be solved directly for that class of features based on these feature values and the corresponding labels. We sorted the WMINS under each joint in ascending order according to their p-values and selected the features whose p-values were less than 0.02. We finally obtained 88 data features and replaced the original motion data with these features to represent the motion state of each subject. These 88 features were used as the original feature data set for building the classification model.

3.2. Sample Balancing

As there was an imbalance in the sample data between non-acrophobia and acrophobia samples, which could lead to a decrease in model generalization and prediction accuracy, K-means smote was used to oversample the sample data [32]. K-means smote is an improved oversampling method used to address the issue of class imbalance in classification problems. K-means smote uses a weight scheme based on the K-nearest neighbor algorithm to control the weights between different samples to avoid oversampling between different clusters. The 88 features were oversampled using K-means smote for each feature dimension, resulting in an increase in the data weight of each feature type from 43 to 56 samples. With 88 feature types representing 1 sample, a total of 56 samples were obtained, with 13 additional acrophobia feature samples.

3.3. Acrophobia Classification Model

We designed a voting classification algorithm specifically for the acrophobia and non-acrophobia movement data. Voting is an ensemble learning-based classification algorithm [33] that uses a majority vote to determine the final classification. We used KNN [34], extra trees, random forest [35], SVM [36], and logistic regression [37] as the five classifiers for model ensemble. First, we divided the dataset into a training set and a test set. During the training phase, we used the training set data to train the five classifiers and obtained five well-trained models. Finally, we input the test set data into each well-trained model for classification, obtaining the classification results of each model. Then, we aggregated the classification results of each model and used hard voting to determine the final classification result. Specifically, for each test sample, we counted the classification results of the five classifiers, and selected the class with the most votes as the final classification result. The voting classification algorithm we designed combines the advantages of five machine learning models and has better classification performance for acrophobia and non-acrophobia movement data.

3.4. Evaluation Metrics

In order to comprehensively evaluate the classification performance of our algorithm models, we used four quantitative evaluation parameters: accuracy, recall, precision, and F1-score [38]. Accuracy measures the proportion of correctly predicted samples to the total number of samples, and is a good indicator of model performance when the proportion of positive and negative samples is relatively balanced. Recall measures the proportion of correctly predicted positive samples to the actual number of positive samples, reflecting the model's ability to detect positive samples. Precision measures the proportion of correctly predicted positive samples to the total number of predicted positive samples, indicating the proportion of true positives in the predicted positive samples. F1-score is the harmonic mean of precision and recall, providing a comprehensive measure of the model's performance. The higher the F1-score, the better the performance of the classification model.

$$\text{precision} = \frac{TP}{TP + FP} \quad (1)$$

$$\text{recall} = \frac{TP}{TP + FN} \quad (2)$$

$$\text{accuracy} = \frac{TP + TN}{TP + FP + TN + FN} \quad (3)$$

$$F1 = \frac{2 \times \text{Precision} \times \text{Recall}}{\text{Precision} + \text{Recall}} \quad (4)$$

These evaluation metrics can be defined by Equation (1) to Equation (4). Here, TP represents the number of positive samples that the model predicted as positive, FP represents the number of negative samples that the model predicted as positive, FN represents the number of positive samples that the classification model predicted as negative, and TN represents the number of negative samples that the classification model predicted as negative.

3.5. Experimental Platform and Model Setting

We employed a computer equipped with an Intel Core i5-11400 processor, 16 GB of RAM, and an NVIDIA GeForce RTX 3060 GPU for machine learning training and testing. The computer runs on the Windows 10 operating system and utilizes the Python programming language for algorithm implementation.

We utilized leave-one-out cross validation (LOOCV) to evaluate the classification performance of our model, a commonly used cross-validation method in machine learning. LOOCV trains and evaluates the model on each sample as the validation set, ultimately averaging the evaluation results for each sample, providing an objective evaluation of model performance. After dividing the dataset into training and test sets using the leave-one-out method, we standardized the training set by normalizing its feature data, ensuring that the distribution of feature data had similar scales and ranges to improve the model's accuracy. The standardization formula (Equation (5)) was applied to the training set and then to the test set to normalize the test set in the same manner.

$$x_n = \frac{x_n - \text{mean}(x)}{\text{std}(x)} \tag{5}$$

To verify the effectiveness of our designed voting method, we constructed 10 typical machine learning models, including k-nearest neighbor, decision tree, random forest, naive Bayes classifier, support vector machine, ensemble learning, logistic regression, multi-layer perceptron, Gaussian naive Bayes, and extremely randomized trees. We used the leave-one-out cross-validation method to evaluate each model, ensuring the most objective evaluation of model performance.

3.6. Results and Analysis

We evaluated our designed voting algorithm using feature sample data that had not undergone K-means smote through 43 cycles of leave-one-out cross validation, resulting in a final classification accuracy of 88.37%. Additionally, we fed the same data processed through the same procedure into the other 10 machine learning models we built [39]. Figure 7a displays the final classification accuracies of these 11 models, with our voting ensemble model achieving the highest accuracy of 88.37%, followed by ET and GNB at around 86%. Conversely, DT, RF, and AdaBoost had poor classification performance, with an accuracy of only approximately 60%. Among them, AdaBoost was the worst-performing classification model with an accuracy of only 55.8%.

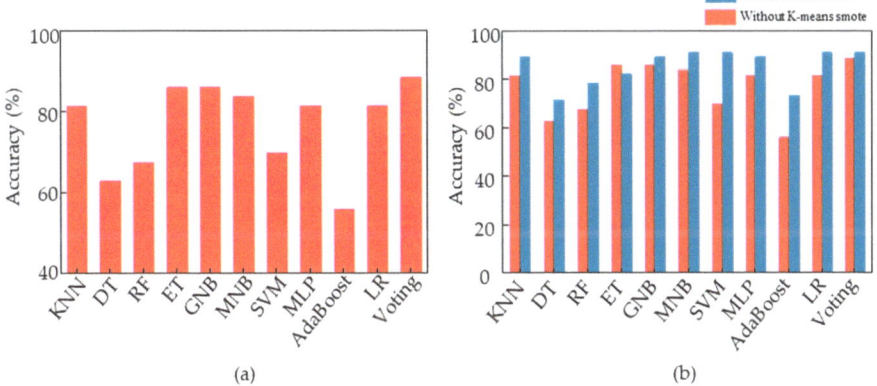

Figure 7. Classification accuracy of each model. (**a**) Classification accuracy of each model obtained by training the sample data without K-means smote. (**b**) Comparison of classification accuracy obtained by training the sample data with and without K-means smote.

To improve the performance of the classification model for acrophobia based on motion data, we applied K-means smote to the feature samples, oversampling the data and feeding it into the voting ensemble model we designed. This resulted in a classification accuracy of 91%. To verify the effectiveness of K-means smote, we fed the new feature samples into 10 other models and compared the accuracy with and without oversampling. Figure 7b illustrates the changes, showing that the K-means smote oversampled data are of higher quality and lead to significant improvements in classification accuracy across various models, including LR, MNB, SVM, and voting, with all achieving a classification accuracy of 91%. The most significant improvement was observed in the AdaBoost model, which achieved a final classification accuracy of 73.2%, an increase of 17.4% from the initial accuracy. Therefore, K-means smote can optimize motion data for acrophobia classification and improve the overall performance of the model at the data level.

Building upon this, we explored the impact of different feature combinations on the performance of the model among 88 features. We reordered the 88 features based on their *p*-values in ascending order and selected the top 10, 20, and so on, at intervals of 10. We thus obtained eight new sets of feature combinations and fed them into the voting ensemble model we designed. We obtained the optimal feature set for each feature data combination. As shown in Figure 8a, we compared the classification accuracy of the voting ensemble model under the 88 features with the other 8 data combinations and found that the classification accuracy reached 94.63% for the top 80 feature combination, which was the highest among all combinations. To further validate the results, we selected the top 75 features with the smallest *p*-values for voting ensemble model training and achieved an accuracy of 91.7%. Figure 8b shows the confusion matrix of the voting ensemble model, with a classification accuracy of 96.3% for non-fear-of-heights samples and 93.1% for fear-of-heights samples.

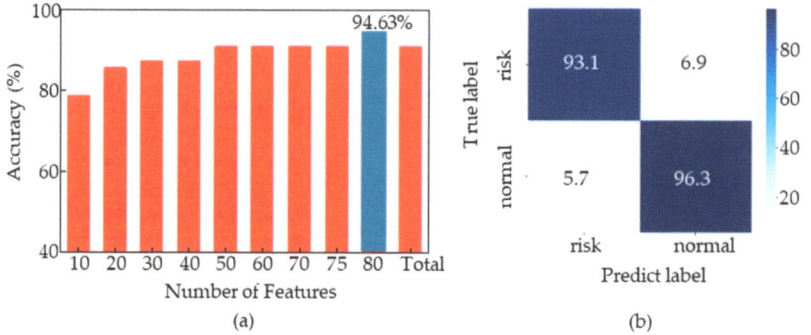

Figure 8. (a) Classification effect of voting model with different combinations of feature data. (b) Confusion matrix with the optimal combination of feature data.

The classification accuracies of the top 80 samples were fed into 10 different machine learning models, and the results are shown in Figure 9. The blue bars represent the classification accuracies of the models using the original top 80 samples without oversampling, while the red line represents the classification accuracies of the models using the top 80 samples after K-means smote oversampling. It can be observed that the Voting algorithm model designed in this study achieved the highest classification accuracy for both the original and oversampled feature samples.

Table 3 presents the specific performance of different algorithm models, and the optimal voting ensemble model achieved a precision of 96.2%, a recall of 92.8%, and an F-score of 94.5%. The precision and F-score were the highest among all models, indicating that the voting ensemble model had the best comprehensive ability to predict non-fearful and fearful samples. In terms of classification accuracy, both the LR model and MNB model had the second-best classification ability, with the same accuracy of 91.07%. However, the precision of the MNB model was 92.59%, while that of the LR model was 88.88%, indicating that the MNB model had a better classification effect on non-fearful samples. In terms of recall, the scores of the 2 models were exactly opposite, with the LR model achieving a score of 92.3% and the MNB model only achieving 89.2%. This indicates that the LR model has a stronger ability to identify non-fearful samples compared to the MNB model.

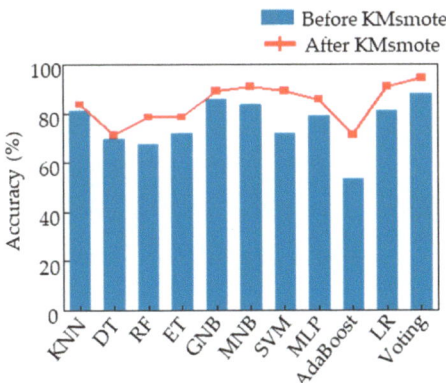

Figure 9. Classification accuracy of each model under Top 80 feature data samples.

Table 3. Classification results of different algorithmic models in top 80 feature sample data.

Models	Accuracy	Precision	Recall	F-Score
KNN	0.83929	0.85185	0.82143	0.83636
DT	0.71429	0.62963	0.73913	0.68000
RF	0.78571	0.74074	0.80000	0.76923
ET	0.78571	0.77778	0.77778	0.77778
GNB	0.89286	0.88889	0.88889	0.88889
MNB	0.91071	0.92593	0.89286	0.90909
SVM	0.89286	0.92593	0.86207	0.89286
MLP	0.85714	0.85185	0.85185	0.85185
AdaBoost	0.71429	0.62963	0.73913	0.68000
LR	0.91071	0.88889	0.92308	0.90566
VOTE	0.94643	0.96296	0.92857	0.94546

In order to explore the optimal combination of human body limb motion features for acrophobia classification and obtain a superior classification model, further analysis was conducted on the top 80 features, as the number of features was too high compared to the final 56 samples. Specifically, we computed the cross-correlation coefficients [40] between different features and determined a series of threshold values to remove redundant features. By setting the correlation coefficient threshold to 0.9, we removed one of the features with a correlation coefficient greater than 0.9, resulting in 48 feature combinations. Figure 10 shows the classification performance of the voting ensemble model trained on these new feature combinations with different correlation coefficient thresholds ranging from 0.4 to 1. We found that the model achieved the best classification accuracy when the correlation coefficient threshold was set to 1, 0.75, 0.7, 0.6, or 0.55, with all accuracies reaching 94.6% through the leave-one-out cross validation. When the threshold was set lower than 0.55, the model's performance began to decline. Therefore, 0.55 was determined as the minimum optimal threshold, resulting in a reduced number of feature types to only 21.

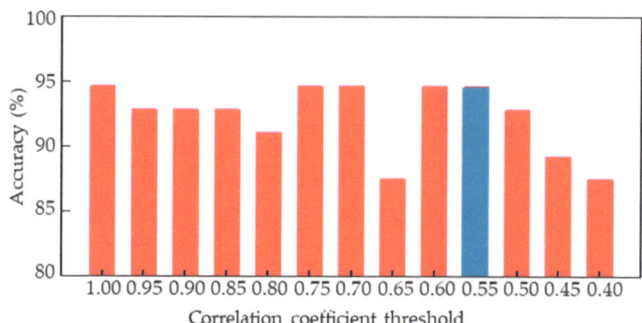

Figure 10. Classification accuracy of feature combinations in voting ensemble models with different correlation coefficient thresholds.

4. Discussion

In this study, a high classification accuracy model of acrophobia and non-acrophobia was developed; on the basis of that, it was revealed that acrophobic and non-acrophobic populations possess different motor patterns under high-altitude conditions. Figure 11 illustrates the selected feature series after feature selection. T3.av_z.fea25 is the feature with the smallest p value, representing feature 25 of the Z-axis angular velocity in the left lower tibia movement data during the T3 task phase. The meanings of the other feature symbols on the horizontal axis are the same as described above. Figure 11a shows the selected left and right wrist features after screening, with a total of 22 features for the left wrist and 21 features for the right wrist. Since the number of features in the left and right wrists is almost equal, the contribution weight of the left and right wrists is equal in the classification of the acrophobia and non-acrophobia labels. Figure 11b shows the selected features in the left and right lower tibia movements, with 26 features for the left lower tibia and 11 features for the right lower tibia, indicating that the left lower tibia contributes more weight to the classification of acrophobia and non-acrophobia due to the asymmetry of lower limb movements in the VR acrophobia experiment. This may be due to the limited walking space on the wooden plank and the subjects' high-altitude state, which causes the lower limbs to move continuously in a single direction, leading to differences in lower tibiae movements. Figure 11b reveals that the left leg is the main force-bearing part, and the right leg is the driven part during walking for both acrophobia and non-acrophobia subjects. Moreover, the T3 task phase is the main walking stage throughout the experiment, which explains why the T3 task phase features account for over 75% of the lower tibiae features, while only about 30% of the wrist features. In conclusion, there are 43 selected features in the wrist and 37 in the lower tibiae, indicating that the upper and lower limbs have a similar weight in acrophobia and non-acrophobia labels.

In addition, we compared our results with those of similar related studies. Hu et al. collected EEG data from subjects in a virtual environment and used deep learning networks to build a 4-classification model, achieving an accuracy of 88.77% [41]. R. Zheng et al. analyzed fear by acquiring EMG, Pupil, and ECG signals from subjects, and built a classification model combining deep learning and machine learning, achieving a binary classification accuracy of 93.93%. Other researchers have either used single or multi-modal physiological signals for subject-related analysis in virtual environments and classified them using ML and DL. These detailed results are shown in Table 4. Among these classifications, our system model achieved the highest classification accuracy of 94.6% using sparse joint data. The 21 selected features significantly reduced the training data required while improving the model's generalizability. This will greatly enhance training efficiency when analyzing large datasets in the future.

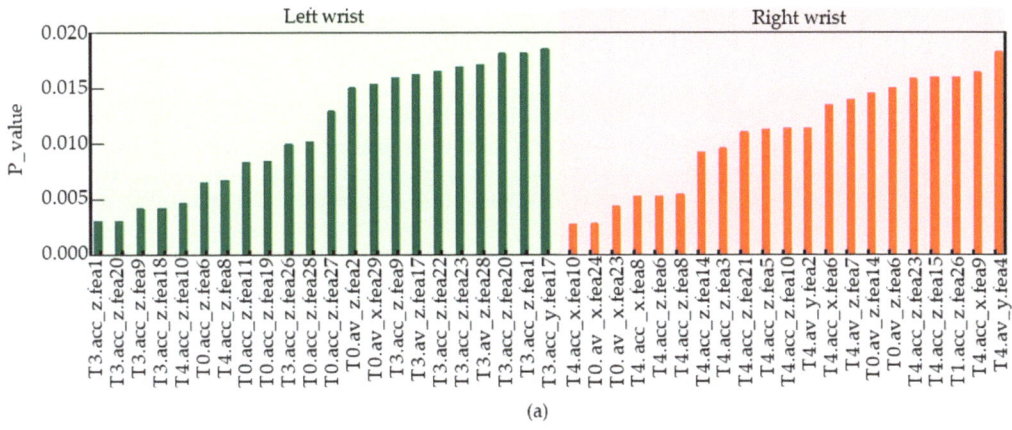

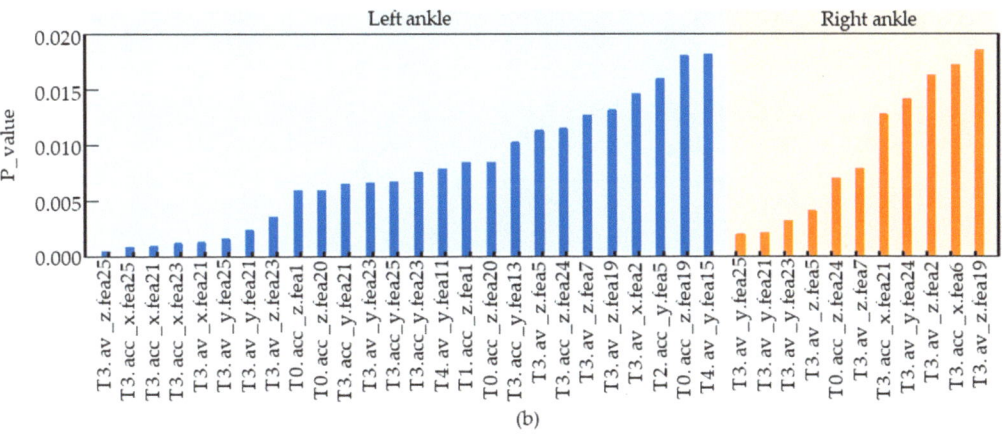

Figure 11. Feature types and corresponding *p* values after feature selection. (**a**) Type of feature selected at the wrist joint and the corresponding *p*-value. (**b**) Type of feature selected at the lower tibiae and the corresponding *p*-value.

Table 4. Comparison with other stress scenario classification models.

Study	Signals	Method	Class	Accuracy
Hu et al. [41]	EEG	DL	4	88.77%
Bălan et al. [39]	EEG, HR, GSR	ML, DL	2-\4-choice scale	89.50%\42.50%
Salkevicius et al. [42]	GSR, BVP, skin temperature	SVM	4	86.30%
Zhang et al. [43]	GSR	BP	2	86.70%
R.Zheng et al. [44]	EMO, Pupil, ECG	ML, DL	2	93.93%
Our method	body motion data	EL	2	94.60%

5. Conclusions

By setting up a series of movement tasks in a virtual reality high-altitude scenario, this study investigated human motion characteristics, and based on these motion characteristics, an in-depth study was conducted to establish a classification model with high accuracy, confirming that there are differences in the movement patterns of people with and without acrophobia in high-altitude environments. This research will enable rapid screening of both acrophobic and non-acrophobic people, and additionally, will provide quantitative profiling that can be used in the future to provide more targeted training guidance for aviation personnel, as well as providing rapid training results.

In the future, we will continue to optimize WMINS to obtain more data on key human joint movements to reconstruct human posture, improve the VR high-altitude scenario while ensuring safety, and explore the contribution of limb movements in different parts of the human body to the correct classification of acrophobia. We will achieve the fine-grained classification of acrophobia and non-acrophobia under different weights, thereby exploring the correlations between limb movement characteristics and human fears.

Author Contributions: All of the authors made significant contributions to this work. H.Y. carried out the whole experimental design and thesis writing guidance, X.C. and B.B. carried out the data analysis algorithm construction and thesis writing, W.C. built the VR scene model, S.L. participated in the experimental process, J.Z. and L.C. designed and built the experimental site. All authors have read and agreed to the published version of the manuscript.

Funding: This research was supported in part by National Key R&D Program of China under Grant 2022YFC2405603 and Key R& D Program of Jiangsu Province under Grant BE2022064-2.

Institutional Review Board Statement: The research was conducted according to the guidelines of the Declaration of Helsinki, and approved by the Institutional Review Board the Suzhou Institute of Biomedical Engineering and Technology, Chinese Academy of Sciences.

Informed Consent Statement: Informed consent was obtained from all subjects involved in the study.

Data Availability Statement: The data are not publicly available due to the relevant project regulations.

Conflicts of Interest: The authors declare no conflict of interest.

References

1. Wilson, R.B.; Vangala, S.; Elashoff, D.; Safari, T.; Smith, B.A. Using wearable sensor technology to measure motion complexity in infants at high familial risk for autism spectrum disorder. *Sensors* **2021**, *21*, 616. [CrossRef] [PubMed]
2. Muñoz-Organero, M.; Powell, L.; Heller, B.; Harpin, V.; Parker, J. Using recurrent neural networks to compare movement patterns in ADHD and normally developing children based on acceleration signals from the wrist and ankle. *Sensors* **2019**, *19*, 2935. [CrossRef]
3. Liu, H.; Gamboa, H.; Schultz, T. Sensor-Based Human Activity and Behavior Research: Where Advanced Sensing and Recognition Technologies Meet. *Sensors* **2023**, *23*, 125. [CrossRef] [PubMed]
4. Hartmann, Y.; Liu, H.; Schultz, T. Interactive and Interpretable Online Human Activity Recognition. In Proceedings of the 2022 IEEE International Conference on Pervasive Computing and Communications Workshops and other Affiliated Events (PerCom Workshops), Pisa, Italy, 21–25 March 2022; IEEE: Piscataway, NJ, USA, 2022; pp. 109–111.
5. Liu, H.; Xue, T.; Schultz, T. On a Real Real-Time Wearable Human Activity Recognition System. In Proceedings of the 16th International Joint Conference on Biomedical Engineering Systems and Technologies, Lisbon, Portugal, 15–17 February 2023; pp. 16–18.
6. Challa, S.K.; Kumar, A.; Semwal, V.B. A multibranch CNN-BiLSTM model for human activity recognition using wearable sensor data. *Vis. Comput.* **2022**, *38*, 4095–4109. [CrossRef]
7. Liu, H.; Schultz, T. How Long Are Various Types of Daily Activities? Statistical Analysis of a Multimodal Wearable Sensor-based Human Activity Dataset. *Healthinf* **2022**, *2022*, 680–688.
8. Xue, T.; Liu, H. Hidden Markov Model and its Application in Human Activity Recognition and Fall Detection: A Review. In *Communications, Signal Processing, and Systems: Proceedings of the 10th International Conference on Communications, Signal Processing, and Systems*; Springer: Berlin/Heidelberg, Germany, 2022; Volume 1, pp. 863–869.
9. Rodrigues, J.; Liu, H.; Folgado, D.; Belo, D.; Schultz, T.; Gamboa, H. Feature-Based Information Retrieval of Multimodal Biosignals with a Self-Similarity Matrix: Focus on Automatic Segmentation. *Biosensors* **2022**, *12*, 1182. [CrossRef]
10. Folgado, D.; Barandas, M.; Antunes, M.; Nunes, M.L.; Liu, H.; Hartmann, Y.; Schultz, T.; Gamboa, H. Tssearch: Time series subsequence search library. *SoftwareX* **2022**, *18*, 101049. [CrossRef]
11. Yi, X.; Zhou, Y.; Xu, F. Transpose: Real-time 3d human translation and pose estimation with six inertial sensors. *ACM Trans. Graph.* **2021**, *40*, 1–13. [CrossRef]
12. Huppert, D.; Grill, E.; Brandt, T. Down on heights? One in three has visual height intolerance. *J. Neurol.* **2013**, *260*, 597–604. [CrossRef]
13. Brandt, T.; Arnold, F.; Bles, W.; Kapteyn, T.S. The mechanism of physiological height vertigo: I. Theoretical approach and psychophysics. *Acta Oto-Laryngol.* **1980**, *89*, 513–523. [CrossRef]
14. Kalantari, S.; Shepley, M. Psychological and social impacts of high-rise buildings: A review of the post-occupancy evaluation literature. *Hous. Stud.* **2021**, *36*, 1147–1176. [CrossRef]

15. Wright, B.; Tindall, L.; Scott, A.J.; Lee, E.; Cooper, C.; Biggs, K.; Bee, P.; Wang, H.I.; Gega, L.; Hayward, E. One session treatment (OST) is equivalent to multi-session cognitive behavioral therapy (CBT) in children with specific phobias (ASPECT): Results from a national non-inferiority randomized controlled trial. *J. Child Psychol. Psychiatry* **2023**, *64*, 39–49. [CrossRef] [PubMed]
16. Heimberg, R.G.; Juster, H.R. Cognitive-behavioral treatments: Literature review. In *Social Phobia: Diagnosis, Assessment, and Treatment*; The Guilford Press: New York, NY, USA, 1995.
17. Giraldy, D.J.; Novaldo, W. A systematic literature review: Acrophobia treatment with virtual reality. *Eng. Math. Comput. Sci. J.* **2022**, *4*, 33–38. [CrossRef]
18. Garcia-Palacios, A.; Hoffman, H.G.; Kwong See, S.; Tsai, A.; Botella, C. Redefining therapeutic success with virtual reality exposure therapy. *CyberPsychol. Behav.* **2001**, *4*, 341–348. [CrossRef] [PubMed]
19. North, M.M.; North, S.M.; Coble, J.R. Effectiveness of virtual environment desensitization in the treatment of agoraphobia. *Int. J. Virtual Real.* **1995**, *1*, 25–34. [CrossRef]
20. Davis, J.R.; Horslen, B.C.; Nishikawa, K.; Fukushima, K.; Chua, R.; Inglis, J.T.; Carpenter, M.G. Human proprioceptive adaptations during states of height-induced fear and anxiety. *J. Neurophysiol.* **2011**, *106*, 3082–3090. [CrossRef]
21. Hüweler, R.; Kandil, F.I.; Alpers, G.W.; Gerlach, A.L. The impact of visual flow stimulation on anxiety, dizziness, and body sway in individuals with and without fear of heights. *Behav. Res. Ther.* **2009**, *47*, 345–352. [CrossRef]
22. Brandt, T.; Kugler, G.; Schniepp, R.; Wuehr, M.; Huppert, D. Acrophobia impairs visual exploration and balance during standing and walking. *Ann. N. Y. Acad. Sci.* **2015**, *1343*, 37–48. [CrossRef]
23. Bzdúšková, D.; Marko, M.; Hirjaková, Z.; Kimijanová, J.; Hlavačka, F.; Riečanský, I. The effects of virtual height exposure on postural control and psychophysiological stress are moderated by individual height intolerance. *Front. Hum. Neurosci.* **2022**, *15*, 824. [CrossRef]
24. Raffegeau, T.E.; Fawver, B.; Young, W.R.; Williams, A.M.; Lohse, K.R.; Fino, P.C. The direction of postural threat alters balance control when standing at virtual elevation. *Exp. Brain Res.* **2020**, *238*, 2653–2663. [CrossRef]
25. Bulić, P.; Kojek, G.; Biasizzo, A. Data transmission efficiency in bluetooth low energy versions. *Sensors* **2019**, *19*, 3746. [CrossRef] [PubMed]
26. Ames, S.L.; Wolffsohn, J.S.; Mcbrien, N.A. The development of a symptom questionnaire for assessing virtual reality viewing using a head-mounted display. *Optom. Vis. Sci.* **2005**, *82*, 168–176. [CrossRef] [PubMed]
27. Steinman, S.A.; Teachman, B.A. Cognitive processing and acrophobia: Validating the Heights Interpretation Questionnaire. *J. Anxiety Disord.* **2011**, *25*, 896–902. [CrossRef] [PubMed]
28. Freeman, D.; Haselton, P.; Freeman, J.; Spanlang, B.; Kishore, S.; Albery, E.; Denne, M.; Brown, P.; Slater, M.; Nickless, A. Automated psychological therapy using immersive virtual reality for treatment of fear of heights: A single-blind, parallel-group, randomised controlled trial. *Lancet Psychiatry* **2018**, *5*, 625–632. [CrossRef]
29. Sreejith, B.; Verma, A.K.; Srividya, A. Fault diagnosis of rolling element bearing using time-domain features and neural networks. In Proceedings of the 2008 IEEE Region 10 and the Third International Conference on Industrial and Information Systems, Kharagpur, India, 8–10 December 2008; IEEE: Piscataway, NJ, USA, 2008; pp. 1–6.
30. Srinivasan, V.; Eswaran, C.; Sriraam, A.N. Artificial neural network based epileptic detection using time-domain and frequency-domain features. *J. Med. Syst.* **2005**, *29*, 647–660. [CrossRef] [PubMed]
31. Andrade, C. The P value and statistical significance: Misunderstandings, explanations, challenges, and alternatives. *Indian J. Psychol. Med.* **2019**, *41*, 210–215. [CrossRef]
32. Douzas, G.; Bacao, F.; Last, F. Improving imbalanced learning through a heuristic oversampling method based on k-means and SMOTE. *Inf. Sci.* **2018**, *465*, 1–20. [CrossRef]
33. Khairy, R.S.; Hussein, A.; ALRikabi, H. The detection of counterfeit banknotes using ensemble learning techniques of AdaBoost and voting. *Int. J. Intell. Eng. Syst.* **2021**, *14*, 326–339. [CrossRef]
34. Guo, G.; Wang, H.; Bell, D.; Bi, Y.; Greer, K. KNN Model-Based Approach in Classification. In *On The Move to Meaningful Internet Systems 2003: CoopIS, DOA, and ODBASE: OTM Confederated International Conferences, CoopIS, DOA, and ODBASE 2003, Catania, Sicily, Italy, November 3–7, 2003. Proceedings*; Springer: Berlin/Heidelberg, Germany, 2003; pp. 986–996.
35. Biau, G.; Scornet, E. A random forest guided tour. *Test* **2016**, *25*, 197–227. [CrossRef]
36. Jakkula, V. Tutorial on support vector machine (svm). *Sch. EECS Wash. State Univ.* **2006**, *37*, 3.
37. Wright, R.E. Logistic regression. In *Reading and Understanding Multivariate Statistics*; American Psychological Association: Washington, WA, USA, 1995.
38. Powers, D.M. Evaluation: From Precision, Recall and F-Measure to ROC, Informedness, Markedness and Correlation. *arXiv* **2020**, arXiv:2010.16061.
39. Bălan, O.; Moise, G.; Moldoveanu, A.; Leordeanu, M.; Moldoveanu, F. An investigation of various machine and deep learning techniques applied in automatic fear level detection and acrophobia virtual therapy. *Sensors* **2020**, *20*, 496. [CrossRef] [PubMed]
40. Hao, T.; Zheng, X.; Wang, H.; Xu, K.; Chen, S. Linear and nonlinear analyses of heart rate variability signals under mental load. *Biomed. Signal Process. Control.* **2022**, *77*, 103758. [CrossRef]
41. Hu, F.; Wang, H.; Chen, J.; Gong, J. Research on the Characteristics of Acrophobia in Virtual Altitude Environment. In Proceedings of the 2018 IEEE International Conference on Intelligence and Safety for Robotics (ISR), Shenyang, China, 24–27 August 2018; IEEE: Piscataway, NJ, USA, 2018; pp. 238–243.

42. Šalkevicius, J.; Damaševičius, R.; Maskeliunas, R.; Laukienė, I. Anxiety level recognition for virtual reality therapy system using physiological signals. *Electronics* **2019**, *8*, 1039. [CrossRef]
43. Zhang, X.; Wen, W.; Liu, G.; Hu, H. In Recognition of public speaking anxiety on the recurrence quantification analysis of GSR signals. In Proceedings of the 2016 Sixth International Conference on Information Science and Technology (ICIST), Dalian, China, 6–8 May 2016; IEEE: Piscataway, NJ, USA, 2016; pp. 533–538.
44. Zheng, R.; Wang, T.; Cao, J.; Vidal, P.-P.; Wang, D. Multi-modal physiological signals based fear of heights analysis in virtual reality scenes. *Biomed. Signal Process. Control* **2021**, *70*, 102988. [CrossRef]

Disclaimer/Publisher's Note: The statements, opinions and data contained in all publications are solely those of the individual author(s) and contributor(s) and not of MDPI and/or the editor(s). MDPI and/or the editor(s) disclaim responsibility for any injury to people or property resulting from any ideas, methods, instructions or products referred to in the content.

Article

Counting Activities Using Weakly Labeled Raw Acceleration Data: A Variable-Length Sequence Approach with Deep Learning to Maintain Event Duration Flexibility

Georgios Sopidis [1,*], Michael Haslgrübler [1] and Alois Ferscha [2]

1. Pro2Future GmbH, Altenberger Strasse 69, 4040 Linz, Austria; michael.haslgruebler@pro2future.at
2. Institute of Pervasive Computing, Johannes Kepler University, Altenberger Straße 69, 4040 Linz, Austria; ferscha@soft.uni-linz.ac.at
* Correspondence: k11851885@students.jku.at or georgios.sopidis@pro2future.at

Abstract: This paper presents a novel approach for counting hand-performed activities using deep learning and inertial measurement units (IMUs). The particular challenge in this task is finding the correct window size for capturing activities with different durations. Traditionally, fixed window sizes have been used, which occasionally result in incorrectly represented activities. To address this limitation, we propose segmenting the time series data into variable-length sequences using ragged tensors to store and process the data. Additionally, our approach utilizes weakly labeled data to simplify the annotation process and reduce the time to prepare annotated data for machine learning algorithms. Thus, the model receives only partial information about the performed activity. Therefore, we propose an LSTM-based architecture, which takes into account both the ragged tensors and the weak labels. To the best of our knowledge, no prior studies attempted counting utilizing variable-size IMU acceleration data with relatively low computational requirements using the number of completed repetitions of hand-performed activities as a label. Hence, we present the data segmentation method we employed and the model architecture that we implemented to show the effectiveness of our approach. Our results are evaluated using the Skoda public dataset for Human activity recognition (HAR) and demonstrate a repetition error of ± 1 even in the most challenging cases. The findings of this study have applications and can be beneficial for various fields, including healthcare, sports and fitness, human–computer interaction, robotics, and the manufacturing industry.

Keywords: artificial intelligence; deep learning; counting; weakly labeled data; variable length size; non-uniform shape data

Citation: Sopidis, G.; Haslgrübler, M.; Ferscha, A. Counting Activities Using Weakly Labeled Raw Acceleration Data: A Variable-Length Sequence Approach with Deep Learning to Maintain Event Duration Flexibility. *Sensors* 2023, 23, 5057. https://doi.org/10.3390/s23115057

Academic Editors: Tanja Schultz, Hui Liu, Hugo Gamboa and Eui Chul Lee

Received: 15 March 2023
Revised: 19 May 2023
Accepted: 22 May 2023
Published: 25 May 2023

Correction Statement: This article has been republished with a minor change. The change does not affect the scientific content of the article and further details are available within the backmatter of the website version of this article.

Copyright: © 2023 by the authors. Licensee MDPI, Basel, Switzerland. This article is an open access article distributed under the terms and conditions of the Creative Commons Attribution (CC BY) license (https://creativecommons.org/licenses/by/4.0/).

1. Introduction

In recent times, people use more and more new technologies, devices, and sensors that generate data to support their daily activities. Researchers can use this sensor data to identify the human body's actions and movements for human activity recognition, or HAR, as it is more commonly known. Various sensor types collect data in those settings, such as the ones that use video and inertial measurement units (IMUs). Sensors provide means to capture data related to human activities, which can be used to develop machine learning models for human activity recognition (HAR) and human behavior recognition (HBR). Achievements have been made in sports and entertainment [1–3], industrial applications [4], and healthcare [5]. Meanwhile, the academic community is actively researching innovative sensor technologies for human activity and behavior recognition, including new sensor designs, applications of traditional sensors, and the usage of non-traditional sensor types [6]. Many studies are currently being conducted to improve existing approaches or solve newly identified problems for the detection and classification of activities with supervised or unsupervised techniques. The majority of research studies and applications in the field of HAR have, up to this point, focused on detecting activities, such as walking, standing, and sitting, as well as other daily living (DL) activities [7], and analyzing their characteristics to generate new insights [8].

Building upon this existing body of work, the scope of our study is to highlight the counting of events that occur in a given period in different activities of daily human life (DHL) or daily work life (DWL). We focus on counting the end of an activity to determine the number of times that activity occurs. By centering our attention on event counting, we aim to provide a comprehensive understanding of the frequency and occurrence of specific actions within the broader context of human activities. In daily activities, such as workouts or sports, it is critical to correctly segment and recognize the type of activity using a sophisticated model [9]; however, mainly classification models can offer such information. Besides that, it is important to acknowledge that different fields exhibit variations in sensor types, signal characteristics, produced by these sensors and face different challenges. With small adjustments, counting with AI and IMU data can be used in the industry to solve a variety of problems. Some examples include: (i) Sensor data analysis to monitor the performance of equipment, detect anomalies, and optimize operations; (ii) Quality control to count the number of defects or errors and improve the quality of products and reduce costs; (iii) Safety monitoring to count the number of incidents to improve safety and reduce the risk of accidents in industrial environments. For example, in an industrial setting where the tasks are more complicated, workers have many repetitive tasks to complete daily, such as screwing activities during assembly processes, which they occasionally miscount or forget to execute [10]. In this regard, we aim to provide people with information and raise awareness about the number of completed activities.

As Kim et al. [11] stated in their work, counting is one ability that humans usually acquire from a young age, and while it appears to be a simple task, young people still need a long period to master it. Comparably, it is challenging to develop a model that can count the number of completed activities (CA) in a time period, based on data from Inertial Measurement Units (IMUs) or similar body-worn sensors. With the term "completed activities", we refer to the repetitive activities that can constitute a single work step in a workflow, e.g., the screwing of one screw, which is complete with the tightening of it.

The analyzed data for our research are sequences of varying length annotated with weak labels that serve as targets for the machine learning models. Traditional approaches for handling time-series data often involve dividing the data into fixed window lengths. However, due to the high variance in activity durations, using a small window for a long activity or a large window for a short activity can result in the loss of important information [12]. Up until recently, one commonly used approach required very large window samples that could fit all activity sizes inside, padding them with a value (typically zeros) and feeding them as input to the networks. Annotating data, on the other hand, is usually laborious and time-consuming, and requires considerable attention and precision.

As an overview of the challenges motivating this work, we concentrate on the following: (i) spread of valuable information across consecutive sequences, (ii) information loss caused by using a single, fixed window size for varying-duration activities, (iii) limited flexibility of models that are more specific for particular data due to manual preprocessing methods, and (iv) the topic of data annotation. To address the aforementioned issues, we propose a model design that works with (i) variable length of data as input, (ii) data that have some form of annotation but is not completely annotated, known as weakly labeled, (iii) raw calibrated data that are normalized but not subjected to any further filtering, to reduce complexity, simplify the preprocessing stage, and develop a more robust model in unprocessed data to achieve counting, (iv) the implemented counting method, integrated within the model's training process rather than simply incrementing the count of correctly classified instances, and (v) fewer training parameters than existing model architectures in the literature to enable suitability for deployment on devices that have limited processing and power resources. Our approach aims to improve the model's ability to accurately count the activities performed by a user, rather than just detect them. To the best of our knowledge, this is the first study that investigates the counting of completed activities and tasks, in a way that goes beyond counting the quantity of previously correctly recognized activities from a classifier, employing an LSTM [13] for counting patterns in a sequence.

Counting is an essential skill that humans employ in their daily lives across a wide range of activities and tasks. Whether it is a simple task or a more complex one, the ability to accurately count holds immense value and can benefit from technological assistance in various fields. According to our literature review, studies in this area mostly used video-capturing sensors and have been conducted in the sports or medical sector.

Fang et al. [14], in their study, explored the possibility of counting the number of items in a display and raise the question, "Can a recurrent neural network learn to count things?", with their findings favoring a positive answer. While they also used an LSTM model, our model takes as input unfiltered accelerometer data relating to human activities in daily life. In a different setting, the authors of [15] proposed to count repetitive activities in a video by sight and sound using an audiovisual model, which differs from our approach among others in the choice of the sensors, since we aim to use body-worn sensors. In ref. [16], the MM-Fit dataset is introduced, which contains data from inertial sensors and ambient video sensors capturing full-body workouts. A single 3D accelerometer worn at the chest is employed in [17] to recognize four types of workouts and count repetitions after the workout is firstly determined and classified by their algorithm. Another study focused on fitness exercises is the one by Ferreira et al. [18], where the authors select camera sensors for realizing their approach to do workout repetition counting. The researchers in [19] designed and implemented a body capacitance-based sensor and employed a residual deep convolutional network that uses dilated convolutions for recognizing and counting gym workouts, while their approach had competitively high counting accuracy, we opt for sensors available in devices of daily use, such as smartwatches or smartphones, and utilize unfiltered data in our approach. In [20], with 91% of the used Cross-fit exercises having an error within a margin of ± 1 repetition, the authors used a vibration signal during their data collection and trained a neural network for counting that relied on whether an input window contains a repetition start. However, our model uses only weak labels as target data for its variable-size input, which requires less human annotation effort than models with dense labels.

Weakly labeled data can be beneficial for deep learning algorithms in certain situations and refer to data that are only partially labeled, meaning that they have some form of annotation, but not all the information is present. This type of data is less expensive and time-consuming to obtain than fully labeled data, and it can be used to train deep learning models in a semi-supervised manner. The authors in [21] proposed an attention-based convolution neural network to process weakly labeled human activities and recognize them. The dataset contains information only about the type of activity that occurred in a sequence of sensor data. A weakly labeled dataset was also included in a Dual Attention Network For Multimodal Human Activity Recognition Using Wearable Sensors in [22], where they blend channel attention and temporal attention on a CNN, for multimodal HAR. The activities that are contained in the dataset are walking, jogging, jumping, going upstairs, or going downstairs, and have a significant difference from the activities that we explore and the way that we create our training dataset. In a related field, for locomotion activities, several studies explored step counting using IMUs or smartphones [23], including approaches that utilize deep learning techniques. One such approach is the attention-based LSTM model by the authors of [24], which has been shown to effectively count steps with high accuracy. However, unlike the continuous and repetitive movements associated with step counting, our approach concentrates on hand-performed activities that involve discrete movements. Furthermore, our model maintains simplicity regarding processing resources, low power consumption, and suitability for edge computing devices.

Raw data as input for the models have the advantage of reducing the need for preprocessing techniques, which can be a time-consuming and resource-intensive task. In our study, we refer to the public dataset's raw calibrated data that have not been subjected to any further preprocessing steps other than normalization. When working with raw data, the model can automatically learn useful features from the data, which can save computational resources and reduce the risk of human error. Important contributions have been made by Shen et al. in [25], where they proposed a workout tracking system that uses smartwatches to accurately and efficiently track both cardio and weightlifting workouts without the need for user input. Their counting strategy begins with detecting and labeling weightlifting sessions, followed by a naive peak detection algorithm based on auto-correlation results. They filter out non-repeating signals and calculate the number of repetitions by counting detected peaks. Likewise, Prabhu et al. in [26] also based their approach on classifying the activities before counting with a peak detector method. Their research aims to identify the most effective artificial intelligence model for repetition counting in LME exercises to be used in wrist-worn rehabilitation programs.

In their work, Taborri et al. [27] implemented the following algorithms, one for recognizing activities based on SVMs and one for counting actions related to workers in the industry. Twenty-three body-worn sensors collected data from the participants, which were divided into windows of 0.6 s and had features such as mean, standard deviation, maximum, and minimum, were computed for each activity. Physical exercises for indoor and outdoor environments were used to recognize the real-time segmentation and classification algorithm in [28]. The method they proposed requires one sample of data for each target exercise; however, once more, the counting relies on accurate classification of the activities. Another algorithm in the context of human activity recognition that segments repetitive motion is the one presented by the authors in [29]. This algorithm was utilized to identify similar location patterns in indoor localization and addresses the problem of subsequence search in univariate and multivariate time series. An automated segmentation way and labeling of single-channel or multimodal biosignal data using a self-similarity matrix (SSM), generated with the feature-based representation of the signals, is proposed by the authors in [30]. Examples of data with the few-shot learning were employed by Nishino et al. [31] to recognize workouts using a wearable sensor including data augmentation and diversification techniques for their data to achieve repetition counting.

In our approach, we leverage deep learning models to extract features and train them to accurately count activities using public datasets that contain raw and calibrated data of human activities performed with hands. As with many machine learning techniques, we normalize the raw calibrated acceleration data before we feed them to the deep learning model. However, we do not apply additional preprocessing steps or filtering to the data. Moreover, the sensor placement described by the dataset's authors in [32] is essential to recreating the study's results. To train the model, we divide the data into segments of variable sizes with weak labels that utilize only the number of repetitions of activities for each sequence as the target value. Hence, the model learns to count activities regardless of the sequences' size, which is important for real-world applications, where activity durations may vary.

2. Materials and Methods

2.1. Counting Approach

Counting repetitions in a sequence is a fundamental problem in various fields, such as speech and image processing, bioinformatics, and cognitive science. Many methods for counting can be deployed, the majority of which require hand-crafted rules, feature extraction and statistical methods, or rule-based systems to manually count objects or events. As was described previously, neural-network-based approaches can be used to count repetitions in a sequence with a combination of CNNs and RNNs or encoder-decoder architectures, trained on a labeled dataset of sequences, to learn a mapping between the input sequence and the number of repetitions. Focusing on a system that can handle weakly labeled data while being less reliant on human intervention and more automated can reduce the complexity of the counting process, make the model more robust, and provide flexibility for applying the method to a wide range of problems.

Weakly labeled data offers a cost-effective and efficient alternative to acquiring fully labeled data, as it is less expensive, time-consuming, and tedious. This type of data enables the use of semi-supervised learning approaches, which can be beneficial when considering the annotation cost associated with large and complex datasets. By leveraging weak labels, the model is encouraged to learn more generalized patterns in the data, leading to improved performance on unseen examples. For our experimental setup, we use IMU acceleration data from daily human activities performed at a quality control checkpoint in a car maintenance scenario that captures activities relevant to the inspection process. Thus, it provides a representation of real-world conditions by addressing the complexity of variable-length data, to develop robust and realistic models for activity counting and recognition tasks.

Despite the fact that counting repetitions in a sequence with variable-length data is more challenging than counting repetitions in a sequence with fixed-length data, this structure is more realistic because data from signals, time series, texts, and other sources have varying length. Using fixed-length tensors for the data can be efficient in certain situations because they simplify the problem, as the model only needs to process a fixed amount of data, regardless of the length of the input sequence. Furthermore, because the libraries and software tools required to build the model are more widely accessible, its implementation and deployment may be simpler. However, using variable length tensors can also be beneficial in many situations. They allow the model to handle input sequences of different length, which is important when dealing with complex real-world data and activities of various length. Additionally, variable-length tensors enable the model to process the entire input sequence at once, rather than only a fixed-length subset of it, which can be valuable when the position of the repetitions is not known in advance.

In our approach for counting, we use data from public datasets that contain data from human activities in the car manufacturing industry, recorded with IMU sensors. We create sequences of data that have a variable size and we obtain a weak label for each sequence. The label shows the executions number of one type of activity observed in the sequences, which is fed into an LSTM regression model built with the ragged tensors. For each sequence that is input to our algorithm, one single count is predicted as the output.

2.2. Dataset

The data used for this study are part of the Skoda Public dataset [32], which includes repetitive activities regarded as single, discrete actions as opposed to continuous activities, such as walking or running. The example signals of the manipulative gestures of the dataset that were performed in a car maintenance scenario, visualized in Figure 1, are "write on notepad", "open hood", "close hood", "check gaps on the front door", "open left front door", "close left front door", "close both left door", "check trunk gaps", "open/close trunk", and "check steering wheel". These activities were recorded for about 3 h by 20 sensors placed on one subject's left and right upper and lower arms. For each sensor, there are acceleration values on the x, y, and z axes that are calibrated in milli-g units (1000 = earth gravity vector, which in S.I. units would be 0.001 g or 0.00981 m/s^2), and the sensor sample rate is approximately 98 Hz, as stated by the dataset's authors.

The objective of this work is to count how many times one activity happened in a period of time, e.g., detect in the data patterns how many times the person closed the hood in the activity "close engine hood". The dataset contains a dense label for each sample, which allows the detection of the end of the activity. The weak label targets are generated by "recording" one repetition for each task completion. Therefore, for the training phase, for every instance where a task is successfully completed, a single repetition sample is marked as "activity end". By using this approach, we create weak annotations that indicate the presence of completed repetitions, allowing the model to learn and recognize the patterns associated with activity completion. We normalize the data with minmaxscaler [33] in a range of [0, 1] and divide it into variable-size sequences. By using an algorithm to generate an array of random numbers, we split the entire dataset into segments, which define the sample length of the sequences, as shown in Figure 2. For example, if we want to generate 20 sequences of variable-length data, the algorithm will create 20 random numbers between 0 and the dataset's maximum index value. Next, we replicate the original data for each activity to provide our model with a larger dataset to train, without using data augmentation techniques for generating variation in the signal's patterns. This expanded dataset introduces greater variability in the unique length of activity sequences and the number of activities contained within them, thereby enhancing the model's robustness.

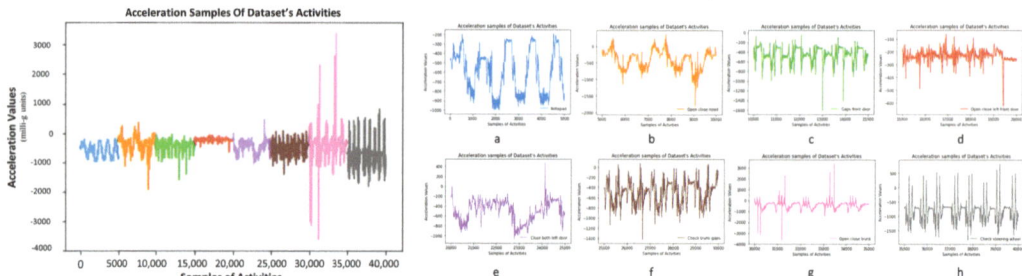

Figure 1. This figure shows examples of the signals that represent each class of the Skoda Public dataset with different colors. Starting from left to right (**a**) Notepad, (**b**) Open close hood, (**c**) Gaps front door, (**d**) Open close left door, (**e**) Close both left door, (**f**) Check trunk gaps, (**g**) Open close trunk, (**h**) Steering wheel. For all the activities, samples are taken from the X-axis accelerometer on the right hand. The activities of the "open hood" and "close hood" as well as the "open left door" and "close left door" are displayed together as "open close hood" and "open close left door", since they are always consecutive. The acceleration is provided in milli-g units.

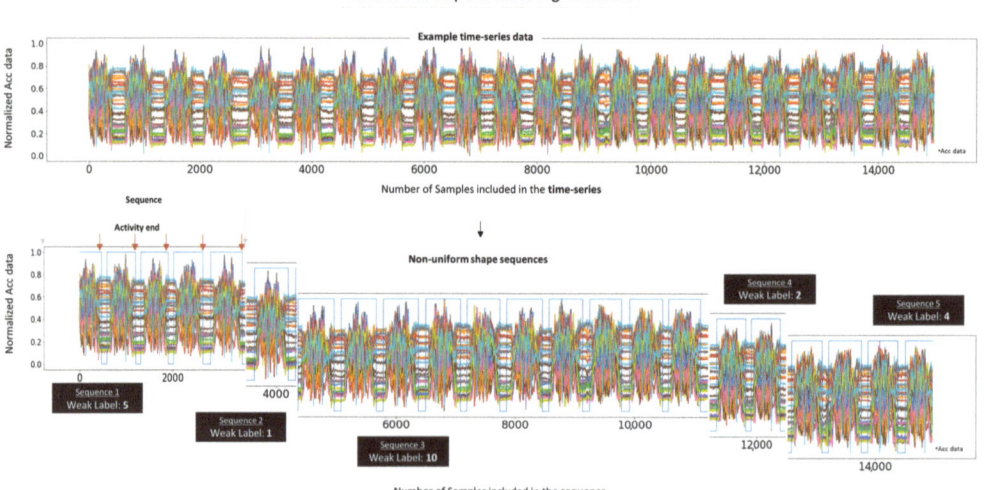

Figure 2. This figure presents the non-uniform shape input for the neural network. We visualize on top of the image time-series data that will be divided into five distinct example sequences of segmented acceleration data with varying duration, number of samples, and weak labels that have a range from 1 to 10 counted activities. The blue squared line shows the start and end of each activity within each sequence. The weak label is generated by the number of spotted endings (red arrow) inside each individual sequence.

The labels for each sequence in this dataset are produced by the number of spotted endings or finished tasks in the sequences, where the last timestamp of each observed activity adds 1 count to the final label of each unique sequence. Figure 2 visualizes the division of a time series into sequences of variable size and how the weak labels are formed. The term "weak labels" in our method denotes the absence of data annotations that map the start and end of an event in the sequence. The number of activities in the sequence is the only information of the sequence that the model utilizes as a target value.

We divided the subsets of activities into 600–900 sequences, where 100 of each type were left as a test dataset, as presented in Table 1, and 10% of each training dataset was used as a validation set. Consecutive activities, such as "open left front door" and "close left front door", were merged into one class, as explained in Figure 1. In this case, the algorithm must count +1 when one of the activities of interest is happening. In the last entry of the table with the label "combined activities", one can see results with 8000 training sequences, for a class that is generated with combined data from all the previous classes in a single one. Every activity that is not a null class will be counted in this scenario, to distinguish between an occurring activity and null class without considering the type of activity. We created 8000 varying-length sequences from all classes, of which 7000 were used to train the network. Despite the more complex approach, employing variable-size sequences allows us to extract the most valuable information from our data without padding.

Table 1. This table lists the overall summary results for accuracy and MAE across all activity classes for the test datasets. As one can see, it contains nine separate datasets of activities. For each activity, the samples of the original dataset, the variable length training sequences created from the data, the range of the number of activities within the sequences, the test sequences, and the results such as the dataset accuracy and mean accuracy in 100 sequences of each test dataset are available. The counts' range shows the maximum number of activity counts contained in 100 sequences of different lengths of the test dataset. The mean percentage accuracy provides an overall assessment of the model's performance by displaying the average deviation of the predicted number of repetitions from the true value across all test sequences, whereas the dataset accuracy assesses the model's ability to predict the precise number of repetitions accurately. The last entry in the table represents a class where all the data from all classes were combined into one class and then split into variable-length sequences so that the model is trained on more complex data. In that case, the model learns a larger variety of patterns from all classes as a single activity class and must identify between the activity class and the null class to perform the counting.

Training A/A	Activity	No of Samples in Original Dataset	Training Seq.	Range Counts in Test Seq.	Test Seq.	Test Dataset Accuracy	MAE	Mean % Accuracy in Test Seq.
1	Steering wheel	51,904	500	0–08	100	60/100	0.4	72.19
2	Check trunk gaps	70,000	500	0–07	100	89/100	0.11	91.66
3	Notepad	74,000	500	0–06	100	92/100	0.08	96.12
4	Open close hood	186,399	800	0–07	100	70/100	0.3	78.58
5	Open close left door	82,000	600	0–12	100	68/100	0.33	80.22
6	Gaps front door	60,000	500	0–09	100	84/100	0.18	90.38
7	Close both left door	72,000	500	0–06	100	75/100	0.25	79.22
8	Open close trunk	95,000	600	0–10	100	74/100	0.26	81.65
9	Combined activities	705,904	7000	0–11	1000	765/1000	0.242	81.29

2.3. Counting Algorithm

A fixed window approach is a commonly used method for segmenting time-series data before using them as input to deep learning models. The segmentation is based on characteristics of the event that we want to identify, such as its periodicity, frequency, and length, among others. Despite their ease of implementation and interpolation with other libraries, fixed-length tensors with a predefined shape have limitations. For example, they are not well suited to handling non-uniform shape data, such as sequences of varying length, without the need for padding or truncation, which can result in additional noise to the data, increase in the computation time, information loss, and storage inefficiency, and it may not always be appropriate. To address the above issues, we used TensorFlow's ragged tensors [34], which support variable-length sequences of samples.

In this study, nine sub-datasets were used with our algorithm to count activities, with nine separate trainings for each subset. Figure 3 shows the architecture of the model that is used for the counting task. The acceleration data are separated into variable-length sequences, each of which comprises several activities and is used as input data, as was previously mentioned. The weak label that the model uses as target data is the total number of activities in each sequence. The model learns to relate acceleration data to the number of activities, so when we feed as input " new unseen" acceleration data of variable length, it outputs the number of spotted activities. The annotation provided no information about the location of activities within the sequence, nor does it provide any additional supporting details to guide the model. A large grid search was deployed to explore the best combination of parameters for the number of layers, learning rate, batch size, optimizer, loss function, and activation functions to use in our network. We experimented with various hyperparameters to achieve the best performance. For the number of layers, we tried configurations ranging from one to three time-distributed layers and one to 4 Lstm layers. For the learning rate, we tested values such as 0.01, 0.001, 0.0001, and 0.00001. We also explored different batch sizes, including 2, 4, 8, 16, 32, and 64. As for the optimizer, we experimented with Adam, RMSprop, and SGD. We evaluated loss functions, such as mean squared error, mean absolute error, and Huber loss. Finally, we tried different activation functions, such as ReLU, tanh, and sigmoid, to achieve the optimal performance on our task.

As shown in Figure 4, two dense layers were used at the beginning of the model to reduce the dimensions of the input data before entering the LSTM. The first dense layer is composed of 60 neurons (number of input signals), and the second consists of 2 neurons with the rectified linear unit (RELU) as an activation function. Two custom layers are then placed after an LSTM layer that outputs a three-dimensional sequence and has one neuron with a linear activation function. The custom masking layer thresholds the signal and converts the output of the LSTM to a more binary format, and xthen, the counting layer counts the regions where the signal value is not zero and summarizes them to one final number, as shown in Figure 3, of the output's graph. After experimenting with several parameter values as mentioned above, a batch size of 2 with a learning rate of 0.0001 and the "Adam" optimizer were selected for optimizing the model. The algorithm's performance is evaluated using the Huber loss as the loss function, which is a combination of the mean squared error (MSE) loss function and the mean absolute error (MAE) loss function. This combination improves the performance of the model when outliers are present in the data, which is possible in our study because the input sequences were generated arbitrarily.

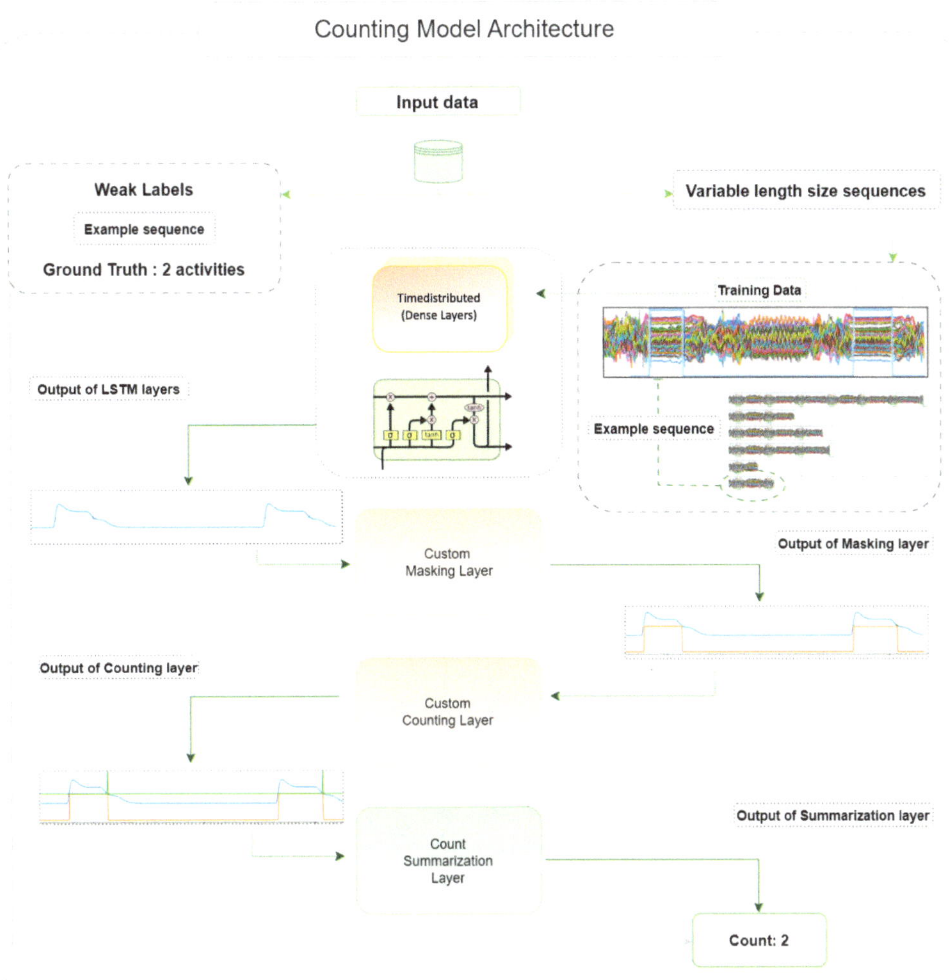

Figure 3. This figure presents an example of the model architecture that was used for the counting of activities. The acceleration data are divided into variable-length sequences and then used as input to the model. For each sequence, there is one weak label that is generated by the number of activities that are included in the sequence. Two time-distributed dense layers process each sensor reading independently before entering an LSTM layer where we get an output for each time step. Since the input data have a variable size, ragged tensors are employed for this task. The output of the LSTM part is inserted into a mask layer that detects values above a threshold and converts the signal into a square form before it continues to a layer that detects the created "edges" of the square shape and gives a final summation of all edges of the sequence to one single number.

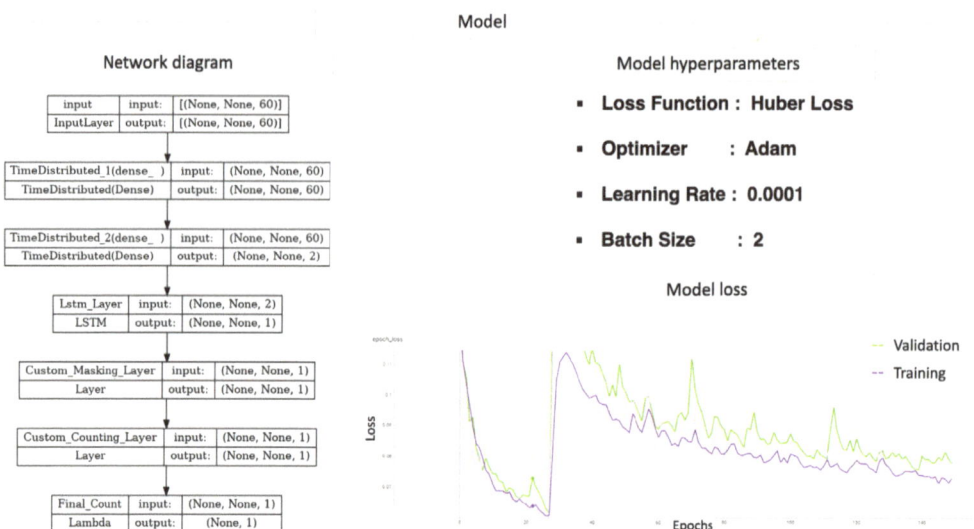

Figure 4. From left to right, this image presents the network's diagram of the counting model with the input (60 acceleration signals) of variable length and output of 1 number. Moreover, the hyperparameters include Huber loss, ADAM optimizer, a learning rate of 0.0001, and a batch size of 2. An example of a learning curve for the training and validation sets demonstrates the model's performance during training. The x-axis represents the number of training epochs, while the y-axis represents the loss metric.

3. Results

In our study, we applied deep learning approaches to acceleration data to count the number of activities in variable-length sequences, as presented in the model architecture.

The ground truth in Figure 3 is two activities in the illustrated example sequence. It is evident that the LSTM outputs a signal with two peaks, which is then converted to a binary format by the masking layer, and we count +1 at the edge of each square area. Two is the final result predicted by the model for the specific input sequence. Figure 4 visualizes the learning curve of the training and a validation loss to present the model's performance during the training of the "notepad" dataset. After each training, we evaluated the model with unseen data sequences of the same class, and the results show that the model can predict very close to the weak label.

For two of the activities of the Skoda Dataset, we visualize example results of the weakest and best cases of the model's predictions in variable-length data sequences. In Figure 5, one can see the graph showing the ground truth and prediction of the model for the activity "writing in notepad". There, we trained the model with 500 sequences of variable size and variety in the number range of contained activities. The findings indicate that the model predicts 92 out of 100 correctly, while the remaining 8 predictions have an error of one activity. Similarly, even though the model predicted less accurately for the dataset's "steering wheel" class, the predictions have a maximum error of one activity, as shown in Figure 6. Table 1 contains information regarding the training data and the results of all dataset activities. The discussion section provides further details about the results.

The table shows in approximation the number of data samples contained in the original dataset for each class, the number of training and test sequences, the range of the number of activities included in different sequences, the accuracy in the test data, the mean absolute error, and the mean % accuracy of test sequences in the test data.

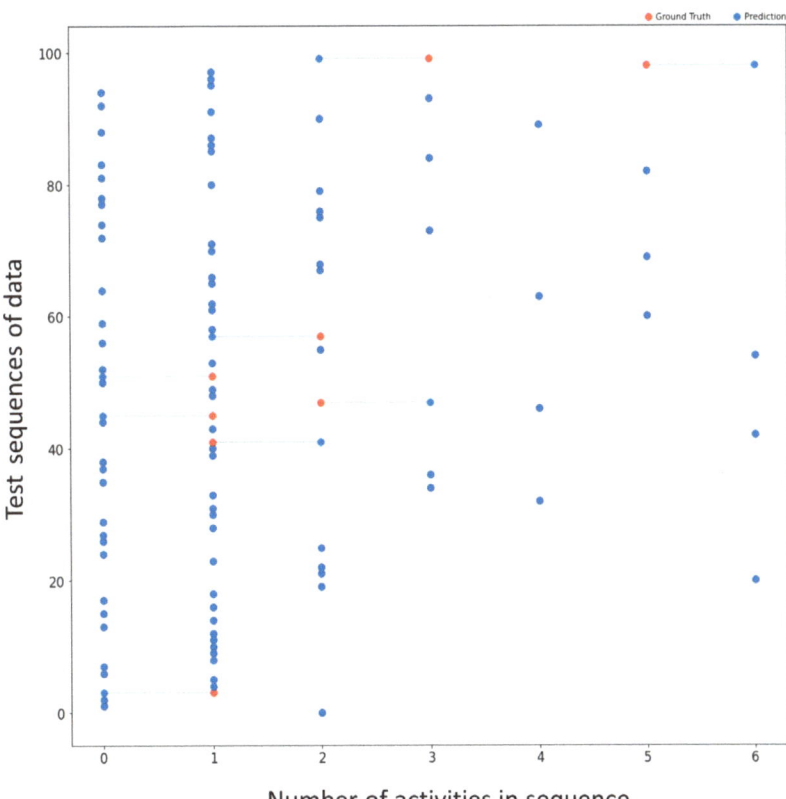

Figure 5. The figure presents the ground truth (red dots) and prediction (blue dots) of the model for 100 unseen sequences of data. The input data are from the notepad writing activity of the Skoda dataset. The algorithm for counting predicted accurately 92 out of 100 activities. A line connecting the two numbers shows the difference in the incorrectly predicted sequences. The largest error per sequence observed in the graph is 1 count.

The "notepad" class has the smallest MAE, 0.08, and the highest mean % accuracy, while the "steering wheel" class has the highest, 0.4, and the lowest mean % accuracy. In the "combined activities" class, the algorithm counts interesting activities in a sequence, regardless of the activity type, in a dataset consisting of all classes combined into a single one. The table lists 8000 training sequences of varying length, of which 7000 were used for the network's training. Randomly, 1000 sequences were kept as test data, and the number of activities in 765 out of 1000 was predicted correctly. From those 1000 sequences, 230 had an error of ±1 counts, 3 of them an error of 2 counts, and 2 of them an error of 3 counts. As shown in the table, the mean accuracy is presented as a percentage, representing the average accuracy of 100 sequences from each test dataset. For example, for the "open close trunk" dataset, the mean accuracy for the 100 test sequences is 81.65%. This means that we found the accuracy of the model for each predicted sequence of this test dataset, and subsequently, provide an average estimate of the accuracy across all the test sequences to evaluate the performance of the model. The dataset accuracy evaluates the model's ability to predict the exact number of repetitions accurately, while the mean accuracy gives an overall measure of the model's performance by indicating the average deviation of the predicted number of repetitions from the true value across all test sequences.

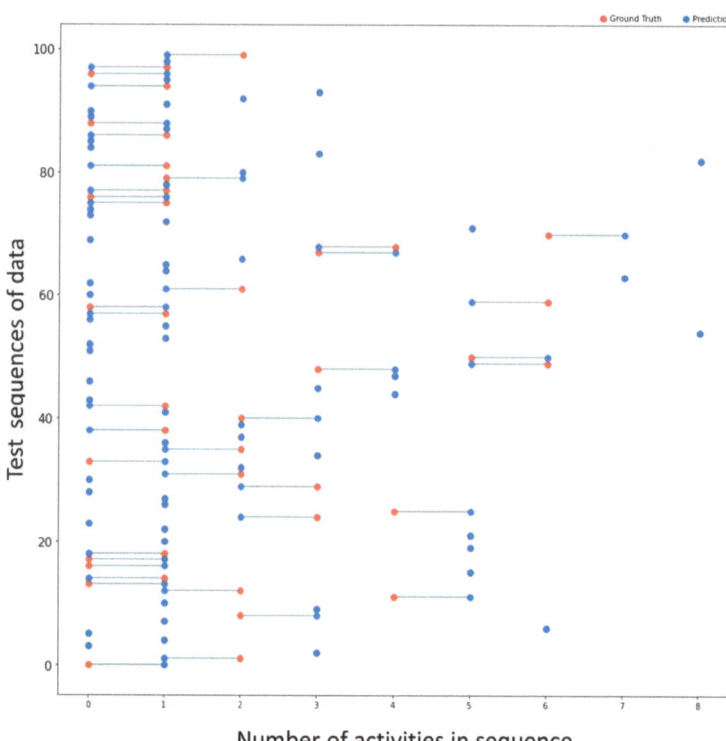

Figure 6. The figure presents the ground truth (red dots) and prediction (blue dots) of the model for 100 unseen sequences of data. The input data are from the steering wheel activity of the Skoda dataset. The algorithm for counting predicted accurately 60 out of 100 activities. A line connecting the two numbers shows the difference in the incorrectly predicted sequences. The largest error per sequence observed in the graph is 1 count.

4. Discussion

The current study confirms that it can count interesting events in time series with more flexibility concerning the size of each input sequence from a model that uses (i) solely normalized data, (ii) weak labels, and (iii) deep learning. According to our method's preliminary findings, when we train the model for specific activities, our algorithm can accurately predict, in most cases, the number of times an event is repeated in a sequence. For some of the activities, the prediction is better than others. For example, activities such as "open close hood" and "open close left door" contain patterns of both opening and closing the object, which can possibly create a larger confusion for the model to recognize the pattern. The lower results were achieved for the steering wheel class. The wheel rotates three times in each direction, clockwise and counterclockwise, before switching. In this case, the orientation for each side within the same sequence may be contributing to the confusion, or the data may not be sufficient for the model's design and a deeper architecture or new data may be needed to capture the dependencies. Likewise, for the "combined activities" class, in each sequence, the model may contain patterns from one or more different activities that need to be counted. However, due to the weak labels, no other information about the activity type is available, except for the total number of events contained in each. The consistently small error, typically within the range of ± 1, demonstrates the effectiveness of our current architecture in accurately counting activities. However, it also highlights the potential for further research and improvements to reduce this error even further and achieve even more precise activity counting results.

The model must be trained on a dataset of labeled sequences where the number of repetitions in each sequence is known, regardless of the architecture being used. To address the diversity in the different activities without padding and by utilizing the entire information of the sequence, weak labels that are less time consuming and variable-size sequences are used. Nevertheless, it is important to note that using weakly labeled data does introduce certain limitations since the data are only partially labeled. A more comprehensive target for the model might be provided, for example, by a second model trained with information on the type of activity occurring, the location, the duration, or even data examples to use for training. In our case, we developed our ragged tensor model using calibrated data, that we normalized on a range of [0,1] to ensure a common scale. We selected the calibrated version because it is in S.I. units and can be replicated by anyone even though the raw data of the public dataset with our model provided comparable results. Additionally, the calibrated data support the use of any sensor that takes readings using the same units, not just the specific sensor that the dataset's authors used in their study. Despite the benefits of ragged tensors, such as efficient storage and easy handling of variable-length data, working with them proved to be challenging, requiring additional effort and consideration of alternative approaches, as some operations, libraries, and software tools outside of the TensorFlow environment are not currently sufficiently supported.

Counting with deep learning can be beneficial for a variety of fields, such as health care, to monitor and track physical activity levels and rehabilitation progress, sports and fitness, to track and analyze athletic performance, robotics applications, to detect and track human movements for use in collaborative robots, identifying actions and events in industrial settings, etc. These are only a few examples of how counting with AI can be used for the analysis and interpretation of IMU data in industrial applications. Although depending on the problem that needs to be addressed and the kind of data that are accessible, a specific application must be selected and modifications and optimizations must be completed. Furthermore, merging AI with technologies such as edge computing, IoT, and cloud computing for data analysis in real time and making decisions based on the results can improve these applications. Initial contributions in this direction have been made in [35], where the authors discuss the transition to real-time models, as well as in [36], where the authors introduce their system that uses wearable sensors to capture online data and perform activity recognition using Hidden Markov Models.

Our next steps include the improvement of the current model to count activities from data that are not collected in laboratory settings to enable better generalization and make it more robust across various sources of IMU data. Counting different activities within one sequence would be another challenge to address. Besides that, it is interesting to investigate the potential application of our method to other time-series data. However, it is important to consider individual differences in movement patterns that may be influenced by factors such as body size, gender, and age. One approach is to use a diverse dataset of individuals with varying body sizes, genders, and ages to train a deep-learning model that can generalize to new individuals and accurately estimate their movements. Additionally, optimizing the placement of the sensors can improve the accuracy of the model and reduce the need for unnecessary sensor data, which can be an obstacle in the learning process.

5. Conclusions

In this paper, we presented a counting method for activities and tasks to identify the end of an activity based on raw calibrated acceleration data that are weakly labeled. The generated sequences from the data have variable sizes instead of a fixed window size, which restricts the system to operate with specific settings and makes it more challenging to fit different types of activities with varying duration. An LSTM model for regression analysis was developed for the task where we tested its performance with data from different classes of the Skoda dataset for HAR. Our research focuses on using raw calibrated data rather than preprocessed or filtered data to build models that are more resistant to changes in their settings and generalize better to different scenarios. Furthermore, when fully labeled data are not available, the use of weak labels reduces the expense of data annotation.

Our results indicate that our method can count instances of activities when it corresponds to a single type and has shown promising results when training input data contain multiple types of activities. This is supported by the findings in the aforementioned cases, which state that the error for the sequences was always within ±1 iteration. Our next stages will involve testing the model using more complex data collected under real-world conditions and making it more robust against outliers and different sources of data. Thorough investigation and application of this method in other domains, such as object detection in videos, can also be included in the scope of the counting problem while adapting the approach to work with the minimum required sensor signals to produce accurate results to create user-friendly solutions that can be used during the daily work life.

Author Contributions: Conceptualization, G.S. and M.H.; methodology, G.S. and M.H.; software, G.S. and M.H.; validation, G.S.; formal analysis, G.S.; investigation, G.S.; resources, G.S.; data curation, G.S.; writing—original draft preparation, G.S.; writing—review and editing, G.S., M.H., and A.F.; visualization, G.S.; supervision, G.S.; project administration, G.S.; funding acquisition, G.S., M.H., and A.F. All authors have read and agreed to the published version of the manuscript.

Funding: Supported by Johannes Kepler Open Access Publishing Fund and the federal state of Upper Austria.

Institutional Review Board Statement: Not applicable.

Informed Consent Statement: Not applicable.

Data Availability Statement: Skoda public dataset for Human activity recognition.

Acknowledgments: This work has been supported by the FFG, Contract No. 881844: Pro²Future is funded within the Austrian COMET Program Competence Centers for Excellent Technologies under the auspices of the Austrian Federal Ministry for Climate Action, Environment, Energy, Mobility, Innovation and Technology, the Austrian Federal Ministry for Digital and Economic Affairs and of the Provinces of Upper Austria and Styria. COMET is managed by the Austrian Research Promotion Agency FFG.

Conflicts of Interest: The authors declare no conflict of interest.

References

1. Azadi, B.; Haslgrübler, M.; Anzengruber-Tanase, B.; Grünberger, S.; Ferscha, A. Alpine skiing activity recognition using smartphone's IMUs. *Sensors* **2022**, *22*, 5922. [CrossRef] [PubMed]
2. Rana, M.; Mittal, V. Wearable sensors for real-time kinematics analysis in sports: A review. *IEEE Sens. J.* **2020**, *21*, 1187–1207. [CrossRef]
3. Qiu, S.; Zhao, H.; Jiang, N.; Wang, Z.; Liu, L.; An, Y.; Zhao, H.; Miao, X.; Liu, R.; Fortino, G. Multi-sensor information fusion based on machine learning for real applications in human activity recognition: State-of-the-art and research challenges. *Inf. Fusion* **2022**, *80*, 241–265. [CrossRef]
4. Mark, B.G.; Rauch, E.; Matt, D.T. Worker assistance systems in manufacturing: A review of the state of the art and future directions. *J. Manuf. Syst.* **2021**, *59*, 228–250. [CrossRef]
5. Filippeschi, A.; Schmitz, N.; Miezal, M.; Bleser, G.; Ruffaldi, E.; Stricker, D. Survey of motion tracking methods based on inertial sensors: A focus on upper limb human motion. *Sensors* **2017**, *17*, 1257. [CrossRef] [PubMed]
6. Liu, H.; Gamboa, H.; Schultz, T. Sensor-Based Human Activity and Behavior Research: Where Advanced Sensing and Recognition Technologies Meet. *Sensors* **2023**, *23*, 125. [CrossRef]
7. Demrozi, F.; Pravadelli, G.; Bihorac, A.; Rashidi, P. Human activity recognition using inertial, physiological and environmental sensors: A comprehensive survey. *IEEE Access* **2020**, *8*, 210816–210836. [CrossRef]
8. Liu, H.; Schultz, T. How Long Are Various Types of Daily Activities? Statistical Analysis of a Multimodal Wearable Sensor-Based Human Activity Dataset. In Proceedings of the HEALTHINF, Online Streaming, 9–11 February 2022; pp. 680–688.
9. Gao, H.; Guo, F.; Zhu, J.; Kan, Z.; Zhang, X. Human motion segmentation based on structure constraint matrix factorization. *Sci. China Inf. Sci.* **2022**, *65*, 119103. [CrossRef]
10. Sopidis, G.; Haslgrübler, M.; Azadi, B.; Anzengruber-Tánase, B.; Ahmad, A.; Ferscha, A.; Baresch, M. Micro-activity recognition in industrial assembly process with IMU data and deep learning. In Proceedings of the 15th International Conference on PErvasive Technologies Related to Assistive Environments, Corfu, Greece, 29 June–1 July 2022; pp. 103–112.

1. Kim, J.Z.; Lu, Z.; Nozari, E.; Pappas, G.J.; Bassett, D.S. Teaching recurrent neural networks to infer global temporal structure from local examples. *Nat. Mach. Intell.* **2021**, *3*, 316–323. [CrossRef]
2. Antar, A.D.; Ahmed, M.; Ahad, M.A.R. Challenges in sensor-based human activity recognition and a comparative analysis of benchmark datasets: A review. In Proceedings of the 2019 Joint 8th International Conference on Informatics, Electronics & Vision (ICIEV) and 2019 3rd International Conference on Imaging, Vision & Pattern Recognition (icIVPR), Spokane, WA, USA, 30 May–2 June 2019; IEEE: Piscataway, NJ, USA, 2019; pp. 134–139.
3. Hochreiter, S.; Schmidhuber, J. Long short-term memory. *Neural Comput.* **1997**, *9*, 1735–1780. [CrossRef]
4. Fang, M.; Zhou, Z.; Chen, S.; McClelland, J. Can a recurrent neural network learn to count things? In Proceedings of the CogSci, Madison, WI, USA, 25–28 July 2018.
5. Zhang, Y.; Shao, L.; Snoek, C.G. Repetitive activity counting by sight and sound. In Proceedings of the IEEE/CVF Conference on Computer Vision and Pattern Recognition, Nashville, TN, USA, 20–25 June 2021; pp. 14070–14079.
6. Strömbäck, D.; Huang, S.; Radu, V. Mm-fit: Multimodal deep learning for automatic exercise logging across sensing devices. *Proc. ACM Interact. Mob. Wearable Ubiquitous Technol.* **2020**, *4*, 168. [CrossRef]
7. Skawinski, K.; Montraveta Roca, F.; Findling, R.D.; Sigg, S. Workout type recognition and repetition counting with CNNs from 3D acceleration sensed on the chest. In Proceedings of the International Work-Conference on Artificial Neural Networks, Gran Canaria, Spain, 12–14 June 2019; Proceedings, Part I 15; Springer International Publishing: New York, NY, USA, 2019.
8. Ferreira, B.; Ferreira, P.M.; Pinheiro, G.; Figueiredo, N.; Carvalho, F.; Menezes, P.; Batista, J. Deep learning approaches for workout repetition counting and validation. *Pattern Recognit. Lett.* **2021**, *151*, 259–266. [CrossRef]
9. Bian, S.; Rey, V.F.; Hevesi, P.; Lukowicz, P. Passive capacitive based approach for full body gym workout recognition and counting. In Proceedings of the 2019 IEEE International Conference on Pervasive Computing and Communications, Kyoto, Japan, 11–15 March 2019; IEEE: Piscataway, NJ, USA, 2019; pp. 1–10. [CrossRef]
10. Soro, A.; Brunner, G.; Tanner, S.; Wattenhofer, R. Recognition and repetition counting for complex physical exercises with deep learning. *Sensors* **2019**, *19*, 714. [CrossRef] [PubMed]
11. Wang, K.; He, J.; Zhang, L. Attention-based convolutional neural network for weakly labeled human activities' recognition with wearable sensors. *IEEE Sens. J.* **2019**, *19*, 7598–7604. [CrossRef]
12. Gao, W.; Zhang, L.; Teng, Q.; He, J.; Wu, H. DanHAR: Dual attention network for multimodal human activity recognition using wearable sensors. *Appl. Soft Comput.* **2021**, *111*, 107728. [CrossRef]
13. Yao, Y.; Pan, L.; Fen, W.; Xu, X.; Liang, X.; Xu, X. A robust step detection and stride length estimation for pedestrian dead reckoning using a smartphone. *IEEE Sens. J.* **2020**, *20*, 9685–9697. [CrossRef]
14. Khan, S.S.; Abedi, A. Step Counting with Attention-based LSTM. *arXiv* **2022**, arXiv:2211.13114.
15. Shen, C.; Ho, B.J.; Srivastava, M. Milift: Efficient smartwatch-based workout tracking using automatic segmentation. *IEEE Trans. Mob. Comput.* **2017**, *17*, 1609–1622. [CrossRef]
16. Prabhu, G.; O'connor, N.E.; Moran, K. Recognition and repetition counting for local muscular endurance exercises in exercise-based rehabilitation: A comparative study using artificial intelligence models. *Sensors* **2020**, *20*, 4791. [CrossRef]
17. Taborri, J.; Bordignon, M.; Marcolin, F.; Donati, M.; Rossi, S. Automatic identification and counting of repetitive actions related to an industrial worker. In Proceedings of the 2019 II Workshop on Metrology for Industry 4.0 and IoT (MetroInd4. 0&IoT), Naples, Italy, 4–6 June 2019; IEEE: Piscataway, NJ, USA, 2019; pp. 394–399. [CrossRef]
18. Ishii, S.; Nkurikiyeyezu, K.; Luimula, M.; Yokokubo, A.; Lopez, G. Exersense: Real-time physical exercise segmentation, classification, and counting algorithm using an imu sensor. In *Activity and Behavior Computing*; Ahad, M.A.R., Inoue, S., Roggen, D., Fujinami, K., Eds.; Springer: Singapore, 2021; pp. 239–255. [CrossRef]
19. Folgado, D.; Barandas, M.; Antunes, M.; Nunes, M.L.; Liu, H.; Hartmann, Y.; Schultz, T.; Gamboa, H. Tssearch: Time series subsequence search library. *SoftwareX* **2022**, *18*, 101049. [CrossRef]
20. Rodrigues, J.; Liu, H.; Folgado, D.; Belo, D.; Schultz, T.; Gamboa, H. Feature-Based Information Retrieval of Multimodal Biosignals with a Self-Similarity Matrix: Focus on Automatic Segmentation. *Biosensors* **2022**, *12*, 1182. [CrossRef]
21. Nishino, Y.; Maekawa, T.; Hara, T. Few-Shot and Weakly Supervised Repetition Counting With Body-Worn Accelerometers. *Fron. Comput. Sci.* **2022**, *4*, 925108. [CrossRef]
22. Zappi, P.; Lombriser, C.; Stiefmeier, T.; Farella, E.; Roggen, D.; Benini, L.; Tröster, G. Activity recognition from on-body sensors: Accuracy-power trade-off by dynamic sensor selection. In Proceedings of the Wireless Sensor Networks: 5th European Conference, EWSN 2008, Bologna, Italy, 30 Januar–1 February 2008; Springer: Berlin/Heidelberg, Germany, 2008; pp. 17–33. [CrossRef]
23. Bisong, E. Introduction to Scikit-learn. In *Building Machine Learning and Deep Learning Models on Google Cloud Platform*; Apress: Berkeley, CA, USA, 2019. [CrossRef]
24. Abadi, M.; Agarwal, A.; Barham, P.; Brevdo, E.; Chen, Z.; Citro, C.; Corrado, G.S.; Davis, A.; Dean, J.; Devin, M.; et al. *TensorFlow: Large-Scale Machine Learning on Heterogeneous Systems*. In Proceedings of the 12th USENIX Symposium on Operating Systems Design and Implementation (OSDI '16), Savannah, GA, USA, 2–4 November 2016.

35. Liu, H.; Xue, T.; Schultz, T. On a Real Real-Time Wearable Human Activity Recognition System. In Proceedings of the 16th International Joint Conference on Biomedical Engineering Systems and Technologies (BIOSTEC 2023), Lisbon, Portugal, 16–18 February 2023.
36. Hartmann, Y.; Liu, H.; Schultz, T. Interactive and Interpretable Online Human Activity Recognition. In Proceedings of the 2022 IEEE International Conference on Pervasive Computing and Communications Workshops and other Affiliated Events (PerCom Workshops), Pisa, Italy, 21–25 March 2022; IEEE: Piscataway, NJ, USA, 2022; pp. 109–111.

Disclaimer/Publisher's Note: The statements, opinions and data contained in all publications are solely those of the individual author(s) and contributor(s) and not of MDPI and/or the editor(s). MDPI and/or the editor(s) disclaim responsibility for any injury to people or property resulting from any ideas, methods, instructions or products referred to in the content.

Article

Decoding Mental Effort in a Quasi-Realistic Scenario: A Feasibility Study on Multimodal Data Fusion and Classification

Sabrina Gado [1,†], Katharina Lingelbach [2,3,*,†], Maria Wirzberger [4,5] and Mathias Vukelić [2]

1. Experimental Clinical Psychology, Department of Psychology, Julius-Maximilians-University of Würzburg, 97070 Würzburg, Germany; sabrina.gado@uni-wuerzburg.de
2. Applied Neurocognitive Systems, Fraunhofer Institute for Industrial Engineering IAO, 70569 Stuttgart, Germany; mathias.vukelic@iao.fraunhofer.de
3. Applied Neurocognitive Psychology Lab, Department of Psychology, Carl von Ossietzky University, 26129 Oldenburg, Germany
4. Department of Teaching and Learning with Intelligent Systems, University of Stuttgart, 70174 Stuttgart, Germany; maria.wirzberger@iris.uni-stuttgart.de
5. LEAD Graduate School & Research Network, University of Tübingen, 72072 Tübingen, Germany
* Correspondence: katharina.lingelbach@iao.fraunhofer.de
† These authors contributed equally to this work.

Abstract: Humans' performance varies due to the mental resources that are available to successfully pursue a task. To monitor users' current cognitive resources in naturalistic scenarios, it is essential to not only measure demands induced by the task itself but also consider situational and environmental influences. We conducted a multimodal study with 18 participants (nine female, M = 25.9 with SD = 3.8 years). In this study, we recorded respiratory, ocular, cardiac, and brain activity using functional near-infrared spectroscopy (fNIRS) while participants performed an adapted version of the warship commander task with concurrent emotional speech distraction. We tested the feasibility of decoding the experienced mental effort with a multimodal machine learning architecture. The architecture comprised feature engineering, model optimisation, and model selection to combine multimodal measurements in a cross-subject classification. Our approach reduces possible overfitting and reliably distinguishes two different levels of mental effort. These findings contribute to the prediction of different states of mental effort and pave the way toward generalised state monitoring across individuals in realistic applications.

Keywords: mental effort; machine learning; multimodal physiological signals; sensor fusion; neuroergonomics; human–machine interaction

1. Introduction

In everyday life, we constantly face situations demanding high stakes for maximum gains; for instance, to succeed in rapidly acquiring complex cognitive skills or making decisions under high pressure. Thereby, a fit between personal skills and the task's requirements determines the quality of outcomes. This fit is vital, especially in performance-oriented contexts such as learning and training, safety-critical monitoring, or high-risk decision-making. A person's performance can be affected by several factors: (1) level of experience and skills, (2) current physical conditions (e.g., illness or fatigue), (3) current psychological conditions (e.g., stress, motivation, or emotions), or (4) external circumstances (e.g., noise, temperature, or distractions; Hart and Staveland [1], Young et al. [2]).

To reliably quantify the mental effort during a particular task, different measures can be used: (1) behavioural (i.e., performance-based), (2) subjective, and (3) neurophysiological measures [3–5]. While performance can be inspected by tracking the user's task-related progress, the actual pattern of invested cognitive resources can only be derived by measuring brain activity with neuroimaging techniques. Coupled with sophisticated signal processing and machine learning (ML), advances in portable neuroimaging techniques have paved the way for studying mental effort and its possible influences from a neuroergonomic perspective [6,7]. Recently, functional near-infrared spectroscopy (fNIRS) has been used to study cognitive and emotional processes with high ecological validity [6,8–10]. fNIRS is an optical imaging technology allowing researchers to measure local oxy-haemoglobin (HbO) and deoxy-haemoglobin (HbR) changes in cortical regions. Higher mental effort is associated with an increase in HbO and a decrease in HbR in the prefrontal cortex (PFC) [11–13]. The PFC is crucial for executive functions like maintaining goal-directed behaviour and suppressing goal-irrelevant distractions [14,15]. In addition to changes in the central nervous system, an increased mental effort also leads to changes in the autonomic nervous system. The autonomic nervous system, as part of the peripheral nervous system, regulates automatic physiological processes to maintain homeostasis in bodily functioning [16,17]. Increased mental effort is associated with decreased parasympathetic nervous system activity and increased sympathetic nervous system activity [18–20]. Typical correlates of the autonomic nervous system for cognitive demands, engagement or mental effort are cardiac activity (e.g., heart rate and heart rate variability), respiration (rate, airflow, and volume), electrodermal activity (skin conductance level and response), blood pressure, body temperature, and ocular measures like pupil dilation, blinks, and eye movements [7,19,21–24].

Not surprisingly, all these measures are, thus, often used as a stand-alone indicator for mental effort (i.e., in a *unimodal approach*). However, a *multimodal approach* has several advantages over using only one measure. It can compensate for specific weaknesses and profit from the strengths of the different complementary measurement methods (performance, subjective experience as well as neuro- and peripheral physiological measures) [25–27]. For instance, (neuro-)physiological measures can be obtained without imposing an additional task [16] and allow for capturing cognitive subprocesses involved in executing the primary task [28]. A multimodal approach, hence, provides a more comprehensive view of (neuro-)physiological processes related to mental effort [4,5,25,29], as it can capture both central and peripheral nervous system processes [21,27]. However, fusing data from different sources remains a major challenge for multimodal approaches. ML methods provide solutions to compare and combine data streams from different measurements. ML algorithms are becoming increasingly popular in computational neuroscience [30,31]. The rationale behind these algorithms is that the relationship between several input data streams and a particular outcome variable, e.g., mental effort, can be estimated from the data by iteratively fitting and adapting the respective models. This allows for data-driven analyses and provides ways to exploratorily identify patterns in the data that are informative [32].

Data-driven approaches can also be advantageous in bridging the disparity between laboratory research and real-world applications. For instance, when specific temporal events (such as a stimulus onset) or the brain correlates of interest, are not precisely known. In contrast to traditional laboratory studies that typically rely on simplified and artificial stimuli and tasks, a naturalistic approach seeks to emulate, to some extent, the intricacy of real-world situations. Hence, these studies can provide insights into how the brain processes information and responds to complex stimuli in the real world [33].

Real-world settings are usually characterised by multiple situational characteristics, including concurrent distractions that affect the allocation of attentional and cognitive resources [34]. According to the working memory model by Baddeley and Hitch [35], performance is notably diminished when distractions deplete resources from the same modality as the primary task. However, Soerqvist et al. [36] propose the involvement of cognitive control mechanisms that result in reduced processing of task-irrelevant information under higher mental effort. To uphold task-relevant cognitive processes, high-level cortical areas, particularly the PFC, which govern top-down regulation and executive functioning, suppress task- or stimulus-irrelevant neural activities by inhibiting the processing of distractions [28]. Consequently, the effects of distractors are mitigated. In light of these considerations, understanding the capacity of a stimulus to capture attention in a bottom-up manner, known as salience, emerges as a crucial aspect. A salient stimulus has the potential to disrupt top-down goal-oriented and intentional attention processes [37] and to impair performance in a primary task [38–40]. Previous studies found that irrelevant, yet intelligible speech exerts such disruptive effects on participants' performance in complex cognitive tasks [41,42]. Consequently, intelligible speech might heighten the salience of a distracting stimulus. Moreover, further studies revealed that the emotional intensity and valence of a stimulus also play a role in influencing its salience [37,43]. Despite their detrimental impact on performance, people frequently experience such salient distractions (such as verbal utterances from colleagues) at work, even in highly demanding safety-relevant tasks. Therefore, gaining an understanding of the underlying cognitive processes in naturalistic scenarios and identifying critical moments that lead to performance decreases in real-world settings are crucial research topics in the field of neuroergonomics.

To decode and predict cognitive states, most research so far focused on subject-dependent classification. These approaches face the challenge of high inter-individual variability in physiological signals when generalising the model to others [44]. Recently, pioneering efforts have been made to develop cross-subject models that overcome the need for subject-specific information during training [45,46]. Solutions to address the challenge of inter-individual variability [47] are crucial for the development of "plug and play" real-time state recognition systems [48] as well as the resource-conserving exploitation of already available large datasets without time-consuming individual calibration sessions. Taking into account the aforementioned considerations and research agenda concerning the decoding of mental effort in naturalistic scenarios, we conducted a feasibility study and developed an ML architecture to decode mental effort across subjects from multimodal physiological and behavioural signals. We used a monitoring task simulating typical work tasks of air traffic controllers. This adapted version of the warship commander task induces mental effort based on a combination of attentional and cognitive processes, such as object perception, object discrimination, rule application, and decision-making [49]. To create a complex close-to-naturalistic scenario, three emotional types of auditory speech-based stimuli with neutrally, positively, and negatively connoted prosody were presented during the task as concurrent distractions [50]. Concurrently, performance-based, brain-related as well as peripheral physiological signals associated with mental effort were recorded.

We hypothesised that a well-designed multimodal voting ML architecture is preferable compared to a classifier based on (a) only one modality (unimodal approach) and (b) a combined, unbalanced feature set of all modalities. We expected that a multimodal voting ML model is capable of predicting subjectively experienced mental effort induced by the task itself but also by the suppression of situational auditory distractions in a complex close-to-realistic environment. Thus, we first investigated whether a combined prediction of various ML models is superior to the prediction of a single model (RQ1) and, second, we explored whether a multimodal classification that combines and prioritises the predictions of different modalities is superior to a unimodal prediction (RQ2). Furthermore, our approach enables a systematic evaluation of the unimodal and multimodal models, assessing their suitability and informativeness of each modality in decoding mental effort. This knowledge provides researchers in the fields of Human Activity Recognition and Behaviour Recognition with references for selecting suitable sensors, as well as a validated multimodal experimental design and ML processing pipeline [51].

2. Materials and Methods

2.1. Participants

Interested volunteers filled in a screening questionnaire that checked eligibility for study participation and collected demographic characteristics. Participants between the ages of 18 and 35 and with normal or corrected-to-normal vision were included in the study. Interested volunteers were excluded if they had insufficient knowledge of the German language or limited colour vision, as these factors could impede their ability to perform the tasks. Additionally, pregnant women, individuals indicating precarious alcohol or drug consumption, and those reporting mental, neurological, or cardiovascular diseases were not included. Due to the data collection period in June 2021 coinciding with the COVID-19 pandemic, individuals belonging to the risk group for severe COVID-19 disease as defined by the Robert Koch Institute, were also not invited to the laboratory. The final sample consisted of 18 participants (nine female, three left-handed, mean age of 25.9 years $SD = 3.8$, range = 21–35 years) who were all tested individually. Before their participation they signed an informed consent according to the recommendations of the Declaration of Helsinki and received monetary compensation for their voluntary participation. The study was approved by the ethics committee of the Medical Faculty of the University of Tübingen, Germany (ID: 827/2020BO1).

2.2. Experimental Task

Participants performed an adapted version of a warship commander task (WCT [52], adapted by Becker et al. [49]). The WCT is a quasi-realistic navy command and control task designed as a basic analogue to a Navy air warfare task [53]. It is suitable to investigate various cognitive processes of human decision-making and action execution [53]. Here, we used a non-military safety-critical task, where participants had to identify two different flying objects on a simulated radar screen around an airport. Objects included either registered drones (neutral, non-critical objects), or non-registered (critical) drones. They had to prevent the non-registered drones, potentially being a safety issue, from entering the airport's airspace. Non-registered drones entering pre-defined ranges close to the airport had to be first warned and then repelled in the next step. A performance score was computed based on participants' accuracy and reaction time. See Becker et al. [49], for a more detailed description of the scoring system, and see Figure 1 for an overview of the interface.

Figure 1. Elements of the WCT interface. Left side of the screen (map): Participants had to monitor the aerial space of the airport. When an unregistered drone entered the yellow area (outer circle), participants had to warn that drone; when an unregistered drone entered the red area (inner circle), participants had to repel it. Right side of the screen (graphical user interface): Participants had to request codes and pictures of unknown flying objects and then classify them as birds, registered drones, or unregistered drones.

During the task, we presented vocal utterances, either spoken in a happy, angry, or neutral way from the Berlin Database of Emotional Speech (Emo-DB [50]). These utterances were combined into different audio files, each one minute long, with speakers and phrases randomly selected and as little repetition as possible within each file. We also included a control condition where no auditory distraction was presented. The task load was manipulated by implementing two difficulty levels in the WCT (low and high). This resulted in a 2 × 4 design with eight experimental conditions. Participants completed two rounds of all conditions in the experiment. Before the respective round, a resting state measurement was conducted (30 s). Each round then consisted of eight blocks, each comprising three 60-s trials of the same experimental condition. The task load condition (operationalised with the difficulty level) was alternated across blocks. Half of the subjects started with a high task load and the other half with a low task load block. Similarly, the concurrent emotional condition (operationalised with different auditory distractions) was randomised and sampled without replacement. Before each block, except for the first, participants completed a baseline condition trial with a very low difficulty level where they had to track six objects, of which three were non-registered drones. In the low task load condition, participants had to track 12 objects, of which six were non-registered drones. In the high task load condition, they had to track 36 objects, of which 17 were non-registered drones. We used different emotional audio files for the trials in one block. Before and after the whole experiment, as well as after each experimental block, participants filled in questionnaires. See Figure 2 for a schematic representation of the whole experimental procedure. Overall, the experiment lasted approximately 120 min, including 30 min of preparation time of the used measurement devices and calibration procedures.

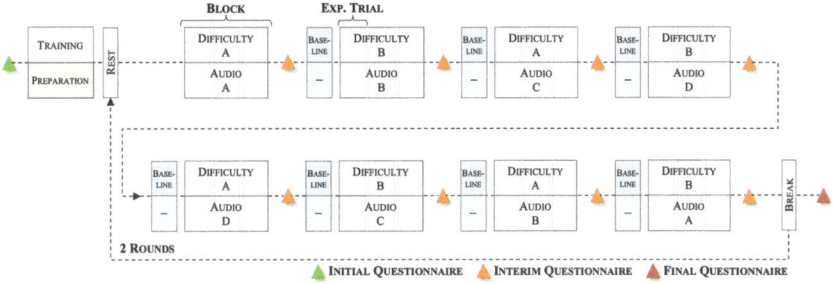

Figure 2. Procedure of the experiment. The presented procedure is exemplary as the task load condition was alternating, and the concurrent emotional condition was pseudo-randomised throughout the different blocks.

2.3. Data Collection

2.3.1. Questionnaires

Subjectively perceived mental effort and affective states were assessed after each experimental block. We used the NASA TLX effort and frustration subscales [1], EmojiGrid [54], and categorical Circumplex Affect Assessment Tool (CAAT [55]). After the experiment, participants answered questionnaires regarding personal traits that might have influenced their performance and behaviour during the study. These questionnaires comprised the short version of the German Big Five Inventory (BFI-K [56]), the German State-Trait-Anxiety Inventory (STAI [57]), the Attention and Performance Self-Assessment (APSA [58]) and the German language version of the Barratt Impulsiveness Scale-11 (BIS [59]). Here, we only used the NASA TLX ratings of mental effort for labelling the (neuro-)physiological data in the ML classification. The other subjective measures were not of interest in this analysis.

2.3.2. Eye-Tracking, Physiology, and Brain Activity

The ocular activity was recorded with the screen-based Tobii Pro Spectrum eye-tracking system, which provides gaze position and pupil dilation data at a sampling rate of 60 Hz. To capture changes in physiological responses, participants were wearing a Zephyr BioHarness™ 3 belt recording electrocardiographic (ECG), respiration, and temperature signals at a sampling rate of 1 Hz. Here, we used automatically computed, aggregated scores for the heart rate, heart rate variability, and respiration rate and amplitude from the device. Physiological as well as behavioural measures were recorded using the iMotions Biometric Research Platform software. Participants' brain activity was recorded with a NIRx NIRSport2 system, which emits light at two wavelengths, 760 and 850 nm. Data were collected with the Aurora fNIRS recording software at a sampling rate of 5.8 Hz. To capture regions associated with mental effort, 14 source optodes and 14 detector optodes were placed over the prefrontal cortex [12,60] using the fNIRS Optodes' Location Decider (fOLD) toolbox [61] (Figure 3, for the montage). Event triggers from the experimental task were sent to iMotions and Aurora using TCP protocols and Lab Streaming Layer (LSL). Signals from the different recording and presentation systems were temporally aligned offline after the data collection.

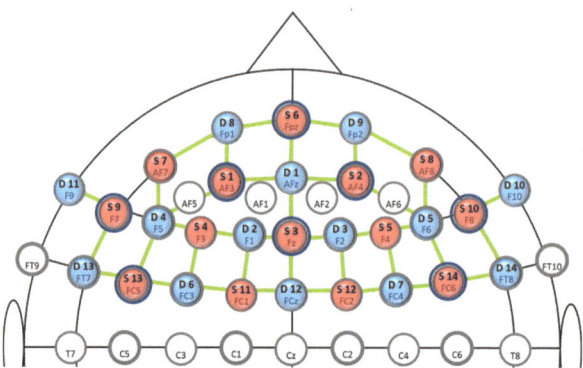

Figure 3. Location of fNIRS optodes. Montage of optodes on fNIRS cap on a standard 10–20 EEG system, red optodes: sources, blue optodes: detectors, green lines: long channels, dark blue circles: short channels. Setup with 41 (source–detector pairs) × 2 (wavelengths) = 82 optical channels of interest.

2.4. Data Preprocessing and Machine Learning

Data preprocessing and ML analyses were performed with custom-written scripts in R (version 4.1.1) and Python™ (version 3.8). Continuous raw data streams were cut into non-overlapping 60-s intervals starting at the onset of each experimental trial (Figure 2). Before feeding the data into the classification pipeline, we applied the following data cleaning and preprocessing steps per modality.

2.4.1. Preprocessing of Eye-Tracking Data

Continuous eye tracker data were preprocessed using the eyetrackingR package in R [62]. Missing values were linearly interpolated and 855 trials with a length of 60 s (on average 47.5 trials per subject, $SD = 0.9$) were extracted. Next, we used the validity index to remove non-consistent data segments from further analysis. The index is provided by the eye tracker and indicates samples in which the eye tracker did not recognise both pupils correctly ("track loss"). A total of 17 trials (1.99%) with a track loss proportion greater than 25% were removed, and 838 trials were left to extract fixations and pupil dilation (on average 46.6 trials per subject, $SD = 2.4$). For the preprocessing of the pupil dilation data, we used the PupillometryR R-package [63]. First, we calculated a simple linear regression of one pupil against the other and vice versa, per subject and trial to smooth out small artefacts [64]. Afterwards, we computed the mean of both pupils and filtered the data using the median of a rolling window with a size of 11 samples. To control for the variance of pupil sizes between participants, we applied a subject-wise z-score normalisation of pupil dilation. For the computation of fixations, we used the saccades R-package [65]. We obtained fixations for 565 trials (on average 31.4 trials per subject, $SD = 1.2$). To control for the variance between participants, we also computed z-scores of the number and the duration of fixations separately for each subject.

2.4.2. Preprocessing of Physiological Data

Epoching in non-overlapping 60-s time windows from the electrocardiographic raw data resulted in 832 trials (on average 46.2 trials per subject, $SD = 1.3$). We applied a correction for the between-participant variance identical to the one described for the eye-tracking data using z-score normalisation.

2.4.3. Preprocessing of fNIRS Data

We used the libraries MNE-Python [66] and its extension MNE-NIRS [67] and guidelines from Yücel et al. [68] to preprocess the fNIRS data. First, we converted the raw data into an optical density measure. A channel pruning was applied using the scalp-coupling index for each channel which is an indicator of the quality of the connection between the optodes and the scalp and looks for the presence of a prominent synchronous signal in the frequency range of cardiac signals across the photo-detected signals [69]. Channels with a scalp-coupling index below 0.5 were marked as bad channels. We further applied a temporal derivative distribution repair accounting for a baseline shift and spike artefacts [70]. Channels marked as bad were interpolated, with the nearest channel providing good data quality. Afterwards, a short-separation regression was used, subtracting short-channel data from the standard long-channel signal to correct for systemic signals contaminating the brain activity measured in the long-channel [68,71]. Next, the modified Beer–Lambert Law was applied to transform optical density into HbO and HbR concentration changes [72] with a partial pathlength factor of 6 [66]. Data were filtered using a fourth-order zero-phase Butterworth bandpass filter to remove instrumental and physiological noise (such as heartbeat and respiration; cut-off frequencies: 0.05 and 0.7 Hz; transition bandwidth: 0.02 and 0.2 Hz). HbO and HbR data was cut into epochs with a length of 60 s and channel-wise z-scored normalised. In total, 730 trials were obtained for the analysis (on average 40.6 trials per subject, $SD = 9.6$).

2.4.4. Feature Extraction

Our feature space comprised brain activity, physiological, ocular and performance-related measures. Table 1 gives an overview of the included features per subject and trial for each modality. We extracted the features of the fNIRS data using the mne-features package [73].

Figure S24 in the supplementary material provide exploratory analyses of the distribution and relationship between behavioural, heart activity, respiration, ocular measures, and the NASA TLX questionnaire scale effort during low and high subjective load. The Supplementary Figures S25 and S26 compare the grand average of the behavioural and physiological measures as well as single fNIRS channels of the prefrontal cortex using bootstrapping with 5000 iterations and 95% confidence intervals (CI) during low and high subjective load.

2.4.5. Ground Truth for Machine Learning

Our main goal was to predict the mental effort experienced by an individual using ML and training data from other subjects (e.g., [74,75]). Since the experimentally manipulated task load was further influenced by situational demands (e.g., inhibiting task-irrelevant auditory emotional distraction), the perceived mental effort might not be fully captured by the experimental condition. Therefore, we explored two approaches to operationalise mental effort as a two-class classification problem: First, based on self-reports using the NASA TLX effort subscale, and second, based on the experimental task load condition.

Table 1. Included features per modality.

Modality	Features
Brain Activity	Mean, standard deviation, peak-to-peak (PTP) amplitude, skewness, and kurtosis of the 82 optical channels
Physiology	
Heart Rate	Mean, standard deviation, skewness, and kurtosis of heart rate
	Mean, standard deviation, skewness, and kurtosis of heart rate variability
Respiration	Mean, standard deviation, skewness, and kurtosis of respiration rate
	Mean, standard deviation, skewness, and kurtosis of respiration amplitude
Temperature	Mean, standard deviation, skewness, and kurtosis of body temperature
Ocular Measures	
Fixations	Number of fixations, total duration and average duration of fixations, and standard deviation of the duration of fixations
Pupillometry	Mean, standard deviation, skewness, and kurtosis of pupil dilation
Performance	Average reaction time and cumulative accuracy

For the mental effort prediction based on subjective perception, we performed a subject-wise median split and categorised values above the threshold as "high mental effort" and below as "low mental effort". Across all subjects, we had a mean median-based threshold of 3.8 ($SD = 3.2$, scale range = 0–20) leading to an average of 23.8 trials per subject with low mental effort ($SD = 6.6$, range = 12–39) and 14.5 trials per subject with high mental effort ($SD = 6.2$, range = 3–21; see Supplementary Figure S1 for a subject-wise distribution of the classes).

In addition, we performed a subject-wise split at the upper quartile of the NASA TLX effort subscale. The upper (or third) quartile is the point below which 75% of the data lies. We introduced this data split to also investigate the prediction and informative features of extremely high perceived mental effort, which may indicate cognitive overload. By performing a quartile split, we had a mean threshold of 6.1 ($SD = 4.3$, scale range = 0–20) across all relevant subjects (excluding subjects 5 and 9 which did not show enough variation to identify these two classes) with an average number of 30.9 low mental effort trials per subject ($SD = 8.0$, range = 16–39) and 6.6 high mental effort trials per subject ($SD = 2.3$, range = 3–9; see Supplementary Figure S2 for a subject-wise distribution of the classes).

At last, we compared the prediction of subjectively perceived mental effort with a prediction of the mental effort induced by the task, that is the experimental condition ("high task load" vs. "low task load"; see Supplementary Figure S15 for a subject-wise comparison of perceived mental effort dependent on the experimental load condition). The comparison allows to further control for confounding effects that are typical for self-reports, e.g., consistency effects or social desirability effects.

2.4.6. Model Evaluation

We fitted six ML approaches: (1) Logistic Regression (LR), (2) Linear Discriminant Analysis (LDA), (3) Gaussian Naïve Bayes Classifier (GNB), (4) K-Nearest Neighbour Classifier (KNN), (5) Random Forest Classifier (RFC), and (6) Support Vector Machine (SVM). They were implemented using the scikit-learn package (version 1.0.1; [76]). Figure 4 shows a schematic representation of our multimodal classification scheme and cross-subject validation procedure using multiple randomised grid search operations.

For the cross-subject classification, we used a leave-one-out (LOO) approach where each subject served as a test subject once (leading to 18 "outer" folds). With this 18-fold cross-subject approach, we simulate a scenario where a possible future system can predict an operator's current mental effort during a task without having seen any data (e.g., collected in a calibration phase) from this person before. This has the advantage that the model learns to generalise across individuals and allows the exploitation of already collected datasets from a similar context as training sets.

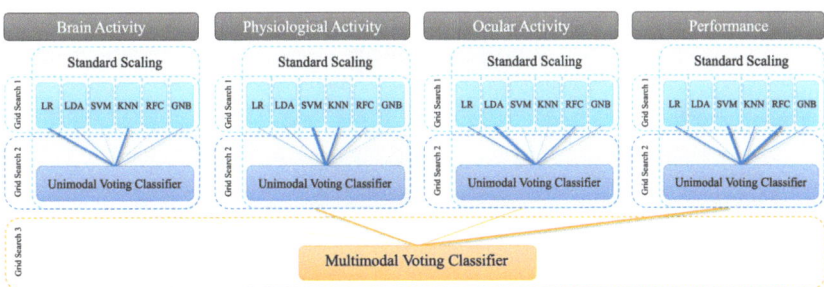

Figure 4. Classification procedure with cross-validated randomised grid searches (maximum number of 100 iterations) and a validation set consisting of one or two subjects. The first grid search optimises the hyperparameters for the different individual and unimodal classifiers. The second grid search optimises the weights as well as voting procedure (soft or hard) for the unimodal voting classifier. The third grid search optimises the weights as well as the voting procedure (soft or hard) for the multimodal voting classifier.

Our multidimensional feature space consisted of four modalities: (1) brain activity, (2) physiological activity, (3) ocular measures, and (4) performance measures. All features were z-standardised (Figure 4). This scaling ensured, that for each feature the mean is zero and the standard deviation is one, thereby, bringing all features to the same magnitude. We then trained the six classifiers (LR, LDA, GNB, KNN, RFC, and SVM) separately for each modality. Hyperparameters for each classifier were optimised by means of a cross-validated randomised grid search with a maximum number of 100 iterations and a validation set consisting of either one or two subjects. We tested both sizes of the validation set to find an optimal compromise between the robustness of the model and the required computing power. While cross-validation with two subjects counteracts the problem that the models highly adapt to an individual's unique characteristics, cross-validation with only one subject leads to a lower number of necessary iterations and a computationally more efficient approach. Due to our cross-subject approach, the selected hyperparameters varied for each predicted test subject.

Afterwards, we combined these classifiers using a voting classifier implemented in the mlxtend package (version 0.19.0 [77]). The ensemble classifier makes predictions based on aggregating the predictions of the previously trained classifiers by assigning weights to each of them. Here, we are interested in whether an ensemble approach achieves higher prediction accuracy than the best individual classifier in the ensemble. An ensemble approach has the advantage that, even if each classifier is a weak learner (meaning it does only slightly better than random prediction), the ensemble could still be a strong learner (achieving high accuracy). The voting either follows a "soft" or a "hard" voting strategy. While hard voting is based on a majority vote combining the predicted classes, soft voting considers the predicted probabilities and selects the class with the highest probability across all classifiers. The weights, as well as the voting procedure (soft or hard voting), were optimised using a third cross-validated randomised grid search with a maximum number of 100 iterations. We restricted the weights to a maximum value of 2 (range = 0–2).

With this procedure, we were able to compare the predictions of the single unimodal classifiers to a weighted combination of all classifiers of one modality.

For the multimodal approach, the voting predictions of each modality were combined into a final multimodal prediction of mental effort using a second voting classifier. This second voting classifier also assigned weights to the different modality-specific classifiers and was optimised in the same manner as the unimodal approach. We report the average F_1 score and a confusion matrix of the training set and the test subject to evaluate model performance. The F_1 score can be interpreted as a weighted average or "harmonic mean" of precision and recall (1—good to 0—bad performance). Precision refers to the number of samples predicted as positive that are positive (true positives). Recall measures how many of the actual positive samples are captured by the positive predictions (also called sensitivity). The F_1 score balances both aspects – identifying all positive, i.e., "high mental effort" cases, but also minimising false positives.

To compare the classification performance of different models, we calculated the bootstrapped mean and its CI over cross-validation folds with 5000 iterations per classification model. We corrected the 95% CI for multiple comparisons using the Bonferroni Method. Significant differences can be derived from non-overlapping notches of the respective boxes, which mark the upper and lower boundaries of bootstrapped 95% CI of the mean F_1 score. The upper CI limit of a dummy classifier represents an empirical chance level estimate (dashed grey line in all subplots of Figure 5). A dummy classifier considers only the distribution of the outcome classes for its prediction. For a prediction to be better than chance (at a significance level below 0.05), the bootstrapped mean of a classifier must not overlap with this grey line [78]. For a significance level below 0.01, the lower CI boundary of a classifier's mean must not overlap with this grey line [78]. This criterion can also be applied when statistically comparing different models, regardless of the chance level.

3. Results

We compare the results for a mental effort prediction based on a subject-wise (1a) median and (1b) upper quartile split of the Nasa TLX effort scale as well as based on the (2) experimentally induced task load. Further, we compare two sizes of the validation set (one subject and two subjects).

3.1. Unimodal Predictions

The performance of the different modalities and classifiers is visualised in Figure 5. We do not see substantially better performance when using a larger validation set of two subjects, neither for the median split (compare Figure 5 and Supplementary Figure S3) nor for the upper quartile split (compare Supplementary Figures S7 and S11) or the prediction of the experimentally induced task load (Supplementary Figures S16 and S20). We will, therefore, focus on the models fitted with a validation set of one subject, as this is more time- and resource-efficient. In this case, we estimated the CI's upper boundary of the mean empirical chance level for predicting subjectively perceived mental effort to be 0.444 ($M = 0.368$, 95% CI [0.284; 0.444]). This estimate now serves as a reference for determining significant performance above the chance level.

Figure 5 and Table 2 show the performance in a median-split-based unimodal approach (Figure 5A,B,D,E) as well as for the multimodal approach (Figure 5C; elaborated on in Section Multimodal Predictions). Regarding the unimodal classifications, we see the highest predictions of the subjectively perceived mental effort for performance data (Figure 5E compared with ocular, physiological, or brain activity measures; Figure 5A,B,D; see Table 2). Except for the performance-based model, we observe overfitting indicated by the large deviation between training and test performance (Figure 5A,B,D). None of the brain activity-based models performs significantly better than the estimated chance level in the test data set (Figure 5A,G, Table 2). When examining the single classification models within each modality, the KNN, RFC, and SVM were more likely to be overfitted, as seen by the good performance in the training set but a significantly worse performance for the test subject. We combined the different classifiers using a voting classifier, of which we ascertained the voting procedure (soft vs. hard voting) and the weights with a randomised grid search. See Figure 6 for an overview of the selected voting procedures and the allocated weights per modality.

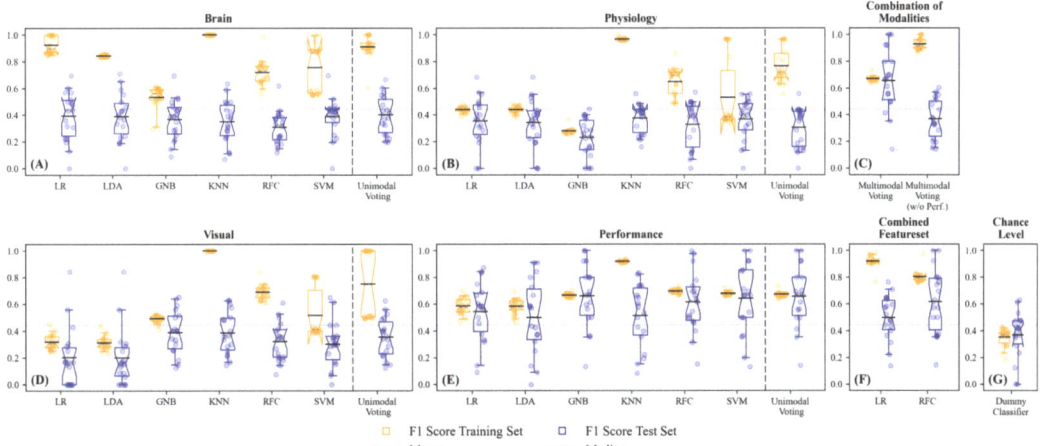

Figure 5. Prediction of the subjectively perceived mental effort based on a median split; validation set: $N = 1$. Classifiers' performance based on (**A**) fNIRS data, (**B**) physiological data (heart rate, respiration, and body temperature, (**C**) a combined, weighted feature set, (**D**) visual data, (**E**) performance data (accuracy and response time), (**F**) a combined, but unweighted feature set, and (**G**) a dummy classifier that considers only the distribution of the outcome classes for its prediction and represents an empirical chance level. Bootstrapped Bonferroni-corrected 95% confidence intervals (CI; 5000 iterations) of the mean F_1 scores for the training set (left, orange) and the test set (right, blue) of the different unimodal and multimodal models. Notches in the boxes of the plot visualise the upper and lower boundary of the CI with the solid line representing the mean and the dashed grey line representing the median. The box comprises 50% of the distribution from the 25th to the 75th quartile. The ends of the whiskers represent the 5th and 95th quartile of the distribution. The continuous grey dashed line shows the upper boundary of the CI of the dummy classifier at 0.444.

Table 2. Bootstrapped Bonferroni-corrected means and 95% CIs of F_1 scores.

	Training Set	Test Set
	Chance Level	
Dummy Classifier	0.351, 95% CI [0.320; 0.379]	0.368, 95% CI [0.284; **0.444**]
	Unimodal Predictions Based on fNIRS	
LR	0.924, 95% CI [0.889; 0.962]	0.392, 95% CI [0.278; 0.500]
LDA	0.845, 95% CI [0.840; 0.850]	0.387, 95% CI [0.271; 0.495]
GNB	0.532, 95% CI [0.469; 0.577]	0.366, 95% CI [0.275; 0.457]
KNN	1.0, 95% CI [1.0; 1.0]	0.348, 95% CI [0.248; 0.451]
RFC	0.721, 95% CI [0.667; 0.786]	0.308, 95% CI [0.235; 0.387]
SVM	0.756, 95% CI [0.644; 0.862]	0.386, 95% CI [0.287; 0.473]
Unimodal Voting	0.911, 95% CI [0.848; 0.954]	0.401, 95% CI [0.314; 0.489]
	Unimodal Predictions Based on Physiology	
LR	0.441, 95% CI [0.426; 0.454]	0.354, 95% CI [0.246; 0.464]
LDA	0.441, 95% CI [0.424; 0.456]	0.341, 95% CI [0.226; 0.448]
GNB	0.279, 95% CI [0.269; 0.297]	0.231, 95% CI [0.141; 0.318]
KNN	0.966, 95% CI [0.962; 0.973]	0.377, 95% CI [0.309; 0.439]
RFC	0.648, 95% CI [0.585; 0.708]	0.327, 95% CI [0.217; 0.425]
SVM	0.532, 95% CI [0.379; 0.715]	0.366, 95% CI [0.262; 0.455]
Unimodal Voting	0.767, 95% CI [0.691; 0.848]	0.305, 95% CI [0.212; 0.396]

Table 2. *Cont.*

	Training Set	Test Set
Unimodal Predictions Based on Visual Measures		
LR	0.318, 95% CI [0.287; 0.355]	0.201, 95% CI [0.080; 0.354]
LDA	0.314, 95% CI [0.285; 0.345]	0.198, 95% CI [0.087; 0.346]
GNB	0.492, 95% CI [0.472; 0.506]	0.390, 95% CI [0.294; 0.485]
KNN	1.0, 95% CI [1.0; 1.0]	0.386, 95% CI [0.291; 0.485]
RFC	0.690, 95% CI [0.658; 0.728]	0.322, 95% CI [0.237; 0.415]
SVM	0.517, 95% CI [0.422; 0.630]	0.301, 95% CI [0.198; 0.401]
Unimodal Voting	0.751, 95% CI [0.589; 0.915]	0.354, 95% CI [0.262; 0.442]
Unimodal Predictions Based on Performance Measures		
LR	0.586, 95% CI [0.549; 0.624]	**0.543**, 95% CI [0.399; 0.676] *
LDA	0.584, 95% CI [0.548; 0.618]	**0.499**, 95% CI [0.325; 0.663] *
GNB	0.667, 95% CI [0.659; 0.677]	0.661, 95% CI [**0.514**; 0.792] **
KNN	0.919, 95% CI [0.914; 0.924]	**0.513**, 95% CI [0.367; 0.650] *
RFC	0.696, 95% CI [0.686; 0.707]	0.616, 95% CI [**0.472**; 0.752] **
SVM	0.679, 95% CI [0.672; 0.689]	0.641, 95% CI [**0.467**; 0.799] **
Unimodal Voting	0.673, 95% CI [0.662; 0.687]	0.656, 95% CI [**0.509**; 0.789] **
Multimodal Predictions		
LR	0.919, 95% CI [0.892; 0.939]	**0.498**, 95% CI [0.412; 0.581] *
RFC	0.803, 95% CI [0.784; 0.830]	0.617, 95% CI [**0.499**; 0.742] **
Multimodal Voting	0.673, 95% CI [0.663; 0.686]	0.658, 95% CI [**0.515**; 0.797] **
Multimodal Voting (w/o Perf.)	0.930, 95% CI [0.905; 0.957]	0.369, 95% CI [0.276; 0.464]

Note. * $p < 0.05$ considering the bootstrapped mean, ** $p < 0.01$ considering the lower CI limit

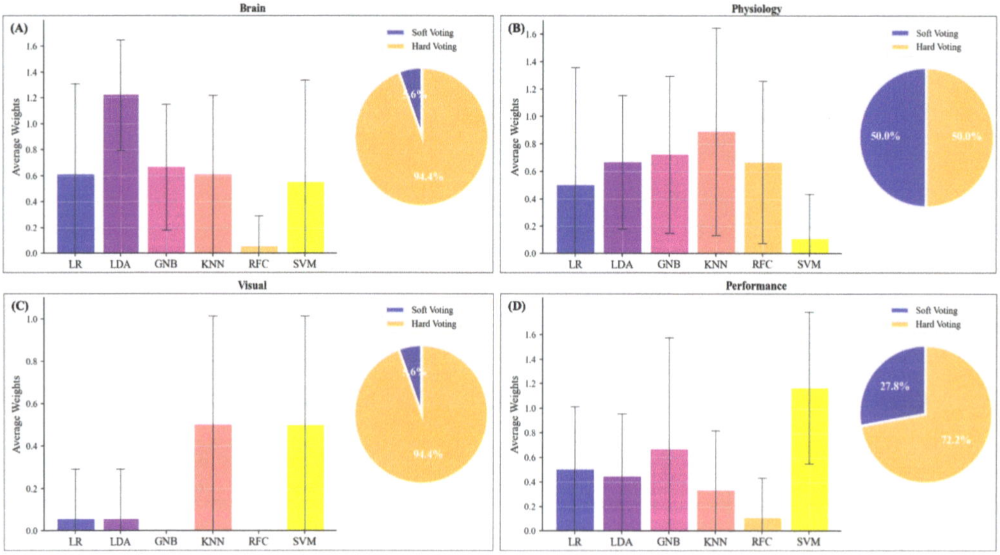

Figure 6. Weights and procedure of an unimodal voting classifier to predict subjectively perceived mental effort based on a median split; validation set: $N = 1$. Allocated weights for an unimodal voting classifier based on (**A**) fNIRS data, (**B**) physiological data (heart rate, respiration, and body temperature), (**C**) visual data, and (**D**) performance data (accuracy and response time). Error bars represent the standard deviation.

Interestingly, for 8 out of 18 participants, we observed high prediction performances with F_1 scores ranging between 0.7 and 1. However, we also identified several subjects whose subjectively perceived mental effort was hard to predict based on the training data of the other subjects. See Tables S1–S3 in the Supplementary Material for a detailed comparison of the classifiers' performances in the different test subjects. Concluding, the results indicate that transfer learning and generalisation over subjects is much more challenging when using the neurophysiological compared with the performance-based features.

3.2. Unimodal Predictions—Brain Activity

The unimodal voting classifiers for brain activity mainly used hard voting (94.4%) and gave the highest weights to the LDA classifier (Figure 6A). However, on average the unimodal voting classifier ($M = 0.40$, 95% CI [0.31; 0.49], Figure 5A) revealed strong overfitting and was neither performing better than the single classifiers nor better than the estimated chance level. We then compared the performance of the classifiers with respect to the percentage of correctly and falsely classified cases in a confusion matrix (Figure 7). Therefore, we used the best-performing classifier for each test subject and then summed over all test subjects. We compared the distribution of the true positives, true negatives, false positives, and false negatives in these classifiers with the respective distribution of the voting classifier. Here (Figure 7A), we see that both distributions indicate a high number of falsely identified "High Mental Effort" cases (False Positives), leading to a recall of 45.6% and precision of only 39.3% for the voting classifier and a recall of 57.5% and precision of 49.8% for best single classifiers.

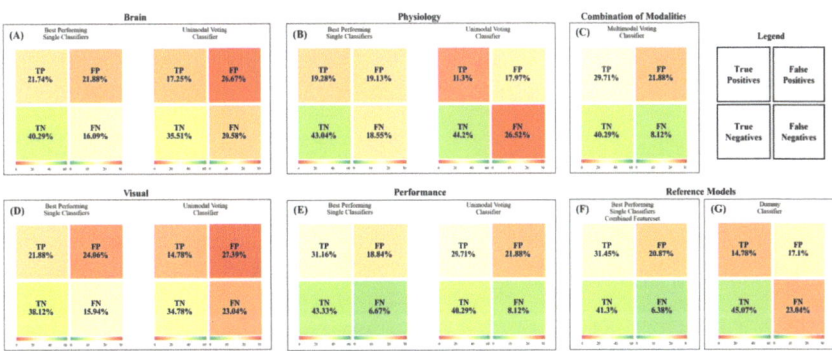

Figure 7. Prediction of the subjectively perceived mental effort (confusion matrix of test set) based on a median split; validation set: $N = 1$. Percentage of correctly and falsely classified perceived mental effort per model across all test subjects: TP = True Positives, TN = True Negatives, FP = False Positives, and FN = False Negatives, with "Positives" representing "High Mental Effort" and "Negatives" representing "Low Mental Effort". For the "Best Performing Single Classifier" we selected the classifier (LDA, LR, SVM, KNN, RFC, or GNB) with the best F_1 score for each subject. Confusion matrices based on (**A**) fNIRS data, (**B**) physiological data (heart rate, respiration, and body temperature), (**C**) a combined, weighted feature set, (**D**) visual data, (**E**) performance data (accuracy and response time), (**F**) a combined, but unweighted feature set, and (**G**) a dummy classifier representing an empirical chance level.

3.3. Unimodal Predictions—Physiological Measures

For classifying subjectively perceived mental effort based on physiological measures such as heart rate, respiration, and body temperature, soft voting was chosen in half of the test subjects. The weighting of the classifiers varied considerably, with the KNN obtaining the highest average weights (Figure 6B). The unimodal voting classifier ($M = 0.31$, 95% CI [0.21; 0.40]; Figure 5B) showed strong overfitting, and its performance in the test subjects was neither significantly better than any of the single classifiers nor better than the estimated chance level. Regarding the percentage of correctly and falsely classified cases (Figure 7B), we see that the distributions for the best-performing single classifiers seem to be slightly better than the distributions of the voting classifier. The latter had difficulties in correctly identifying the conditions with low mental effort as can be seen in the high number of false negatives. When comparing the recall and precision of both approaches, we have a recall of only 29.9% for the voting classifier (precision: 38.6%) and an average recall of 51.0% for the best single classifiers (precision: 50.2%).

3.4. Unimodal Predictions—Ocular Measures

For subjectively perceived mental effort classification based on ocular measures such as pupil dilation and fixations, the split of soft vs. hard voting was 5.6% for soft voting and 94.4% for hard voting. KNN and SVM were weighted highest (Figure 6C). The F_1 score of the unimodal voting classifier ($M = 0.35$, 95% CI [0.26; 0.44]; Figure 5D) did not show a significant above-chance-level classification performance. The percentage of correctly and falsely classified cases (Figure 7D) was similar to the brain models, with a recall of 39.1% for the voting classifier (precision: 35.1%) and an average recall of 57.9% for the best single classifiers (average precision: 47.6%).

3.5. Unimodal Predictions—Performance

At last, we predicted subjectively perceived mental effort based on performance (accuracy and speed). 27.8% of the test subjects had voting classifiers using soft voting, and 72.2% used hard voting with SVM being weighted highest (Figure 6D). The models GNB ($M = 0.66$, 95% CI [0.51; 0.79]), RFC ($M = 0.62$, 95% CI [0.47; 0.75]), and SVM ($M = 0.64$, 95% CI [0.47; 0.80]) showed all a significant above-chance-level performance ($p < 0.01$). The other models (LR, LDA, KNN) revealed also above-chance level performances but with smaller differences ($p \approx 0.05$; Table 2). The performance of the unimodal voting classifier ($M = 0.66$, 95% CI [0.51; 0.79]) was also significantly better than the estimated chance level. The percentage of correctly and falsely classified cases (Figure 7E) reveals superior classification performance compared with the brain-, physiological- and ocular-based models. However, the voting classifier still had many falsely identified "High Mental Effort" cases (False Positives), leading to a recall of 78.5% and a precision of 57.6%. The best-performing single classifiers have an average recall of 82.4% and an average precision of 62.3%.

3.6. Unimodal Predictions Based on the Upper Quartile Split

To identify informative measures for very high perceived mental effort potentially reflecting cognitive overload, we also performed predictions based on the subject-wise split at the upper quartile. Compared with the median-split-based results, we observed decreased classifiers' performance even below dummy classifier performance (Supplementary Figure S7). This might be explained by the fact that we reframed a binary prediction problem with evenly distributed classes into an outlier detection problem. Using the upper quartile split, we created imbalanced classes regarding the number of the respective samples, which made the reliable identification of the less well-represented class in the training set more difficult (reflected in the recall; Supplementary Figure S9).

3.7. Unimodal Predictions Based on the Experimental Condition

We further fitted models to predict the experimentally induced task load instead of the subjectively perceived mental effort. The prediction of mental effort operationalised by the task load was substantially more successful than the prediction of subjectively perceived mental effort. All modalities, including brain activity and physiological activity, revealed at least one classifier that was able to predict the current task load above the chance level (Supplementary Figure S16). The unimodal voting classifiers (Brain: $M = 0.59$, 95% CI [0.55; 0.62], Physiology: $M = 0.66$, 95% CI [0.62; 0.72], Visual: $M = 0.69$, 95% CI [0.62; 0.77], Performance: $M = 0.97$, 95% CI [0.90; 1.0]) were all significantly better than a dummy classifier ($M = 0.51$, 95% CI [0.47; 0.55]). Best unimodal voting classifications were obtained based on performance measures. Interestingly, other classification models were favoured in the unimodal voting, and the distribution between soft- and hard voting differed compared with the subjectively based approach, with soft voting being used more often (Supplementary Figure S17).

3.8. Multimodal Predictions Based on the Median Split

In the final step, we combined the different modalities into a multimodal prediction. Figures 5C and 7C show the performance of the multimodal voting classifier, and Figure 8A the average allocated weights to the different modalities. To compare the rather complex feature set construction of the multimodal voting with a simpler approach, we also trained two exemplary classifiers (LR without feature selection and RFC with additional feature selection) on the whole feature set without a previous splitting into the different modalities (Figure 5F).

In most test subjects (55.6%), soft voting was selected to combine the predictions for the different modalities; 44.4% used hard voting. In line with the results outlined above, the multimodal classifier relied on performance measures to predict subjectively perceived mental effort (Figure 8A), thereby turning it into a unimodal classifier. The voting classifier (Figure 5C, $M = 0.66$, 95% CI [0.52; 0.80]) led to a significantly better classification than the estimated chance level. The multimodal classifier exhibited an equivalent percentage of correctly and falsely classified cases (Figure 7C) compared with the performance-based classifier, demonstrating an average recall of 78.5% and an average precision of 57.6%. On average, it performed better than the classifiers trained with the combined whole feature set, which showed substantial overfitting.

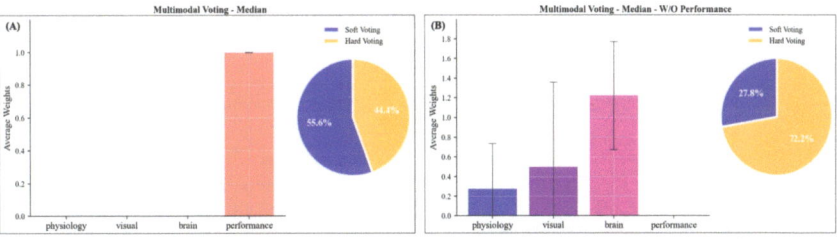

Figure 8. Weights and procedure of a multimodal voting classifier to predict subjectively perceived mental effort based on a median split; validation set: $N = 1$. (**A**) shows the allocation of weights when all modalities are included in the multimodal classification. (**B**) shows the allocation of weights when performance measures are not included in the multimodal classification. Error bars represent the standard deviation.

In order to assess the performance of the multimodal classifier without incorporating performance-based information such as speed and accuracy, we constrained the classifier to utilise only (neuro-)physiological and visual measures. This approach is especially relevant for naturalistic applications where obtaining an accurate assessment of behavioural performance is challenging or impossible within the critical time window. For the multimodal prediction without performance, brain activity was weighted highest (Figure 8B). However, classifiers revealed strong overfitting during the training, and the average performance was decreased to chance level ($M = 0.37$, 95% CI [0.28; 0.46], average recall: 40.6% and average precision: 38.3%; Figure 5C).

3.9. Multimodal Predictions Based on the Upper Quartile Split

With the upper quartile split, we observed a fundamentally different allocation of weights. High weights were assigned to brain and ocular activity (Figure 8B), while performance received only minimal weights. Hence, the exclusion of performance-based measures had minimal impact on the allocation of weights (Supplementary Figure S10B) and the overall performance of the multimodal classifiers remained largely unaffected (Supplementary Figure S7C). Among the eighteen test subjects, the multimodal classification demonstrated the highest performance in two cases (Supplementary Table S2). However, on average, the multimodal classification based on an upper quartile split ($M = 0.19$, 95% CI [0.08; 0.33]; average recall: 21.9% and average precision: 18.5%) did not demonstrate superiority over the unimodal classifiers (Brain: $M = 0.18$, 95% CI [0.08; 0.30], Physiology: $M = 0.24$, 95% CI [0.14; 0.34], Visual: $M = 0.16$, 95% CI [0.06; 0.28], Performance: $M = 0.20$, 95% CI [0.09; 0.34]). It further did not significantly outperform the dummy classifier ($M = 0.20$, 95% CI [0.12; 0.28]) or classifiers trained on a feature set of simply combined modalities without weight assignment (LR: $M = 0.26$, 95% CI [0.16; 0.36], RFC: $M = 0.28$, 95% CI [0.18; 0.39]; Supplementary Figure S7).

3.10. Multimodal Predictions Based on the Experimental Condition

Similar to the multimodal voting classifier based on a subject-wise median split of perceived mental effort, classifiers predicted the experimentally induced task load solely using the performance measures. The average prediction performance was exceptionally high ($M = 0.97$, 95% CI [0.91; 1.0]; average recall: 99.7% and average precision: 91.3%), significantly outperforming a dummy classifier ($M = 0.51$, 95% CI [0.47; 0.55]), and comparable to the performance of the classifiers trained on the combined feature set (LR: $M = 0.95$, 95% CI [0.90; 0.99], RFC: $M = 0.96$, 95% CI [0.91; 1.0]; Supplementary Figure S16). When we only allowed (neuro-)physiological and visual measures as features, visual measures were weighted highest (Supplementary Figure S19B). In this case, the average performance of the multimodal classifiers ($M = 0.69$, 95% CI [0.64; 0.74]) was also significantly above the chance level, with an average recall of 82.7% and precision of 58.7%, indicating a successful identification of mental effort based on neurophysiological, physiological, and visual measures (Supplementary Figure S16C).

4. Discussion

The purpose of our study was to test the feasibility of multimodal voting in a ML classification for complex close-to-realistic scenarios. We used both—the subjectively experienced and experimentally induced mental effort—as ground truths for a cross-subject classification. Our approach represents a crucial investigation for the practical application of mental state decoding under real-world conditions. Our study aims to fill the existing gap in the literature and address the need for online accessible naturalistic data sets by providing a multimodal voting ML architecture along with the dataset to decode mental effort across subjects in a quasi-realistic experiment simulating a real-world monitoring task. This serves as the foundation for enabling the adaptation of systems to users' current mental resources and efforts. By incorporating adaptive systems, individuals can enhance their performance by operating within an optimal level of demand, allowing them to perform at their best. In tasks involving high-security risks, it is crucial for system engineers to make every effort to prevent individuals from being overwhelmed or bored, as such states can increase the likelihood of errors. In our analyses, we employed multimodal voting cross-subject classification and evaluated the model performance using a leave-one-out approach. We contribute to the existing body of knowledge on mental state decoding by systematically evaluating the selection and informativeness of sensors including neurophysiological, physiological, visual, and behavioural measures for the classification of subjective and experimentally induced mental effort. Moreover, our results show which classifier models perform best for each modality. We further observed that in certain modalities, the combination of ML models outperformed predictions made by individual ones. For each modality, we found a different set of classifiers that were better performing in the prediction and, thus, also considered more informative in the unimodal voting.

4.1. Using Subjectively Perceived Mental Effort as Ground Truth

When predicting subjectively perceived mental effort, LDA and LR performed best and were weighted highest in the classifications based on brain activity (Figure 5). Whereas, in physiological activity, the highest weights were assigned to KNN, RFC, and SVM.

Regarding visual activity, the GNB and KNN revealed high classification performance among the test subjects. However, these models aiming to predict subjectively perceived mental effort based on brain activity, physiological activity, and visual measures were still strongly overfitted, and their performances in the test subjects were not significantly better than the dummy classifier. In performance-related measures, the GNB, RFC, and SVM performed significantly better than the dummy classifier when predicting subjective mental effort based on a median split. Using the upper quartile split for performance-related measures, the KNN and SVM showed the highest, but still, chance-level-like performances. Regarding the unimodal voting predictions of subjectively perceived mental effort, we see that a weighted combination of classifiers (LR, LDA, GNB, KNN, RFC, and SVM) was not superior to single classifiers neither when using the median nor the upper quartile split. When we combined the different modalities into a joined prediction of subjectively perceived mental effort, only the performance modality was considered. Hence, our multimodal classification might rather be considered a unimodal (performance-based) prediction. Removing the performance information from the multimodal voting classifier increases overfitting and drops the average classification performance. However, a more detailed investigation of the upper quartile split classification revealed that performance was less predictive in identifying cases of exceptionally high perceived mental effort and potential "cognitive overload" (Figure 7). In the upper quartile split classification, higher multimodal voting weights were assigned to neurophysiological and visual measures compared with performance measures. This seems to imply that subjects were more heterogeneous in their performance under exceptionally high perceived mental effort, and classifiers rather exploited correlates from neurophysiological and visual measures than from performance to predict subjectively perceived mental effort. In summary, our findings indicate that when utilizing only unimodal voting classification, the best prediction of subjectively perceived mental effort was achieved through performance-based measures. Additionally, the inclusion of the performance-based classification model is essential in our multimodal voting classification approach to address potential overfitting in predicting mental effort (median-based split). These findings suggest that further research is necessary to investigate the dependence and variability of mental effort in cross-subject classification.

4.2. Using Experimentally Induced Mental Effort as Ground Truth

For the classification of the experimentally induced task load, all modalities were able to predict mental effort with high performances already on a single classifier level. GNB, KNN, and SVM performed above the chance level and were assigned the highest weights in the unimodal voting based on brain activity (Supplementary Figure S16). For the physiological activity, all classifiers—except the KNN—reached above-chance level performance. The highest average weight in the unimodal voting was assigned to the SVM. For visual and performance measures, we did not see substantial differences between the classification performance of the single models, with all performing above the chance level. A unimodal weighted combination of these classifiers was not superior to the single classifiers in any modality. Performance exhibited the highest predictive capability for task load (Supplementary Figure S16). As a result, the multimodal classifier transitioned again back into unimodal voting, as it relied solely on the performance modality. When excluding the performance-based features, the multimodal prediction based on neurophysiological, physiological, and ocular activity was still significantly above the chance level estimated by a dummy classifier. These findings suggest that it is feasible to differentiate between various mental effort states, represented by experimentally induced task load, by utilizing neurophysiological, physiological, and visual data obtained in a close-to-realistic environment through a cross-subject classification approach. However, it was not possible to replicate these results for subjectively perceived mental effort. The discrepancies observed between these two ground truth approaches could potentially be attributed to the retrospective nature of self-reports. Self-reports rely on an individual's perception, reasoning, and subjective introspection [79]. They are, therefore, vulnerable to various perceptual and response biases like social desirability [21,80]. These post hoc evaluation processes might not be adequately reflected in and could be learned from (neuro-)physiological and visual measures during the task itself.

4.3. Generalisation across Subjects

For all classification approaches, we observed substantial variation in the performance of classifiers between the test subjects. Some individuals had F_1 scores above 0.8 (Supplementary Table S1). Other individuals demonstrated deviations in their neurophysiological reactions, diverging significantly from the patterns learned from the subjects included in the training set. These results are in line with the findings by Causse et al. [81]. The authors concluded that it is quite challenging to identify mental states based on haemodynamic activity across individuals because of the major structural and functional inter-individual differences. For instance, in the context of brain–computer interfaces, a phenomenon called BCI illiteracy describes the inability to modulate sensorimotor rhythms in order to control a BCI observed in approximately 20–30% of subjects [82]. Our findings underscore the importance of developing appropriate methods to address two key aspects. First, identifying subjects who may pose challenges in prediction due to their heterogeneity compared with the training set. Second, enabling transfer learning for these individuals by implementing techniques such as standardisation and transformation of correlates into a unified feature space [83].

4.4. Limitations and Future Research

We acknowledge that certain aspects of this study can be further improved and serve as opportunities for future advancements. One area for improvement is the complexity of the measurement setup used in this study, which required a substantial amount of time for the preparation and calibration of the involved devices. It is important to consider the potential impact on participants' intrinsic engagement and explore ways to further streamline the process during soft- and hardware development. Furthermore, it is worth noting that our feasibility study sample was relatively small and homogeneous in terms of socio-demographic characteristics, consisting predominantly of young individuals with a high level of education. The small sample size likely had a negative impact on the statistical power of our study. Combined with the homogeneity of the sample, it could also limit the generalisability of our results to more diverse populations. While it may seem intuitive to increase the sample size to address the issue of heterogeneity, there is a debate surrounding the relationship between sample size and its impact on classifier performance. With adding more and more samples, the dataset is supposedly at some point large enough to enable the classifier to find more generic and universal predictive patterns and achieve better performance again. Some argue that it is necessary to train ML models with large training datasets, including edge cases, to achieve good generalisability and attain good prediction accuracy on an individual-level Bzdok and Meyer-Lindenberg [84], Dwyer et al. [85]. Nevertheless, as emphasised by Cearns et al. [86], it is worth noting that ML classifiers demonstrate exceptional performance primarily in relatively small datasets. Consequently, the heterogeneity of a large dataset might present a significant challenge for learning. Thus, it may be more reasonable to train separate,specialised models for each homogeneous cluster, rather than attempting to construct a single model that explains the entire variance but yields less accurate predictions. Orrù et al. [87], for example, suggests the use of simple classifiers or ensemble learning methods instead of complex neural networks. Cearns et al. [86] highlight the importance of suitable cross-validation methods. Especially in the case of physiological datasets, one might also identify subjects that are very predictive for the patterns of a specific subgroup and remove subjects from the training set that show unusual patterns in neurophysiological reactions [85]. One interesting idea to address this problem is data augmentation [88,89]. This can be done by artificially generating new samples from existing samples to extend a dataset. For example, using Generative Adversarial Networks (GANs), one could simulate data to create more homogeneous and "prototypic" training datasets and increase the performance and stability of respective ML models [90]. Another suggested method to improve generalisability across subjects might be multiway canonical correlation analysis (MCCA). An approach that allows combining multiple data sets into a common representation and, thereby, achieves the denoising of data, and dimensionality reduction, based on shared components across subjects [83]. Advancements in these methods play a crucial role in enhancing the comparability and potential combinability of datasets, which is a shared objective within the research community [91].

To further increase classification performance, additional artefact analyses [92], or the implementation of inclusion criteria on a subject-, trial-, and channel-level could be explored in order to improve poor signal-to-noise ratios. Friedman et al. [93], who used an XGBoost classifier on EEG data, applied extensive and rigorous trial and subject selection criteria. For example, they did not include trials where participants failed to solve the task because they assumed that the mental effort shown by participants answering incorrectly did not reflect the true level of load (also [94]). Although this bears the risk of a major data loss, these rigid removal criteria might reflect an efficient solution to ensure that the measured neurophysiological signals truly reflect the cognitive processes of interest. Future research is necessary to a) define such exclusion and inclusion criteria depending on the investigated cognitive processes and b) develop standardised evaluation methods to decide which preprocessing step is beneficial and adequate.

A final limitation relates to the arrangement of the fNIRS optodes. Based on previous research (e.g., [12]), we decided to choose a montage solely covering the prefrontal cortex in order to reduce preparation time and facilitate transfer into close-to-realistic applications. However, we probably would have profited from a larger brain coverage that also covers parietal, temporal, and occipital brain areas [94]. Integrating these regions allows identifying features for the classification from larger functional networks that might play a crucial role in distinguishing mental states and cognitive control mechanisms [36,95]. Increased activity in the frontoparietal network is, for example, associated with task-related working memory (WM) processes (e.g., [96,97]), whereas increased connectivity between frontal and sensory areas are linked to the suppression of distractors [95].

4.5. Feature Selection and Data Fusion in Machine Learning

A crucial aim of this study was the selection and fusion of informative sources for cross-subject mental effort prediction. We integrated data from different modalities comprising brain activity as assessed with fNIRS, physiological activity (cardiac activity, respiration, and body temperature), ocular measures (pupil dilation and fixations), as well as behavioural measures of performance (accuracy and speed). However, this selection was naturally not exhaustive. Other measures, such as electroencephalography or electrodermal activity [98], could provide useful information about cognitive and physiological processes related to mental effort. In addition, one could also explore more behaviour-related measures such as speech [99] or gaze [100]. These measures might also provide the possibility to detect predictive patterns without significantly interfering with the actual task. When conducting applied research and incorporating mental state decoding in real-world settings (e.g., healthcare, entertainment, gaming, industry, and lifestyle; [101]), it is crucial to utilise sensors that are unobtrusive, seamlessly integrated into the environment, mobile, and user-friendly. For this purpose, further validation studies are warranted to evaluate the quality and suitability of smart wearables such as smartwatches or fitness trackers [101,102], mobile neurophysiological sensors [103,104], and mobile eye-tracking [105,106]. Our results indicate that performance-based measures as well as a multimodal approach including neurophysiological, physiological, and visual measures are successful in decoding experimentally induced mental effort. Visual features, followed by physiological measures, were particularly informative in the multimodal approach. This insight enables researchers to optimise their sensor setup by prioritizing measures and streamlining their data collection process.

To combine the data streams obtained from the different measurement methods, we implemented data fusion on two levels: (1) the feature level and (2) the classification level.

First, we aggregated our raw data, mainly time series, into informative features. We used standard statistical features like the mean, standard deviation, skewness, and kurtosis. Friedman et al. [93] explored more sophisticated features such as connectivity and complexity metrics, which have the potential to capture additional information about relationships within and between neuronal networks. Further investigations are required to assess the predictive quality of these aggregated features. Additionally, future research can explore the added value of feature selection and wrapping methods, which aim to reduce the complexity of the feature space without compromising the predictive information [107,108]. Such methods, e.g., sequential feature forward selection, might be a way to improve classifiers' performance by keeping only the most informative features. Another approach could be the use of continuous time-series data which provide insights into differences in the experience and processing of mentally demanding tasks separately for the different neurophysiological modalities. Hence, some researchers implemented deep learning methods like convolutional or recurrent neural networks to derive classifications based on multidimensional time-series data [45,109,110]. Nevertheless, these algorithms require that all data streams are complete (no missing data points) and have the same length and sampling frequency. These requirements are often difficult to fulfil in naturalistic settings with multimodal measurement methods using different measurement devices.

Once the feature space is defined, the research focus shifts towards developing strategies for selecting, merging, combining, and weighting multiple classifier models and modalities at the classification level. These strategies are still the subject of ongoing research and exploration. In this context, it is important to strike a balance between computational power, dataset size, and the benefits of finely tuned combinations of optimally stacked or voted classifiers. The exploration of early and late fusion approaches, as commonly employed in the field of robotics, could provide valuable insights. Early fusion involves the early combination of all data points and the fitting of classifiers to multidimensional data. On the other hand, late fusion involves a more fine-grained pipeline, where several classifiers are fitted to different proportions of the dataset and subsequently combined at a later stage. In this study, we implemented a late-fusion approach where we first combined different classifiers for each modality. In a subsequent step, we combined classifiers to create a unified prediction. Exploring early and late fusion strategies is especially important when one wants to account for temporal dynamics in the different measures or the realisation of real-time mental state monitoring. The review of Debie et al. [27] provides a comprehensive overview of the different fusion stages when identifying mental effort based on neurophysiological measures.

5. Practical Implications and Conclusions

Our proposed multimodal voting classification approach contributes to the ecologically valid distinction and identification of different states of mental effort. It paves the way toward generalised state monitoring across individuals in realistic applications. Interestingly, the choice of ground truth had a fundamental influence on the classification performance. The prediction of subjectively perceived mental effort operationalised through self-reports, is most effectively achieved by incorporating performance-based measures. On the other hand, the experimentally induced task load can be accurately predicted not only from performance-based measures but also by incorporating neurophysiological, physiological and visual measures. Our findings provide valuable guidance for researchers and practitioners in selecting appropriate methods based on their specific research questions or application scenarios, taking into account limited resources or environmental constraints. The capacity to predict subjectively perceived and experimentally induced mental effort on an individual level makes this architecture an integral part of future research and development of user-centred applications such as adaptive assistance systems.

Supplementary Materials: The following supporting information can be downloaded at https://osf.io/2vxs5. Figure S1: Distribution of low and high subjectively perceived mental effort for each subject after a subject-wise median split; Figure S2: Distribution of low and high subjectively perceived mental effort for each subject after a subject-wise upper quartile split; Figure S3: Prediction of the subjectively perceived mental effort (F_1 scores in training and test set) based on a median split; validation set: $N = 2$; Figure S4: Weights and procedure of a unimodal voting classifier to predict subjectively perceived mental effort based on a median split; validation set: $N = 2$; Figure S5: Prediction of the subjectively perceived mental effort (confusion matrix of test set) based on a median split; validation set: $N = 2$; Figure S6: Weights and procedure of a multimodal voting classifier to predict subjectively perceived mental effort based on a median split; validation set: $N = 2$; Figure S7: Prediction of the subjectively perceived mental effort (F_1 scores in training and test set) based on an upper quartile split; validation set: $N = 1$; Figure S8: Weights and procedure of a unimodal voting classifier to predict subjectively perceived mental effort based on an upper quartile split; validation set: $N = 1$; Figure S9: Prediction of the subjectively perceived mental effort (confusion matrix of test set) based on an upper quartile split; validation set: $N = 1$; Figure S10: Weights and procedure of a multimodal voting classifier to predict subjectively perceived mental effort based on an upper quartile split; validation set: $N = 1$; Figure S11: Prediction of the subjectively perceived mental effort (F_1 scores in training and test set) based on an upper quartile split; validation set: $N = 2$; Figure S12: Weights and procedure of a unimodal voting classifier to predict subjectively perceived mental effort based on an upper quartile split; validation set: $N = 2$; Figure S13: Prediction of the subjectively perceived mental effort (confusion matrix of test set) based on an upper quartile split; validation set: $N = 2$; Figure S14: Weights and procedure of a multimodal voting classifier to predict subjectively perceived mental effort based on an upper quartile split; validation set: $N = 2$; Figure S15: Distribution of low and high subjectively perceived mental effort for each subject separately for the two experimental task load conditions ("High Task Load" and "Low Task Load"); Figure S16: Prediction of the experimental task load condition (F_1 scores in training and test set); validation set: $N = 1$; Figure S17: Weights and procedure of a unimodal voting classifier to predict the experimental task load condition; validation set: $N = 1$; Figure S18: Prediction of the experimental task load condition (confusion matrix of test set); validation set: $N = 1$; Figure S19: Weights and procedure of a multimodal voting classifier to predict the experimental task load condition; validation set: $N = 1$; Figure S20: Prediction of the experimental task load condition (F_1 scores in training and test set); validation set: $N = 2$; Figure S21: Weights and procedure of a unimodal voting classifier to predict the experimental task load condition; validation set: $N = 2$; Figure S22: Prediction of the experimental task load condition (confusion matrix of test set); validation set: $N = 2$; Figure S23: Weights and procedure of a multimodal voting classifier to predict the experimental task load condition; validation set: $N = 2$; Figure S24: Distribution and correlations; Figure S25: Bootstrapped grand averages and 95% confidence intervals of the physiological and behavioural measurements for low and high subjective load; Figure S26: Bootstrapped grand averages and 95% confidence intervals of relative HbO and HbR concentration in single fNIRS channels. Table S1: Comparison of Subjectively Perceived Mental Effort Classification Performance between Test Subjects (F_1 Score)–Median Split; Table S2: Comparison of Subjectively Perceived Mental Effort Classification Performance between Test Subjects (F_1 Score)–Quartile Split; Table S3: Comparison of Experimentally Induced Mental Effort Classification Performance between Test Subjects (F_1 Score).

Author Contributions: Conceptualization, K.L.; Data curation, S.G. and K.L.; Formal analysis, S.G.; Funding acquisition, M.W.; Methodology, K.L.; Project administration, K.L.; Supervision, K.L. and M.V.; Visualization, S.G.; Writing—original draft, S.G.; Writing—review and editing, S.G., K.L., M.W. and M.V. All authors have read and agreed to the published version of the manuscript.

Funding: The reported research was supported by the Ministry of Economic Affairs, Labour and Tourism, Baden-Wuerttemberg, Germany: KI-Fortschrittszentrum Lernende Systeme und Kognitive Robotik and the Federal Ministry of Science, Research, and the Arts Baden-Württemberg and the University of Stuttgart as part of the Research Seed Capital funding scheme (grant number Az. 33-7533.-30-10/75/3).

Institutional Review Board Statement: This study was approved by the ethics committee of the Medical Faculty of the University of Tübingen, Germany (ID: 827/2020BO1).

Informed Consent Statement: Participants were informed that their participation was voluntary and that they could withdraw at any time during the experiment. They signed an informed consent according to the recommendations of the Declaration of Helsinki.

Data Availability Statement: The datasets analysed for this study as well as the code can be found in a publicly accessible OSF repository: https://osf.io/9dbcj/, accessed on 16 July 2023.

Acknowledgments: We would like to thank Ron Becker, Alina Schmitz-Hübsch, Michael Bui, and Sophie Felicitas Böhm for their contribution to the experimental environment, technical set-up, data collection and data preparation.

Conflicts of Interest: The authors declare that the research was conducted in the absence of any commercial or financial relationships that could be construed as a potential conflict of interest.

Abbreviations

The following abbreviations are used in this manuscript:

fNIRS	functional Near-Infrared Spectroscopy
ECG	Electrocardiography
HbO	Oxy-Haemoglobin
HbR	Deoxy-Haemoglobin
PFC	Prefrontal Cortex
WCT	Warship Commander Task
SD	Standard Deviation
CI	Confidence Interval
ML	Machine Learning
LR	Logistic Regression
LDA	Linear Discriminant Analysis
GNB	Gaussian Naïve Bayes Classifier
KNN	K-Nearest Neighbour Classifier
RFC	Random Forest Classifier
SVM	Support Vector Machine Classifier

References

1. Hart, S.G.; Staveland, L.E. Development of NASA-TLX (Task Load Index): Results of empirical and theoretical research. In *Advances in Psychology*; Hancock, P.A., Meshkati, N., Eds.; Elsevier: Amsterdam, The Netherlands, 1988; Volume 52, pp. 139–183. [CrossRef]
2. Young, M.S.; Brookhuis, K.A.; Wickens, C.D.; Hancock, P.A. State of science: Mental workload in ergonomics. *Ergonomics* **2015**, *58*, 1–17. [CrossRef] [PubMed]
3. Paas, F.; Tuovinen, J.E.; Tabbers, H.; Van Gerven, P.W.M. Cognitive load measurement as a means to advance cognitive load theory. *Educ. Psychol.* **2003**, *38*, 63–71. [CrossRef]
4. Chen, F.; Zhou, J.; Wang, Y.; Yu, K.; Arshad, S.Z.; Khawaji, A.; Conway, D. *Robust Multimodal Cognitive Load Measurement*; Human–Computer Interaction Series; Springer International Publishing: Cham, Switzerland, 2016. [CrossRef]
5. Zheng, R.Z. *Cognitive Load Measurement and Application: A Theoretical Framework for Meaningful Research and Practice*; Routledge: New York, NY, USA, 2017. [CrossRef]
6. von Lühmann, A. Multimodal Instrumentation and Methods for Neurotechnology out of the Lab. Fakultät IV—Elektrotechnik und Informatik. Ph.D. Thesis, Technische Universität Berlin, Berlin, Germay, 2018. [CrossRef]
7. Charles, R.L.; Nixon, J. Measuring mental workload using physiological measures: A systematic review. *Appl. Ergon.* **2019**, *74*, 221–232. [CrossRef] [PubMed]
8. Curtin, A.; Ayaz, H. The age of neuroergonomics: Towards ubiquitous and continuous measurement of brain function with fNIRS. *Jpn. Psychol. Res.* **2018**, *60*, 374–386. [CrossRef]
9. Benerradi, J.; Maior, H.A.; Marinescu, A.; Clos, J.; Wilson, M.L. Mental workload using fNIRS data from HCI tasks ground truth: Performance, evaluation, or condition. In *Proceedings of the Halfway to the Future Symposium, 19–20 November 2019*; Association for Computing Machinery: Nottingham, UK, 2019. [CrossRef]
10. Midha, S.; Maior, H.A.; Wilson, M.L.; Sharples, S. Measuring mental workload variations in office work tasks using fNIRS. *Int. J. Hum.-Comput. Stud.* **2021**, *147*, 102580. [CrossRef]
11. Izzetoglu, K.; Bunce, S.; Izzetoglu, M.; Onaral, B.; Pourrezaei, K. fNIR spectroscopy as a measure of cognitive task load. In Proceedings of the 25th Annual International Conference of the IEEE Engineering in Medicine and Biology Society, Cancun, Mexico, 17–21 September 2003; Volume 4, pp. 3431–3434. [CrossRef]
12. Ayaz, H.; Shewokis, P.A.; Bunce, S.; Izzetoglu, K.; Willems, B.; Onaral, B. Optical brain monitoring for operator training and mental workload assessment. *NeuroImage* **2012**, *59*, 36–47. [CrossRef]
13. Herff, C.; Heger, D.; Fortmann, O.; Hennrich, J.; Putze, F.; Schultz, T. Mental workload during n-back task—Quantified in the prefrontal cortex using fNIRS. *Front. Hum. Neurosci.* **2014**, *7*, 935. [CrossRef]

14. Miller, E.K.; Freedman, D.J.; Wallis, J.D. The prefrontal cortex: Categories, concepts and cognition. *Philos. Trans. R. Soc. Lond.* **2002**, *357*, 1123–1136. [CrossRef]
15. Dehais, F.; Lafont, A.; Roy, R.; Fairclough, S. A neuroergonomics approach to mental workload, engagement and human performance. *Front. Neurosci.* **2020**, *14*, 268. [CrossRef]
16. Babiloni, F. Mental workload monitoring: New perspectives from neuroscience. In *Proceedings of the Human Mental Workload: Models and Applications*; Longo, L., Leva, M.C., Eds.; Springer International Publishing: Cham, Switzerland, 2019; Volume 1107, pp. 3–19. [CrossRef]
17. Matthews, R.; McDonald, N.J.; Trejo, L.J. Psycho-physiological sensor techniques: An overview. In *Foundations of Augmented Cognition*; Schmorrow, D.D., Ed.; CRC Press: Boca Raton, FL, USA, 2005; Volume 11, pp. 263–272. [CrossRef]
18. Wierwille, W.W. Physiological measures of aircrew mental workload. *Hum. Factors* **1979**, *21*, 575–593. [CrossRef]
19. Kramer, A.F. Physiological metrics of mental workload: A review of recent progress. In *Multiple-Task Performance*; Damos, D.L., Ed.; CRC Press: London, UK, 1991; pp. 279–328. [CrossRef]
20. Backs, R.W. Application of psychophysiological models to mental workload. *Proc. Hum. Factors Ergon. Soc. Annu. Meet.* **2000**, *44*, 464–467. [CrossRef]
21. Dirican, A.C.; Göktürk, M. Psychophysiological measures of human cognitive states applied in human computer interaction. *Procedia Comput. Sci.* **2011**, *3*, 1361–1367. [CrossRef]
22. Dan, A.; Reiner, M. Real time EEG based measurements of cognitive load indicates mental states during learning. *J. Educ. Data Min.* **2017**, *9*, 31–44. [CrossRef]
23. Tao, D.; Tan, H.; Wang, H.; Zhang, X.; Qu, X.; Zhang, T. A systematic review of physiological measures of mental workload. *Int. J. Environ. Res. Public Health* **2019**, *16*, 2716. [CrossRef]
24. Romine, W.L.; Schroeder, N.L.; Graft, J.; Yang, F.; Sadeghi, R.; Zabihimayvan, M.; Kadariya, D.; Banerjee, T. Using machine learning to train a wearable device for measuring students' cognitive load during problem-solving activities based on electrodermal activity, body temperature, and heart rate: Development of a cognitive load tracker for both personal and classroom use. *Sensors* **2020**, *20*, 4833. [CrossRef]
25. Uludağ, K.; Roebroeck, A. General overview on the merits of multimodal neuroimaging data fusion. *NeuroImage* **2014**, *102*, 3–10. [CrossRef]
26. Zhang, Y.D.; Dong, Z.; Wang, S.H.; Yu, X.; Yao, X.; Zhou, Q.; Hu, H.; Li, M.; Jiménez-Mesa, C.; Ramirez, J.; et al. Advances in multimodal data fusion in neuroimaging: Overview, challenges, and novel orientation. *Inf. Fusion* **2020**, *64*, 149–187. [CrossRef]
27. Debie, E.; Rojas, R.F.; Fidock, J.; Barlow, M.; Kasmarik, K.; Anavatti, S.; Garratt, M.; Abbass, H.A. Multimodal fusion for objective assessment of cognitive workload: A review. *IEEE Trans. Cybern.* **2021**, *51*, 1542–1555. [CrossRef]
28. Klimesch, W. Evoked alpha and early access to the knowledge system: The P1 inhibition timing hypothesis. *Brain Res.* **2011**, *1408*, 52–71. [CrossRef]
29. Wirzberger, M.; Herms, R.; Esmaeili Bijarsari, S.; Eibl, M.; Rey, G.D. Schema-related cognitive load influences performance, speech, and physiology in a dual-task setting: A continuous multi-measure approach. *Cogn. Res. Princ. Implic.* **2018**, *3*, 46. [CrossRef]
30. Lemm, S.; Blankertz, B.; Dickhaus, T.; Müller, K.R. Introduction to machine learning for brain imaging. *Multivar. Decod. Brain Read.* **2011**, *56*, 387–399. [CrossRef] [PubMed]
31. Vu, M.A.T.; Adalı, T.; Ba, D.; Buzsáki, G.; Carlson, D.; Heller, K.; Liston, C.; Rudin, C.; Sohal, V.S.; Widge, A.S.; et al. A shared vision for machine learning in neuroscience. *J. Neurosci.* **2018**, *38*, 1601. [CrossRef] [PubMed]
32. Herms, R.; Wirzberger, M.; Eibl, M.; Rey, G.D. CoLoSS: Cognitive load corpus with speech and performance data from a symbol-digit dual-task. In *Proceedings of the 11th International Conference on Language Resources and Evaluation*, Miyazaki, Japan, 7–12 May 2018; European Language Resources Association: Miyazaki, Japan, 2018.
33. Ladouce, S.; Donaldson, D.I.; Dudchenko, P.A.; Ietswaart, M. Understanding minds in real-world environments: Toward a mobile cognition approach. *Front. Hum. Neurosci.* **2017**, *10*, 694. [CrossRef] [PubMed]
34. Lavie, N. Attention, distraction, and cognitive control under load. *Curr. Dir. Psychol. Sci.* **2010**, *19*, 143–148. [CrossRef]
35. Baddeley, A.D.; Hitch, G. Working Memory. In *Psychology of Learning and Motivation*; Bower, G.H., Ed.; Academic Press: Cambridge, MA, USA, 1974; Volume 8, pp. 47–89. [CrossRef]
36. Soerqvist, P.; Dahlstroem, O.; Karlsson, T.; Rönnberg, J. Concentration: The neural underpinnings of how cognitive load shields against distraction. *Front. Hum. Neurosci.* **2016**, *10*, 221. [CrossRef]
37. Anikin, A. The link between auditory salience and emotion intensity. *Cogn. Emot.* **2020**, *34*, 1246–1259. [CrossRef]
38. Dolcos, F.; Iordan, A.D.; Dolcos, S. Neural correlates of emotion–cognition interactions: A review of evidence from brain imaging investigations. *J. Cogn. Psychol.* **2011**, *23*, 669–694. [CrossRef]
39. D'Andrea-Penna, G.M.; Frank, S.M.; Heatherton, T.F.; Tse, P.U. Distracting tracking: Interactions between negative emotion and attentional load in multiple-object tracking. *Emotion* **2017**, *17*, 900–904. [CrossRef]
40. Schweizer, S.; Satpute, A.B.; Atzil, S.; Field, A.P.; Hitchcock, C.; Black, M.; Barrett, L.F.; Dalgleish, T. The impact of affective information on working memory: A pair of meta-analytic reviews of behavioral and neuroimaging evidence. *Psychol. Bull.* **2019**, *145*, 566–609. [CrossRef]
41. Banbury, S.; Berry, D.C. Disruption of office-related tasks by speech and office noise. *Br. J. Psychol.* **1998**, *89*, 499–517. [CrossRef]

42. Liebl, A.; Haller, J.; Jödicke, B.; Baumgartner, H.; Schlittmeier, S.; Hellbrück, J. Combined effects of acoustic and visual distraction on cognitive performance and well-being. *Appl. Ergon.* **2012**, *43*, 424–434. [CrossRef] [PubMed]
43. Vuilleumier, P.; Schwartz, S. Emotional facial expressions capture attention. *Neurology* **2001**, *56*, 153–158. [CrossRef] [PubMed]
44. Waytowich, N.R.; Lawhern, V.J.; Bohannon, A.W.; Ball, K.R.; Lance, B.J. Spectral transfer learning using information geometry for a user-independent brain–computer interface. *Front. Neurosci.* **2016**, *10*, 430. [CrossRef] [PubMed]
45. Lawhern, V.J.; Solon, A.J.; Waytowich, N.R.; Gordon, S.M.; Hung, C.P.; Lance, B.J. EEGNet: A compact convolutional neural network for EEG-based brain–computer interfaces. *J. Neural Eng.* **2018**, *15*, 056013. [CrossRef]
46. Lyu, B.; Pham, T.; Blaney, G.; Haga, Z.; Sassaroli, A.; Fantini, S.; Aeron, S. Domain adaptation for robust workload level alignment between sessions and subjects using fNIRS. *J. Biomed. Opt.* **2021**, *26*, 1–21. [CrossRef] [PubMed]
47. Lotte, F.; Bougrain, L.; Cichocki, A.; Clerc, M.; Congedo, M.; Rakotomamonjy, A.; Yger, F. A review of classification algorithms for EEG-based brain–computer interfaces: A 10 year update. *J. Neural Eng.* **2018**, *15*, 031005. [CrossRef]
48. Liu, Y.; Lan, Z.; Cui, J.; Sourina, O.; Müller-Wittig, W. EEG-based cross-subject mental fatigue recognition. In Proceedings of the International Conference on Cyberworlds 2019, Kyoto, Japan, 2–4 October 2019; pp. 247–252. [CrossRef]
49. Becker, R.; Stasch, S.M.; Schmitz-Hübsch, A.; Fuchs, S. Quantitative scoring system to assess performance in experimental environments. In Proceedings of the 14th International Conference on Advances in Computer-Human Interactions, Nice, France, 18–22 July 2021; ThinkMind: Nice, France, 2021; pp. 91–96.
50. Burkhardt, F.; Paeschke, A.; Rolfes, M.; Sendlmeier, W.; Weiss, B. A database of German emotional speech. In Proceedings of the 9th European Conference on Speech Communication and Technology, Lisbon, Portugal, 4–8 September 2005; Volume 5, p. 1520. [CrossRef]
51. Liu, H.; Gamboa, H.; Schultz, T. Sensor-Based Human Activity and Behavior Research: Where Advanced Sensing and Recognition Technologies Meet. *Sensors* **2022**, *23*, 125. [CrossRef]
52. The Pacific Science Engineering Group. *Warship Commander 4.4*; The Pacific Science Engineering Group: San Diego, CA, USA, 2003.
53. St John, M.; Kobus, D.A.; Morrison, J.G. *DARPA Augmented Cognition Technical Integration Experiment (TIE)*; Technical Report ADA420147; Pacific Science and Engineering Group: San Diego, CA, USA, 2003.
54. Toet, A.; Kaneko, D.; Ushiama, S.; Hoving, S.; de Kruijf, I.; Brouwer, A.M.; Kallen, V.; van Erp, J.B.F. EmojiGrid: A 2D pictorial scale for the assessment of food elicited emotions. *Front. Psychol.* **2018**, *9*, 2396. [CrossRef]
55. Cardoso, B.; Romão, T.; Correia, N. CAAT: A discrete approach to emotion assessment. In *Proceedings of the Extended Abstracts on Human Factors in Computing Systems*; Association for Computing Machinery: Paris, France, 2013; pp. 1047–1052. [CrossRef]
56. Rammstedt, B.; John, O.P. Kurzversion des Big Five Inventory (BFI-K). *Diagnostica* **2005**, *51*, 195–206. [CrossRef]
57. Laux, L.; Glanzmann, P.; Schaffner, P.; Spielberger, C.D. *Das State-Trait-Angstinventar*; Beltz: Weinheim, Germany, 1981.
58. Bankstahl, U.; Görtelmeyer, R. *APSA: Attention and Performance Self-Assessment— deutsche Fassung*; Fragebogen; Elektronisches Testarchiv, ZPID (Leibniz Institute for Psychology Information)–Testarchiv: Trier, Germany, 2013.
59. Hartmann, A.S.; Rief, W.; Hilbert, A. Psychometric properties of the German version of the Barratt Impulsiveness Scale, Version 11 (BIS-11) for adolescents. *Percept. Mot. Ski.* **2011**, *112*, 353–368. [CrossRef]
60. Scheunemann, J.; Unni, A.; Ihme, K.; Jipp, M.; Rieger, J.W. Demonstrating brain-level interactions between visuospatial attentional demands and working memory load while driving using functional near-infrared spectroscopy. *Front. Hum. Neurosci.* **2019**, *12*, 542. [CrossRef] [PubMed]
61. Zimeo Morais, G.A.; Balardin, J.B.; Sato, J.R. fNIRS Optodes' Location Decider (fOLD): A toolbox for probe arrangement guided by brain regions-of-interest. *Sci. Rep.* **2018**, *8*, 3341. [CrossRef] [PubMed]
62. Dink, J.W.; Ferguson, B. eyetrackingR: An R Library for Eye-Tracking Data Analysis. 2015. Available online: http://www.eyetrackingr.com (accessed on 4 March 2022).
63. Forbes, S. PupillometryR: An R package for preparing and analysing pupillometry data. *J. Open Source Softw.* **2020**, *5*, 2285. [CrossRef]
64. Jackson, I.; Sirois, S. Infant cognition: Going full factorial with pupil dilation. *Dev. Sci.* **2009**, *12*, 670–679. [CrossRef] [PubMed]
65. von der Malsburg, T. saccades: Detection of Fixations in Eye-Tracking Data. 2015. Available online: https://github.com/tmalsburg/saccades (accessed on 4 March 2022).
66. Gramfort, A.; Luessi, M.; Larson, E.; Engemann, D.A.; Strohmeier, D.; Brodbeck, C.; Parkkonen, L.; Hämäläinen, M.S. MNE Software for Processing MEG and EEG Data. *NeuroImage* **2014**, *86*, 446–460. [CrossRef]
67. Luke, R.; Larson, E.D.; Shader, M.J.; Innes-Brown, H.; Van Yper, L.; Lee, A.K.C.; Sowman, P.F.; McAlpine, D. Analysis methods for measuring passive auditory fNIRS responses generated by a block-design paradigm. *Neurophotonics* **2021**, *8*, 025008. [CrossRef]
68. Yücel, M.A.; Lühmann, A.v.; Scholkmann, F.; Gervain, J.; Dan, I.; Ayaz, H.; Boas, D.; Cooper, R.J.; Culver, J.; Elwell, C.E.; et al. Best practices for fNIRS publications. *Neurophotonics* **2021**, *8*, 012101. [CrossRef]
69. Pollonini, L.; Olds, C.; Abaya, H.; Bortfeld, H.; Beauchamp, M.S.; Oghalai, J.S. Auditory cortex activation to natural speech and simulated cochlear implant speech measured with functional near-infrared spectroscopy. *Hear. Res.* **2014**, *309*, 84–93. [CrossRef]
70. Fishburn, F.A.; Ludlum, R.S.; Vaidya, C.J.; Medvedev, A.V. Temporal Derivative Distribution Repair (TDDR): A motion correction method for fNIRS. *NeuroImage* **2019**, *184*, 171–179. [CrossRef]
71. Saager, R.B.; Berger, A.J. Direct characterization and removal of interfering absorption trends in two-layer turbid media. *J. Opt. Soc. Am. A* **2005**, *22*, 1874–1882. [CrossRef]

72. Beer, A. Bestimmung der Absorption des rothen Lichts in farbigen Flüssigkeiten. *Annalen der Physik und Chemie* **1852**, *86*, 78–88. [CrossRef]
73. Schiratti, J.B.; Le Douget, J.E.; Le Van Quyen, M.; Essid, S.; Gramfort, A. An ensemble learning approach to detect epileptic seizures from long intracranial EEG recordings. In Proceedings of the IEEE International Conference on Acoustics, Speech and Signal Processing 2018, Calgary, AB, Canada, 15–20 April 2018; pp. 856–860. [CrossRef]
74. Keles, H.O.; Cengiz, C.; Demiral, I.; Ozmen, M.M.; Omurtag, A. High density optical neuroimaging predicts surgeons's subjective experience and skill levels. *PLOS ONE* **2021**, *16*, e0247117. [CrossRef] [PubMed]
75. Minkley, N.; Xu, K.M.; Krell, M. Analyzing relationships between causal and assessment factors of cognitive load: Associations between objective and subjective measures of cognitive load, stress, interest, and self-concept. *Front. Educ.* **2021**, *6*, 632907. [CrossRef]
76. Pedregosa, F.; Varoquaux, G.; Gramfort, A.; Michel, V.; Thirion, B.; Grisel, O.; Blondel, M.; Prettenhofer, P.; Weiss, R.; Dubourg, V. Scikit-learn: Machine Learning in Python. *J. Mach. Learn. Res.* **2011**, *12*, 2825–2830.
77. Raschka, S. MLxtend: Providing machine learning and data science utilities and extensions to Python's scientific computing stack. *J. Open Source Softw.* **2018**, *3*, 638. [CrossRef]
78. Cumming, G.; Finch, S. Inference by eye: Confidence intervals and how to read pictures of data. *Am. Psychol.* **2005**, *60*, 170–180. [CrossRef]
79. Ranchet, M.; Morgan, J.C.; Akinwuntan, A.E.; Devos, H. Cognitive workload across the spectrum of cognitive impairments: A systematic review of physiological measures. *Neurosci. Biobehav. Rev.* **2017**, *80*, 516–537. [CrossRef]
80. Matthews, G.; De Winter, J.; Hancock, P.A. What do subjective workload scales really measure? Operational and representational solutions to divergence of workload measures. *Theor. Issues Ergon. Sci.* **2020**, *21*, 369–396. [CrossRef]
81. Causse, M.; Chua, Z.; Peysakhovich, V.; Del Campo, N.; Matton, N. Mental workload and neural efficiency quantified in the prefrontal cortex using fNIRS. *Sci. Rep.* **2017**, *7*, 5222. [CrossRef]
82. Allison, B.Z.; Neuper, C. Could anyone use a BCI? In *Brain-Computer Interfaces: Applying our Minds to Human-Computer Interaction*; Tan, D.S., Nijholt, A., Eds.; Springer: London, UK, 2010; pp. 35–54. [CrossRef]
83. de Cheveigné, A.; Di Liberto, G.M.; Arzounian, D.; Wong, D.D.; Hjortkjær, J.; Fuglsang, S.; Parra, L.C. Multiway canonical correlation analysis of brain data. *NeuroImage* **2019**, *186*, 728–740. [CrossRef]
84. Bzdok, D.; Meyer-Lindenberg, A. Machine learning for precision psychiatry: Opportunities and challenges. *Biol. Psychiatry Cogn. Neurosci. Neuroimaging* **2018**, *3*, 223–230. [CrossRef] [PubMed]
85. Dwyer, D.B.; Falkai, P.; Koutsouleris, N. Machine learning approaches for clinical psychology and psychiatry. *Annu. Rev. Clin. Psychol.* **2018**, *14*, 91–118. [CrossRef] [PubMed]
86. Cearns, M.; Hahn, T.; Baune, B.T. Recommendations and future directions for supervised machine learning in psychiatry. *Transl. Psychiatry* **2019**, *9*, 271. [CrossRef] [PubMed]
87. Orrù, G.; Monaro, M.; Conversano, C.; Gemignani, A.; Sartori, G. Machine learning in psychometrics and psychological research. *Front. Psychol.* **2020**, *10*, 2970. [CrossRef]
88. Lashgari, E.; Liang, D.; Maoz, U. Data augmentation for deep-learning-based electroencephalography. *J. Neurosci. Methods* **2020**, *346*, 108885. [CrossRef]
89. Bird, J.J.; Pritchard, M.; Fratini, A.; Ekárt, A.; Faria, D.R. Synthetic biological signals machine-generated by GPT-2 improve the classification of EEG and EMG through data augmentation. *IEEE Robot. Autom. Lett.* **2021**, *6*, 3498–3504. [CrossRef]
90. Zanini, R.A.; Colombini, E.L. Parkinson's disease EMG data augmentation and simulation with DCGANs and Style Transfer. *Sensors* **2020**, *20*, 2605. [CrossRef]
91. Abrams, M.B.; Bjaalie, J.G.; Das, S.; Egan, G.F.; Ghosh, S.S.; Goscinski, W.J.; Grethe, J.S.; Kotaleski, J.H.; Ho, E.T.W.; Kennedy, D.N.; et al. A standards organization for open and FAIR neuroscience. The international neuroinformatics coordinating facility. *Neuroinformatics* **2021**, *20*, 25–36 [CrossRef]
92. von Lühmann, A.; Boukouvalas, Z.; Müller, K.R.; Adalı, T. A new blind source separation framework for signal analysis and artifact rejection in functional near-infrared spectroscopy. *NeuroImage* **2019**, *200*, 72–88. [CrossRef]
93. Friedman, N.; Fekete, T.; Gal, K.; Shriki, O. EEG-based prediction of cognitive load in intelligence tests. *Front. Hum. Neurosci.* **2019**, *13*, 191. [CrossRef] [PubMed]
94. Unni, A.; Ihme, K.; Jipp, M.; Rieger, J.W. Assessing the driver's current level of working memory load with high density functional near-infrared spectroscopy: A realistic driving simulator study. *Front. Hum. Neurosci.* **2017**, *11*, 167. [CrossRef]
95. García-Pacios, J.; Garcés, P.; del Río, D.; Maestú, F. Tracking the effect of emotional distraction in working memory brain networks: Evidence from an MEG study. *Psychophysiology* **2017**, *54*, 1726–1740. [CrossRef] [PubMed]
96. Curtis, C. Prefrontal and parietal contributions to spatial working memory. *Neuroscience* **2006**, *139*, 173–180. [CrossRef] [PubMed]
97. Martínez-Vázquez, P.; Gail, A. Directed interaction between monkey premotor and posterior parietal cortex during motor-goal retrieval from working memory. *Cereb. Cortex* **2018**, *28*, 1866–1881. [CrossRef] [PubMed]
98. Vanneste, P.; Raes, A.; Morton, J.; Bombeke, K.; Van Acker, B.B.; Larmuseau, C.; Depaepe, F.; Van den Noortgate, W. Towards measuring cognitive load through multimodal physiological data. *Cogn. Technol. Work* **2021**, *23*, 567–585. [CrossRef]
99. Yap, T.F.; Epps, J.; Ambikairajah, E.; Choi, E.H. Voice source under cognitive load: Effects and classification. *Speech Commun.* **2015**, *72*, 74–95. [CrossRef]

100. Marquart, G.; Cabrall, C.; de Winter, J. Review of eye-related measures of drivers' mental workload. *Procedia Manuf.* **2015**, *3*, 2854–2861. [CrossRef]
101. Zhang, S.; Li, Y.; Zhang, S.; Shahabi, F.; Xia, S.; Deng, Y.; Alshurafa, N. Deep learning in human activity recognition with wearable sensors: A review on advances. *Sensors* **2022**, *22*, 1476. [CrossRef]
102. Lo, J.C.; Sehic, E.; Meijer, S.A. Measuring mental workload with low-cost and wearable sensors: Insights into the accuracy, obtrusiveness, and research usability of three instruments. *J. Cogn. Eng. Decis. Mak.* **2017**, *11*, 323–336. [CrossRef]
103. Tsow, F.; Kumar, A.; Hosseini, S.H.; Bowden, A. A low-cost, wearable, do-it-yourself functional near-infrared spectroscopy (DIY-fNIRS) headband. *HardwareX* **2021**, *10*, e00204. [CrossRef]
104. Niso, G.; Romero, E.; Moreau, J.T.; Araujo, A.; Krol, L.R. Wireless EEG: A survey of systems and studies. *NeuroImage* **2023**, *269*, 119774. [CrossRef]
105. Mantuano, A.; Bernardi, S.; Rupi, F. Cyclist gaze behavior in urban space: An eye-tracking experiment on the bicycle network of Bologna. *Case Stud. Transp. Policy* **2017**, *5*, 408–416. [CrossRef]
106. Ahmadi, N.; Sasangohar, F.; Yang, J.; Yu, D.; Danesh, V.; Klahn, S.; Masud, F. Quantifying workload and stress in intensive care unit nurses: Preliminary evaluation using continuous eye-tracking. *Hum. Factors* **2022**, 00187208221085335. [CrossRef] [PubMed]
107. Gottemukkula, V.; Derakhshani, R. Classification-guided feature selection for NIRS-based BCI. In Proceedings of the 5th International IEEE/EMBS Conference on Neural Engineering 2011, Cancun, Mexico, 27 April–1 May 2011; pp. 72–75. [CrossRef]
108. Aydin, E.A. Subject-Specific feature selection for near infrared spectroscopy based brain–computer interfaces. *Comput. Methods Programs Biomed.* **2020**, *195*, 105535. [CrossRef]
109. Chakraborty, S.; Aich, S.; Joo, M.i.; Sain, M.; Kim, H.C. A multichannel convolutional neural network architecture for the detection of the state of mind using physiological signals from wearable devices. *J. Healthc. Eng.* **2019**, *2019*, 5397814. [CrossRef]
110. Asgher, U.; Khalil, K.; Khan, M.J.; Ahmad, R.; Butt, S.I.; Ayaz, Y.; Naseer, N.; Nazir, S. Enhanced accuracy for multiclass mental workload detection using long short-term memory for brain–computer interface. *Front. Neurosci.* **2020**, *14*, 584. [CrossRef] [PubMed]

Disclaimer/Publisher's Note: The statements, opinions and data contained in all publications are solely those of the individual author(s) and contributor(s) and not of MDPI and/or the editor(s). MDPI and/or the editor(s) disclaim responsibility for any injury to people or property resulting from any ideas, methods, instructions or products referred to in the content.

Article

Empowering Participatory Research in Urban Health: Wearable Biometric and Environmental Sensors for Activity Recognition

Rok Novak [1,2,*], Johanna Amalia Robinson [1,2,3], Tjaša Kanduč [1], Dimosthenis Sarigiannis [4,5,6], Sašo Džeroski [2,7] and David Kocman [1]

1. Department of Environmental Sciences, Jožef Stefan Institute, 1000 Ljubljana, Slovenia; johanna.robinson@acs.si (J.A.R.); tjasa.kanduc@ijs.si (T.K.); david.kocman@ijs.si (D.K.)
2. Ecotechnologies Programme, Jožef Stefan International Postgraduate School, 1000 Ljubljana, Slovenia; saso.dzeroski@ijs.si
3. Centre for Research and Development, Slovenian Institute for Adult Education, 1000 Ljubljana, Slovenia
4. Environmental Engineering Laboratory, Department of Chemical Engineering, Aristotle University of Thessaloniki, 54124 Thessaloniki, Greece; denis@eng.auth.gr
5. HERACLES Research Centre on the Exposome and Health, Centre for Interdisciplinary Research and Innovation, 57001 Thessaloniki, Greece
6. Environmental Health Engineering, Department of Science, Technology and Society, University School of Advanced Study IUSS, 27100 Pavia, Italy
7. Department of Knowledge Technologies, Jožef Stefan Institute, 1000 Ljubljana, Slovenia
* Correspondence: rok.novak@ijs.si

Abstract: Participatory exposure research, which tracks behaviour and assesses exposure to stressors like air pollution, traditionally relies on time-activity diaries. This study introduces a novel approach, employing machine learning (ML) to empower laypersons in human activity recognition (HAR), aiming to reduce dependence on manual recording by leveraging data from wearable sensors. Recognising complex activities such as smoking and cooking presents unique challenges due to specific environmental conditions. In this research, we combined wearable environment/ambient and wrist-worn activity/biometric sensors for complex activity recognition in an urban stressor exposure study, measuring parameters like particulate matter concentrations, temperature, and humidity. Two groups, Group H (88 individuals) and Group M (18 individuals), wore the devices and manually logged their activities hourly and minutely, respectively. Prioritising accessibility and inclusivity, we selected three classification algorithms: k-nearest neighbours (IBk), decision trees (J48), and random forests (RF), based on: (1) proven efficacy in existing literature, (2) understandability and transparency for laypersons, (3) availability on user-friendly platforms like WEKA, and (4) efficiency on basic devices such as office laptops or smartphones. Accuracy improved with finer temporal resolution and detailed activity categories. However, when compared to other published human activity recognition research, our accuracy rates, particularly for less complex activities, were not as competitive. Misclassifications were higher for vague activities (resting, playing), while well-defined activities (smoking, cooking, running) had few errors. Including environmental sensor data increased accuracy for all activities, especially playing, smoking, and running. Future work should consider exploring other explainable algorithms available on diverse tools and platforms. Our findings underscore ML's potential in exposure studies, emphasising its adaptability and significance for laypersons while also highlighting areas for improvement.

Keywords: wearable sensors; particulate matter; activity recognition; machine learning; low-cost sensors; participatory research

Citation: Novak, R.; Robinson, J.A.; Kanduč, T.; Sarigiannis, D.; Džeroski, S.; Kocman, D. Empowering Participatory Research in Urban Health: Wearable Biometric and Environmental Sensors for Activity Recognition. *Sensors* **2023**, *23*, 9890. https://doi.org/10.3390/s23249890

Academic Editors: Hui Liu, Loris Nanni, Hugo Gamboa and Tanja Schultz

Received: 19 October 2023
Revised: 20 November 2023
Accepted: 15 December 2023
Published: 18 December 2023

Copyright: © 2023 by the authors. Licensee MDPI, Basel, Switzerland. This article is an open access article distributed under the terms and conditions of the Creative Commons Attribution (CC BY) license (https://creativecommons.org/licenses/by/4.0/).

1. Introduction

Exposure studies often rely on participants or subjects to provide information about their movements and activities relevant to the study. Time use diaries, or time activity diaries (TADs), have been extensively used to record specific activities and their relation to economic or health factors [1]. While TADs have been mostly paper-based, in the last decade, activity tracking has transitioned to smartphone apps and web-based applications, improving diary data quality [2]. On the other hand, there are indications that using smartphone apps can increase nonresponse levels due to several factors, e.g., not owning a smartphone and unfamiliarity with digital tools, though there are options available to overcome some of these issues [3].

Analysing the everyday activities of individuals can present a useful way to compartmentalise human behaviour and subsequently assess exposure to stressors, such as pollution or noise. Strong evidence exists that different activities increase exposure to stressors, e.g., elevated levels of airborne particulate matter when dusting, folding clothes, making a bed [4], smoking cigarettes [5], vaping [6], or walking/vacuuming on carpeted flooring [7,8], and increased exposure to noise on public transport [9]. Manually recording activities by a large group of individuals can be imprecise or require more resources [10]. An important constraint is temporal resolution, which has to be suited to participants/subjects' availability and responsiveness. When individuals self-report activities, there is little control over how precise the reports are, especially when taking into account recall bias and reliability [11–13]. In this context, participatory research, a collaborative approach that involves stakeholders, particularly those affected by the issue being studied, in all aspects of the research process, from defining the problem to collecting and analysing data, may offer a more integrated and accurate method of data collection. Reviewing TAD data in this study, in collaboration with the participants, showed a possible error rate of up to 5–10% for each activity. To reduce the probability of human error, different approaches are employed, e.g., user-cantered study design to construct better TADs [14] and using GPS and other variables as activity identifiers to reduce manual input [15–17].

Different classification algorithms developed over the past decades could potentially classify different activities by using data recorded with sensors as learning data. Equipping each individual with low-cost sensors would provide data about their movement, physiology, and environment. Relying solely on movement data or environmental data does not necessarily provide enough information to predict complex activities. This study utilised machine learning methods for classification, in combination with sensor and activity data, to provide a proof of concept for an alternative to manually recording complex activities. Furthermore, the approach was centred on analysing the usefulness of these tools to non-expert users involved in participatory research. To this end, two groups of participants, equipped with biometric and environmental sensors, recorded their activities with different temporal resolutions. The collected data were used to learn three different classification algorithms, observe how accurately each of them classifies simple and complex activities, and determine the role of different temporal resolutions. Overall, the aims of this study are:

(i) To evaluate the effectiveness of using a combined dataset from environmental and biometric sensors for the recognition of complex individual activities, utilizing classifiers selected for their transparency, interpretability, and accessibility to laypersons. This involves comparing the predictive performance of different classifiers to ascertain their suitability in a participatory-based urban health stressor context.

(ii) To investigate how different temporal resolutions of the collected data influence the predictive performance of each classifier, thereby determining the optimal data granularity for accurate activity recognition in urban health studies.

(iii) To assess the individual contributions and overall value of the environmental and biometric sensors used in the study, particularly focusing on their role in enhancing the accuracy of human activity recognition for complex urban activities.

(iv) To assess the role of these classifiers and sensors in empowering lay individuals for participatory urban health research. This involves enhancing the accessibility and understandability of human activity recognition technology, thereby enabling more effective community involvement in urban environmental health studies.

1.1. Air Quality and Environmental Data from Personal Monitors

Low-cost personal sensors and monitors that measure ambient conditions are becoming increasingly popular. Particulate matter concentrations, temperature, relative humidity, and various gases are just a few of the parameters that can be measured with low-cost personal sensors and monitors. These devices have certain drawbacks, mainly the uncertainty of their results [18,19], although they have been improving in the past years [20] and are "cautiously encouraged" for monitoring indoor air quality [21].

Certain complex activities, e.g., cooking, cleaning, and smoking, are characterised by distinct environmental conditions (smoke, resuspension of particles, high humidity), which could potentially provide enough data for a classification algorithm to identify them. Environmental sensors for temperature, humidity, and light have been successfully used to aid in activity recognition [22], including using humidity, CO_2, and temperature to classify specific (simple) activities in a single room [23]. To the best of our knowledge, individual-level particulate matter concentration, in combination with biometric data, has not been used for HAR.

1.2. Human Activity Recognition

Human activity recognition (HAR) methods have notably evolved, enabling the distinction of various activities using low-cost sensors without manual input. Recent advancements have shown the potential of HAR in applications such as mobile healthcare, smart homes, and fitness monitoring [24]. While earlier research focused on utilising sensors like accelerometers, compasses, gyroscopes, barometers, magnetometers, and GPS present in smartphones to predict specific actions, events, or activities such as walking, running, and falling [25,26], dedicated activity trackers (usually worn on the wrist) are more sophisticated and provide more accurate data [27,28]. On the other hand, even as these devices are better at HAR, they still have a high error rate (mean absolute percentage error (MAPE) of >50%) for more complex activities (dusting, cleaning, playing cards, etc.), but provide heart rate measurements with lower mean errors, i.e., between −3.57 bpm and 4.21 bpm, and MAPE of up to 16% [29,30]. In some instances, they have shown accuracy >90% for certain complex activities, e.g., smoking [31], by utilising hand gestures. Sensors measuring ambient conditions have been deployed in various HAR-oriented research projects.

1.2.1. HAR Challenges and HAR Pipeline

Several technical challenges persist in HAR, according to Chen et al. [32]:

1. Difficult feature extraction is due to activities having similar characteristics.
2. The high cost and time-intensive nature of activity data collection leads to annotation scarcity.
3. Person-dependent activity patterns, temporal variability of activity concepts, and diverse sensor layouts in individuals result in sensory data heterogeneity.
4. Composite or complex activities encompass several actions, making them more difficult to classify. Concurrent and multi-occupant activities, where an individual performs multiple activities simultaneously or with multiple people, add to the complexity.
5. A high computational cost is associated with HAR systems that have to provide instant responses and fit into portable devices.
6. The privacy and interpretability of the collected data have to be considered.

A recent HAR research pipeline has been introduced, offering a structured approach to human activity recognition [33]. This pipeline consists of nine interconnected components, guiding research from equipment selection to real-world application. It emphasises the importance of data acquisition, segmentation, annotation, signal processing, and feature extraction. The final phase integrates the research into practical scenarios [33]. In this research, the proposed pipeline has been implemented into the workflow. The equipment used was chosen for its relevance in capturing data across a range of activities in variable environmental conditions in Ljubljana, Slovenia. The data acquisition phase was focused on quality data capture amidst environmental variations. The subsequent segmentation and annotation phases were critical for accurately categorising diverse activities from indoor to outdoor settings. In signal processing, we processed the data to filter out noise. These methodologies were applied in the scope of this work, with an emphasis on real-world applicability and data interpretation. Each stage addressed specific challenges, including data quality and environmental adaptability.

1.2.2. Deep Learning Models for HAR

Deep learning models, such as those based on the self-supervised learning framework SimCLR, have showcased competitive performance in HAR using ambient sensor data [34]. In smart homes, the use of ambient sensors has become crucial due to the increasing demand for applications that can recognise activities in real-time [35]. Transformer-based filtering networks combined with LSTM-based early classifiers have been proposed to address the challenges posed by unrefined data in real-time HAR [35]. Cross-house human activity recognition is another area of interest, aiming to use labelled data from available houses (source domains) to train recognition models for unlabelled houses (target domains) [36]. Wearable devices have also played a pivotal role in HAR, supporting the detection of human activities without the need for environmental sensors like cameras [37]. These devices offer the advantage of not constraining users to remain in controlled environments. A deep understanding of the situations in which activities are performed is essential for applications in domains like safety, security, surveillance, and health monitoring [37].

1.2.3. Classification and Shallow Algorithms for HAR

While deep learning has shown high accuracy, usefulness, and advancements in HAR [38], shallow machine learning algorithms continue to demonstrate effectiveness in this field. Classification specifically is one of the major tasks in machine learning, where an algorithm learns, from examples in the training data, how to assign a specific class to the testing data. A task of this kind is to classify emails as "spam" or "not spam". In its most rudimentary form, an algorithm would check in a labelled training dataset which words or phrases are associated with a spam email and which are not. With these learned associations, the algorithm would be used on a new (testing) dataset, classifying new emails into "spam" and "not-spam". This is an illustrative task of binary classification, whereas activity classification usually requires multi-class classification, where there are several different classes, such as walking, running, cleaning, smoking, and so on. A variety of algorithms can be used for classification, including kNN [39], decision trees [40], Naïve Bayes [41], random forests [42], gradient boosting [43], support vector machines [44], etc.

Classifiers have been used in various HAR applications that use smartphones and low-cost activity trackers or other mobility sensors, in some cases with accuracy >98% [45], and in most cases >80% [46,47]. Combining these data points with ambient conditions, such as temperature and relative humidity measured with a smartphone (which has certain drawbacks [48]) or with static sensors, has shown up to 99.96% of correctly classified activities, such as walking, sitting, cycling, running, and other similar, less complex activities [49,50]. Specifically, decision tree classifiers (DTC), random forest classifiers (RFC), and K-nearest neighbours (KNN) have been utilised to recognise activities such as walking upstairs, walking downstairs, and walking normally, among others. The RFC model exhibited superior performance, achieving an accuracy score of 97.67% [51]. A proposed framework for HAR using smartphone sensors employed random forest, decision tree, and k-nearest neighbour classifiers, achieving an accuracy of 93.10% [52].

These approaches have utilised an array of different classifiers with varying results, where some algorithms, such as Naïve Bayes, achieve an average accuracy of 43.29% as compared to random forests, with the accuracy of 99.96% [50] and 99.86% [49]. Fewer studies have attempted to identify more complex activities, such as cooking, cleaning, gardening, playing, smoking, and others, as they are more difficult to characterise or distinguish from each other. Dernbach et al. [53] report over 90% accuracy for simpler activities for all classifiers (except Naïve Bayes with ~74%), while the accuracy for more complex activities was ~50% (only for K-star, otherwise between 35% and 50%). As complex and simple activities are broad terms, it is useful to define them. Sousa Lima et al. [54] provide a good explanation to delimit these two types of activities: "Simple or low-level activities are those activities that can only be recognized by analysing data from one or more sensors in a short period of time (e.g., walking and running). While complex or high-level activities can be seen as a set of low-level activities that can be recognized over a long period of time (e.g., work and shopping)." In this study, the scope of definitions is broadened to encompass various sensor types essential for HAR, including movement, biometric, and environmental sensors. Simple activities such as running, sleeping, resting, and sports typically require only a single device for effective monitoring. Conversely, complex activities like cleaning, cooking, playing, and smoking often demand additional data on ambient conditions. The research primarily investigates complex activities while also considering some simpler ones.

To promote the wider adoption of machine learning and classification methods, particularly in participatory and citizen science initiatives where individuals actively contribute data and collaborate in research, it is preferential to prioritize the explainability and accessibility of the algorithms and tools employed. In this context, WEKA, a collection of visualisation tools and algorithms designed for data analysis and predictive modelling, is identified as a promising tool [55]. Combining data from different environmental and biometric sensors and devices could provide enough information to distinguish different complex activities and somewhat resolve the listed challenges. Employing explainable algorithms available in accessible and user-friendly environments could lead to a wider adoption of machine learning and classification for HAR in participatory-based research.

2. Methodology

2.1. Data Collection

Two sets of data were used for this research, collected from participants living in Ljubljana, Slovenia:

The first set (group H—hourly data) was collected as part of the ICARUS H2020 project from 88 participants [56]. The participants were involved in the winter (February to March 2019) and summer (April to June 2019) seasons of the campaign for approximately 7 days and were equipped with two sensor devices: a smart activity tracker (SAT) and a portable particulate matter (PM) measuring device (PPM). Basic personal information was obtained from each participant (age, body mass, sex, etc.). All participants had to fill out a TAD, where information about their activities was provided for each hour. They could select their hourly activity from 7 indoor activities (resting, sleep, playing, sports, cooking, smoking, cleaning) and 2 outdoor activities (running, sports). The activities chosen to be included in the TAD were based on the criteria developed within the ICARUS project, based on available research and activity pattern databases [57]. Sensor data, collected with a 1-min resolution, were aggregated to a 1-h resolution by calculating the mean value. A detailed description of the sampling campaigns was published by Robinson [58].

The second set (group M—minute data) was collected from September to November of 2020 from 18 participants. They were equipped with the same devices as the first group. An important distinction was that they (a) had more activities to choose from and (b) had to log activity data on the scale of minutes, not hours. The activities used for group M were modified activities from the initial ICARUS TAD. Sensor data with a 1-min resolution were used as-is.

All participants involved in the study provided their informed consent. Ethical approval for the ICARUS project in Slovenia was obtained from the National Medical Ethics Committee of the Republic of Slovenia (approval nr. 0120-388/2018/6 on 22 August 2018). The data in this paper were selected only from participants in Slovenia, and all methods were performed in accordance with the relevant guidelines and regulations.

A graphic representation of the methodology and dataflows used in this work is shown in Figure 1.

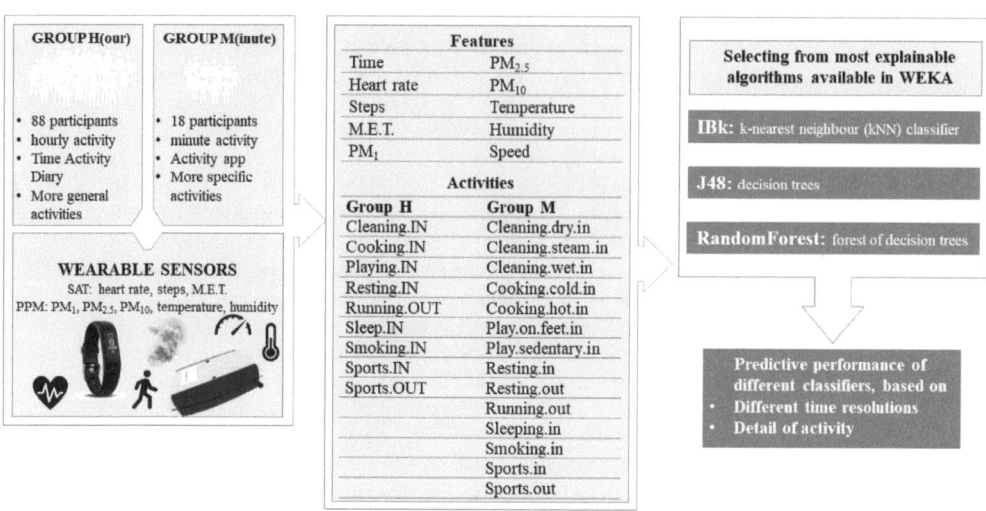

Figure 1. Schematic representation of the overall methodology and data flows used in this work.

2.1.1. Smart Activity Tracker

A Garmin (Garmin, Olathe, KS, USA) Vivosmart 3 activity tracker was strapped to each participant's wrist for the entire duration of the data collection period, except for two hours when the device had to be recharged. Information about the participant (sex, age, body mass, height, etc.) was logged into the device before deployment. The temporal resolution of the data was one minute. Data for average minute heart rate [beats per minute], steps [number of steps], and metabolic equivalent of task (M.E.T.) [between 0.01 and 45.60] was collected from each participant. The SAT provided several other variables, though they were not relevant to the scope of this research. Raw data, e.g., accelerometer, was not accessible.

The device measured heart rate using photoplethysmography (PPG) [59], with the wrist as the preferred location for its cost-effectiveness and convenience. According to information disclosed by Garmin, this specific device contains four sensors: a Garmin Elevate™ write heat rate monitor, a barometric altimeter, an accelerometer, and an ambient light sensor [60]. Validation studies confirm the Garmin Vivosmart's accuracy in capturing relevant data, including in older adults [61–63], though caution is advised for energy expenditure and high-intensity activities [64,65].

2.1.2. Portable Particulate Matter Sensing Device

This low-cost PPM device was developed for the ICARUS project by IoTECH (IoTECH Telecommunications, Thessaloniki, Greece) using a Plantower pms5003 sensor (Nanchang Panteng Technology Co., Ltd., Nanchang, China). This sensor uses the optical particle counting principle to measure particle size and mass concentration in real time. A fan draws particles into a beam of light, illuminating each particle as it passes through, with scattered light being recorded on a photodetector and converted into an electrical signal. The device provided data at a one-minute resolution. Participants carried it with them the entire period, strapped to their clothes, handbags, backpacks, or something similar. When sedentary, they were instructed to "have it in the same room, as close as possible", as the device needed to be recharged every six to seven hours or continuously plugged into a power source. The pms5003 sensor consistently demonstrates accuracy in both short-term and long-term evaluations, exhibiting moderate to high correlation with reference instruments in various settings [66–68]. While some studies suggest minimal drift over time [68,69], others recommend regular calibration, especially in high-humidity environments [67]. Additionally, the Plantower sensor recorded ambient temperature and humidity, and the ancillary GPS component provided data on speed of movement. Group M was provided with a power bank with a 10.000 mAh capacity, which prolonged the use of the device to ~24 h. The PPM provided minute-resolution data for PM_1, $PM_{2.5}$, PM_{10}, temperature [°C], relative humidity [%], and speed [km/h]. Some other variables were also provided by the PPM, though they were out of the scope of this research. The PPM was validated by co-location with reference research-grade sensors [70].

2.1.3. Activity Recording

Group H was provided with 7 blank daily time activity diaries (TADs), where they were able to fill in circles for each activity they did for every hour of the day. These files were collected and digitalized. Information about all indoor and outdoor activities was used. An example of a TAD can be found as supplemental information in Novak et al. [71].

Group M installed the Clockify app [72] on their smartphone, which had activities already pre-set by the research team on the online portal. Several activity-tracking apps were tested and reviewed, and though the Clockify app was generally meant as a time-tracking app for work and projects, it had the functionalities that were needed for this research. Each participant selected the activity they were beginning to perform, and the timer would start. After they finished the activity, they would select the next activity, which would automatically finish the first one. The time stamps had date, hour, minute, and second information. While the activity data technically had a 1-s resolution, it was rounded to the nearest minute. The reasoning was fourfold: (1) a few instances of activities with a duration of <1 min, (2) the compiled dataset would be unnecessarily large, (3) the changes between activities included in the analysis are not relevant at <1-min resolution, and (4) all sensor data had a 1-min resolution. The recorded data were exported from the Clockify portal in csv format.

Recording more than one activity simultaneously was deemed out of the scope of this research. In the case of Group M, recording more than one activity at once was not possible in the app. On the other hand, there were some instances of participants recording more than one activity per hour in Group M. Generally, these were not the same variables as those used in this research. In the few cases where this overlap happened, the first activity from the right side of the TAD was selected.

All annotations of data and any segmentation were based on the activity data provided by the participants.

2.2. Dataset Overview

Sensor, TAD, and Clockify portal data were harmonised and compiled into two datasets: group H with 228,267 instances (per-minute recordings of all variables), and group M with 70,139 instances. Each instance was associated with 10 variables (time, PM_1, $PM_{2.5}$, PM_{10}, temperature, humidity, speed, heart rate, steps, M.E.T., activity):

- time—indicating a time of day or a specific hour when the measurement took place (from 0 to 23)
- PM_1, $PM_{2.5}$, and PM_{10}—particulate matter concentrations in three size classes, recorded as non-negative integer values (PPM)
- temperature, humidity—ambient temperature and humidity, recorded as a float value (PPM)
- speed—calculated based on GPS module data, recorded as a float value (PPM).
- heart rate—heart rate per minute, recorded as a positive integer value (SAT).
- steps—number of steps per minute, recoded as a non-negative integer value (SAT).
- M.E.T.—a non-negative integer value (SAT).
- activity—recorded on TAD or in the Clockify app.

The minimum requirement for each instance to be included in the dataset was an activity label and at least one non-empty variable. Data preparation was done in R [73].

The number of included instances of each activity for each dataset (group M and group H) is listed in Table 1. All activities were capped at 5000 instances for each group. When >5000 instances were available, a random selection was made from the dataset. Resting and sleeping were capped for group M. The ceiling was determined based on several iterations, which showed a considerably longer time to build the models without having an impact on the overall performance.

Table 1. Activities and number of instances of each activity in each of the two datasets.

Group M		Group H	
Activity/Task	Nr.	Activity/Task	Nr.
Cleaning.dry.in	438	Cleaning.in	5000
Cleaning.steam.in	416	Cooking.in	5000
Cleaning.wet.in	516	Playing.in	5000
Cooking.cold.in	387	Resting.in	5000
Cooking.hot.in	1923	Running.out	5000
Play.on.feet.in	85	Sleep.in	5000
Play.sedentary.in	469	Smoking.in	5000
Resting.in	5000	Sports.in	5000
Resting.out	225	Sports.out	5000
Running.out	80		
Sleeping.in	5000		
Smoking.in	769		
Sports.in	361		
Sports.out	774		

Basic statistics for all numeric variables in the final datasets are presented in Table 2. All values were within expected limits. All PM variables had a ceiling fixed at 180 µg/m^3 as the highest possible value; otherwise, the mean, median, and quartile values are as expected. The mean and median values for speed are low, as >20 km/h values were removed, as there are no activities included in this research where speed could be >20 km/h. PM statistics are similar between the groups. There are some differences in max and min temperature and relative humidity, which is due to a larger and more diverse dataset for group H and data from two seasons. Speed, heart rate, steps, and M.E.T. do not show wide discrepancies. As individuals, on average, spend most of their time in a sedentary or stationary position, the low median, mean, and quartile values reflect this. Similarly, the highest values reflect vigorous movement, e.g., running, producing >200 steps per minute, and an M.E.T. of 15. These results show that the two datasets are quite similar when observed through basic statistics, which was a key aim that was set when collecting data for group M. The two groups should have the same general characteristics and differ only in the temporal resolution of the data collected to facilitate an accurate comparison of the classification results. The mean values of all variables for each activity are available in the Supplementary Information. These show that the values are in line with expectations as there are generally higher concentrations of PM outdoors than indoors, though this is somewhat dependent on season, time of day, and specific activity [74,75].

Table 2. Basic statistics for all numeric variables in the dataset.

	Median		Mean		st. dev.		Max		Min		1stQ		3rdQ	
Variable	H	M	H	M	H	M	H	M	H	M	H	M	H	M
PM_1 [µg/m³]	10.0	9.0	18.7	15.4	29.9	26.1	180.0	180.0	0.0	0.0	5.0	4.0	19.0	19.0
$PM_{2.5}$ [µg/m³]	13.0	13.0	25.7	21.8	37.5	30.4	180.0	180.0	0.0	0.0	7.0	6.0	28.0	29.0
PM_{10} [µg/m³]	15.0	14.0	28.7	24.1	39.1	31.8	180.0	180.0	0.0	0.0	8.0	7.0	31.0	31.0
Temperature [°C]	24.0	23.8	23.6	23.6	3.3	3.2	34.3	34.6	5.9	8.5	22.4	22.4	25.4	25.0
Relative humidity [%]	32.2	39.6	32.8	39.9	8.3	6.9	76.5	67.2	6.7	19.7	27.2	35.4	38.0	44.4
Speed [km/h]	0.6	0.0	1.5	0.1	2.4	0.6	20.0	10.6	0.0	0.0	0.0	0.0	2.0	0.0
Avg. heart rate [bpm]	83.0	69.0	86.2	73.2	22.4	20.0	195.0	177.0	38.0	38.0	70.0	58.0	98.0	85.0
Steps [nr.]	0.0	0.0	16.7	6.1	37.5	20.0	245.0	157.0	0.0	0.0	0.0	0.0	0.0	0.0
M.E.T. [mL O2/kg/min]	0.1	0.1	0.3	0.5	0.5	0.5	15.0	6.1	0.0	0.0	0.1	0.1	0.2	1.0

2.3. Classifiers Used

In this research, shallow algorithms were opted for over deep learning techniques, given their inherent advantages tailored to the objectives of this research. While deep learning methods frequently offer high accuracy, they come with increased computational demands. This becomes a challenge when classifications are executed on devices with limited computational power, such as smartphones, wearable sensors, or standard office laptops. Although these demands can be mitigated through sophisticated feature extraction methods [76], it might render the approach less intuitive for those not well-versed in machine learning. Importantly, certain shallow algorithms are recognised for their transparency and interpretability. For instance, tree-based algorithms, like decision trees and random forests, are visually representable, making them more comprehensible to laypersons [77]. The k-nearest neighbours (kNN) algorithm, with its principle of similarity, is also straightforward in its logic. In this study, shallow algorithms not only facilitate easier explanations for research participants but also ensure that the data analysis remains accessible to researchers. As artificial intelligence becomes increasingly integrated into research and policy making, the emphasis on the explainability of these algorithms grows [78,79]. Notably, the trade-off between model accuracy and interpretability in AI has been a focal point in recent research, with a survey paper offering an in-depth analysis of explainable AI methodologies and suggesting future research avenues to optimize this balance [80].

The classifiers chosen for the tasks outlined in this research were selected based on the requirements outlined in Sections 1 and 2.3. These requirements can be summarised into four criteria:

1. All selected algorithms must be appropriate for this task, based on their use in existing literature and proof-of-concept cases, and show promising results in terms of accuracy in published research.
2. They should be considered (easily) explainable to laypersons, with the processes used being transparent and understandable.
3. Accessibility must be considered, i.e., the algorithms must be available in a user-friendly, GUI-based experimental environment allowing access to laypersons, e.g., WEKA.
4. The algorithms used should be executable on devices with limited computational power, such as smartphones or office laptops, proving results in a reasonable time frame, as per the aims of specific research.

Based on the outlined criteria, the following classifiers were chosen: kNN, decision trees, and random forests. All analyses were performed using the WEKA 3.8.3 [55] "Explorer" application. The specific classifiers within WEKA that were used in this research are listed in Table 3, which also contains short descriptions of each of the classifiers.

Table 3. Classifiers in WEKA used for this research, with short descriptions.

Classifier	Description
IBk	Instance-based learner [39], otherwise known as the k-nearest neighbour (kNN) classifier; kNN takes the k closest examples (typically according to a Euclidean distance) to the given instance in the feature space and counts how many of the k belong to each class. The new instance object is classified by plurality vote.
J48	J48 is a Java implementation of the C4.5 decision tree algorithm developed by Ross Quinlan [40]. It can be used for classification and allows a high number of attributes. Deemed as a "machine learning workhorse", ranked no. 1 in the Top 10 Algorithms in Data Mining [81]. To classify data from a testing set, each sample from the data are propagated through the tree (according to the conditions satisfied by its attribute values). When an example reaches a leaf node, it is assigned the class value of that node.
Random Forest	Constructs a forest of decision trees in a randomized manner. Developed by Leo Breiman [42]. The Random Forest (RF) method is an ensemble learning method for classification that constructs a forest of decision trees in a randomised fashion. Each tree is constructed from a different randomly selected subset of the dataset (bootstrap/sample), with a subset of (randomly chosen) features considered to select a split at each step of tree construction. When the forest is applied to a new instance, each tree votes for one class. The output is the class that gets the most votes from the individual trees.

2.4. Parameter Settings for the Classifiers

The settings for all classifiers were at their WEKA defaults. IBk used 1 nearest neighbour for classification and did not perform distance weighting. J48 trees were pruned. The RF contained 100 trees.

2.5. Feature Ranking Using the Relief Approach

Not all attributes in the dataset are necessarily useful for classification models, and some can be omitted. In turn, this can reduce the time and computational cost of building the model. The features in these datasets were ranked using the Relief approach, i.e., the Relief Attribute Evaluator in WEKA, with 10-fold cross-validation. Relief "evaluates the worth of an attribute by repeatedly sampling an instance and considering the value of the given attribute for the nearest instance of the same and different class" [82].

2.6. Performance Metrics

There are many measures of the performance of classifiers, typically defined for each class value (and then averaged across the different class values). These include true positive and false positive rates, precision, recall, F-measure, and others. Classification accuracy is defined as the percentage of instances that have been classified correctly. It is the most commonly used indicator of performance. Another performance metric is Kappa (K), which allows for direct comparison between models as it shows how closely the classified instances match the labelled data while also considering random chance (agreement with a random classifier).

Performance metrics are typically calculated based on the entries of the confusion matrix for a given classifier C and a given dataset D. The entry in row x and column y specifies the number of instances from D that actually belong to class x but have been classified as class y by the classifier C. The diagonal entries of the matrix specify the numbers of correctly classified instances. The confusion matrices for all three classifiers and the two groups are provided in the supplementary information.

For a given class x, the diagonal entry corresponds to the number of true positives (TP) for class x. The sum of all non-diagonal entries in column x corresponds to the number of all instances incorrectly classified as x (false positives). Based on TP and FP, the precision for x is calculated as:

$$\text{Precision}_x = \frac{\text{TP}}{\text{TP} + \text{FP}}$$

Recall (sensitivity) for class x is defined as the ratio of true positives to the total number of instances that truly belong to class x:

$$\text{Recall}_x = \frac{\text{TP}}{\text{TP} + \text{FN}}$$

The F-measure is a metric that combines precision and recall into a single score, providing a balanced measure of a classifier's performance:

$$F_x = \frac{2 * \text{Precision}_x * \text{Recall}_x}{\text{Precision}_x + \text{Recall}_x}$$

A receiver operating characteristic curve, or ROC curve, is a graphical plot that illustrates the diagnostic ability of a probabilistic binary classifier, and the area under the ROC curve (ROC-AUC) is also often used as a performance metric. The ROC curve plots the true positive rate (TPR), also known as sensitivity or recall, against the false positive rate (FPR) at various threshold settings. FPR is calculated as:

$$\text{FPR} = \frac{\text{FP}}{\text{FP} + \text{TN}}$$

These metrics and the ROC curve provide valuable insights into the performance of classifiers by considering both the ability to correctly classify instances (precision, recall) and the ability to evaluate a model's performance across different decision thresholds (ROC curve)."

The performance of all three classifiers on unseen cases was estimated by using the 10-fold cross-validation procedure. Cross-validation reduces the variance in the performance estimates by averaging over different partitions of the dataset. The dataset is divided into 10-subsets (folds), which in turn are used as testing sets, while all remaining instances are used as training instances. This procedure ensures that every instance from the dataset appears in the test set exactly once.

3. Results and Discussion

3.1. Feature Importance and Ranking

In the analysis of feature importance and ranking, the focus was on identifying significant features influencing activity recognition in groups H and M. This was necessary for understanding activity patterns in relation to environmental exposures.

Time (hour of day) was the top-ranked feature for both groups, aligning with the expectation that certain activities are time-specific. Heart rate, humidity, and temperature were also important, with their order of importance varying between groups.

Notable differences between the groups were observed beyond the top four features. For group M, Table 4 indicates that PM_{10} and $PM_{2.5}$ were ranked 5th and 6th, with PM1 at an average rank of 9.2. This variation may be due to different temporal resolutions in recording sensor and activity data. In group H, as shown in Table 5, steps, speed, and metabolic equivalent of task (M.E.T.) ranked higher, followed by particulate matter (PM) variables with the lowest average merit.

Table 4. Feature merits (importance scores) and ranks for group M.

Group M		
Average Merit	Average Rank	Attribute
0.127 ± 0.001	1 ± 0	Time
0.052 ± 0.001	2 ± 0	Heart rate
0.036 ± 0	3 ± 0	Humidity
0.03 ± 0	4 ± 0	Temperature
0.028 ± 0.001	5 ± 0	PM_{10}
0.02 ± 0	6 ± 0	$PM_{2.5}$
0.017 ± 0	7 ± 0	Speed
0.016 ± 0	8 ± 0	Steps
0.014 ± 0	9.2 ± 0.4	PM_1
0.013 ± 0.001	9.8 ± 0.4	M.E.T.

Table 5. Feature merits (importance scores) and ranks for group H.

Group H		
Average Merit	Average Rank	Attribute
0.193 ± 0.001	1 ± 0	Time
0.017 ± 0.001	2.5 ± 0.5	Humidity
0.017 ± 0	2.5 ± 0.5	Heart rate
0.016 ± 0	4	Temperature
0.015 ± 0	5	Steps
0.007 ± 0	6	Speed
0.001 ± 0.001	7	M.E.T.
−0.006 ± 0	8	PM_1
−0.008 ± 0.001	9	PM_{10}
−0.011 ± 0	10	$PM_{2.5}$

Due to the small dataset size, no attributes were excluded. Therefore, models were trained using a comprehensive set of features: PM_1, $PM_{2.5}$, PM_{10}, humidity, temperature, speed, heart rate, steps, M.E.T., and time.

The analysis showed expected patterns, such as the high ranking of time, and less intuitive differences between the groups in the ranking of PM variables and physical activity indicators. These findings emphasise the complexity of human activity patterns in relation to environmental exposures and the need to consider a range of features in such studies.

3.2. Overall Predictive Performance of Classifiers

Table 6 shows a comparison of the most relevant metrics for all the classifiers used in this research for groups H and M. Random Forest (RF) shows the highest correctly classified (CC) values for both groups, and IBk shows the lowest CC values. For group H, the percent of correctly classified instances increases gradually from IBk to J48 (Δ11.3%) to RF (Δ6.4%). On the other hand, for group M, the share of correctly classified instances jumps up from IBk to J48 (Δ42.6%), but then only marginally increases from J48 to RF (Δ0.3%). This difference could be a consequence of the different number of instances for each activity between the two groups. Some activities in group M have <500 instances, while all activities in group H have 5000 instances. The trend for Kappa is similar to

CC, though all values for J48 and RF are ~0.06 lower than the correctly classified percent (divided by 100). Though there is some disagreement on the applicability of the Kappa statistic in the context of "The Paradox of Cohen's Kappa" [83,84] and what the guidelines are for evaluating it [85], in this context, a value of >0.7 can be interpreted as moderate to strong agreement. Both J48 and RF fall in this category for group M. For group H, all Kappa values are <0.5. Importantly, the difference between J48 and RF for group M is only 0.01.

Table 6. Summary of results for all models for both groups.

Classifier/Metric	CC [%]		Kappa		TP		FP		Precision		Recall		F-Measure		ROC-AUC	
	H	M	H	M	H	M	H	M	H	M	H	M	H	M	H	M
IBk	35.2	34.3	0.3	0.2	0.4	0.3	0.1	0.1	0.4	NA	0.4	0.3	0.4	NA	0.6	0.6
J48	46.5	76.9	0.4	0.7	0.5	0.8	0.1	0.1	0.5	0.8	0.5	0.8	0.5	0.8	0.8	1
RF	52.9	77.2	0.5	0.7	0.5	0.8	0.1	0.1	0.5	0.8	0.5	0.8	0.5	0.8	0.9	1

Calculated values for TP and precision again follow the CC and Kappa metrics, showing that J48 and RF provide the highest precision. The RF model for group M has the highest ROC-AUC (0.97), indicating the lowest FP and highest TP rate.

An important evaluator that is not included in the table is the time it took to construct each model. For IBk, the time to build the model was <1 s for both groups. This is a positive aspect for IBk, though all the evaluation metrics show that this model is not suited for this type of data in comparison with J48 and RF. For group H, J48 took 4.67 s to build the model and 0.42 s for group M. In contrast, the RF-based model took 121.89 s for group H (more than 57 times as much time as J48) and 24.16 s for group M (26 times as much time as J48). As these two models perform very similarly based on the evaluation metrics, the time it takes to build and cross-validate the model is a relevant factor when considering which one to use. In the case of the group M subset of data, it would be efficient to use the J48 classifier, as the improvement in the correctly classified percent of instances does not offset the time and processing power that have to be allotted. Real-world applications, based on collecting data with personal sensors, experience larger volumes of instances and would have to account for considerably longer run times. When a ML approach is applied to improve the classification of activities and reduce the probability of human errors, time and processing power should be considered. In line with the 5th challenge listed in Section 1.2, reducing the number of unnecessary instances, e.g., for sleeping, and selecting a more fit-for-purpose algorithm would reduce the computational cost associated with activity recognition.

3.3. Predictive Performance Per Group and Activity

Results comparing the predictive performance of the used classifiers for group H (Table 7) show similar results as described in Section 3.2. J48 and RF show an overall higher TP rate, precision, recall, F-measure, and ROC-AUC values compared to IBk. These differences are more obvious in simpler activities, i.e., running, sleeping, and sports, with a 0.3 difference in ROC-AUC between IBk and J48/RF and a ≤0.2 ROC-AUC difference for other, more complex activities. This result highlights the drawbacks of hourly recorded activity data, with less resolution on dynamic changes in more complex activities.

Table 7. Summary for group H, showing TP, FP, precision, recall, F-measure, and ROC-AUC for all classifiers.

Class/Classifier	TP			FP			Precision			Recall			F-measure			ROC-AUC		
	IBk	J48	RF	IBk	J48	RF	IBk	J48	RF	IBk	J48	RF	IBk	J48	RF	IBk	J48	RF
Cleaning.in	0.3	0.4	0.5	0	0.1	0.1	0.5	0.4	0.5	0.3	0.4	0.5	0.4	0.4	0.5	0.6	0.8	0.8
Cooking.in	0.5	0.5	0.4	0.1	0.1	0.1	0.3	0.4	0.5	0.5	0.5	0.4	0.4	0.4	0.4	0.7	0.8	0.8
Playing.in	0.3	0.4	0.4	0.1	0.1	0.1	0.3	0.4	0.5	0.3	0.4	0.4	0.3	0.4	0.5	0.6	0.8	0.8
Resting.in	0.4	0.3	0.3	0.2	0.1	0.1	0.2	0.3	0.3	0.4	0.3	0.3	0.3	0.3	0.3	0.6	0.7	0.7
Running.out	0.3	0.6	0.8	0	0	0	0.6	0.7	0.7	0.3	0.6	0.8	0.4	0.6	0.7	0.7	0.9	1
Sleep.in	0.7	0.8	0.8	0	0	0	0.8	0.7	0.7	0.7	0.8	0.8	0.7	0.8	0.8	0.8	1	1
Smoking.in	0.4	0.3	0.5	0.1	0	0.1	0.3	0.5	0.5	0.4	0.3	0.5	0.3	0.4	0.5	0.6	0.8	0.9
Sports.in	0.2	0.4	0.6	0	0.1	0.1	0.5	0.5	0.6	0.2	0.4	0.6	0.3	0.5	0.6	0.6	0.8	0.9
Sports.out	0.2	0.4	0.5	0	0.1	0.1	0.4	0.4	0.5	0.2	0.4	0.5	0.3	0.4	0.5	0.6	0.8	0.9
Weighted average	0.4	0.5	0.5	0.1	0.1	0.1	0.4	0.5	0.5	0.4	0.5	0.5	0.4	0.5	0.5	0.6	0.8	0.9

The IBk results for group M showed several activities with values of 0 in all evaluation classes, having a considerably worse predictive performance result than J48 and RF classifiers, as evident in Table 8. Running stands out in the ROC-AUC metric for IBk, with a relatively low FP rate compared to its TP rate. Unique characteristics of running, compared to other activities, such as high values for heart rate, speed, steps, intensity, and lower temperatures and PM concentrations, could contribute to better predictive performance. J48 and RF show a ROC-AUC and FP value of 1 and 0, respectively, for running. Results for group M show similar patterns as group H, with simpler activities showing better performance for all classifiers. On the other hand, group M TP and precision values are on average higher compared to group H. Cooking has a TP rate of 0.5 in group H, and while the TP rate for cooking.cold.in shows a value of 0.4, cooking.hot.in had a TP value of 0.8. The latter is associated with specific environmental conditions that can make it more distinguishable from other activities, e.g., higher temperatures and PM concentrations. Contrary to this assumption, the activity of playing does not show the same pattern. Playing.on.feet, associated with dust resuspension and an elevated heart rate, would be more distinctive than playing.sedentary. Although, on average, group M playing activities have better metrics compared to group H.

These conclusions are corroborated by the confusion matrix results available in the SI. Activities with value definitions have more misclassified instances, in contrast to well-defined activities. For example, out of the 5000 instances of resting in group M for J48 (Table S4 in Supplementary Material), 3738 are correctly classified. Out of the incorrectly classified, two-thirds (818) are labelled as sleeping and one-fifth as cooking.hot.in. Moreover, activities often have a high number of their misclassified instances labelled as resting. Out of 769 instances of smoking, 102 are misclassified as resting. On the other hand, 96 instances are misclassified as cooking.hot.in, which would be expected as both activities can show high concentrations of PM.

Furthermore, sleeping and resting have well-defined time intervals, a low heart rate, and no movement. They are consistently indicated by all participants and evenly distributed. Unlike other activities, sleep is uninterrupted for several consecutive hours, resulting in minimally distorted minute values within an hour. For instance, if a person runs for only 20 min but claims it is the main activity for that hour, only 1/3 of the data support this claim, while the remaining 40 min include other activities.

Table 8. Summary for group M, showing TP, FP, precision, recall, F-measure, and ROC-AUC for all classifiers.

Class	TP			FP			Precision			Recall			F-Measure			ROC-AUC		
	IBk	J48	RF	IBk	J48	RF	IBk	J48	RF	IBk	J48	RF	IBk	J48	RF	IBk	J48	RF
Cleaning.dry.in	0	0.5	0.4	0	0	0	0.1	0.6	0.6	0	0.5	0.4	0	0.6	0.5	0.5	1	1
Cleaning.steam.in	0	0.6	0.5	0	0	0	NA	0.6	0.7	0	0.6	0.5	NA	0.6	0.6	0.5	1	1
Cleaning.wet.in	0	0.4	0.5	0	0	0	NA	0.6	0.5	0	0.4	0.5	NA	0.5	0.5	0.5	0.9	1
Cooking.cold.in	0	0.4	0.4	0	0	0	0.6	0.5	0.4	0	0.4	0.4	0	0.4	0.5	0.5	0.9	1
Cooking.hot.in	0.3	0.8	0.7	0.3	0.1	0.1	0.1	0.6	0.6	0.3	0.8	0.7	0.2	0.7	0.7	0.5	0.9	1
Play.on.feet.in	0	0.6	0.4	0	0	0	NA	0.8	0.9	0	0.6	0.4	NA	0.7	0.6	0.5	0.9	1
Play.sedentary.in	0.5	0.9	0.8	0	0	0	0.4	0.9	0.9	0.5	0.9	0.8	0.4	0.9	0.8	0.7	1	1
Resting.in	0.4	0.7	0.8	0.2	0.1	0.1	0.4	0.8	0.7	0.4	0.7	0.8	0.4	0.8	0.8	0.6	0.9	0.9
Resting.out	0.4	0.4	0.5	0	0	0	0.1	1	0.8	0.4	0.4	0.5	0.2	0.6	0.6	0.7	1	1
Running.out	0.6	0.8	0.9	0.1	0	0	0	0.9	0.9	0.6	0.8	0.9	0.1	0.9	0.9	0.8	1	1
Sleeping.in	0.4	0.9	0.9	0.1	0.1	0.1	0.6	0.8	0.9	0.4	0.9	0.9	0.5	0.9	0.9	0.7	1	1
Smoking.in	0.6	0.7	0.7	0	0	0	0.5	0.9	1	0.6	0.7	0.7	0.5	0.8	0.8	0.7	1	1
Sports.in	0	0.2	0.4	0	0	0	NA	0.8	0.6	0	0.2	0.4	NA	0.4	0.4	0.5	0.9	1
Sports.out	0	0.8	0.9	0	0	0	1	0.9	0.9	0	0.8	0.9	0.1	0.9	0.9	0.5	1	1
Weighted Average	0.3	0.8	0.8	0.1	0.1	0.1	NA	0.8	0.8	0.3	0.8	0.8	NA	0.8	0.8	0.6	1	1

Resting is also somewhat characterised by longer, consecutive time intervals without interruptions. It also has a high FP rate and low precision in both groups. It is the second most frequent activity chosen by participants (behind sleeping) in the study, frequently overlapping with other activities. Resting could be understood as a "default" activity, chosen when no other activity fits the description. It is a vaguely defined activity and open to interpretation. Participants tend to include various activities under this term, e.g., reading a book, playing board or computer games, watching television, chatting with friends, taking a leisurely walk, napping, having a dinner party, etc. All of these activities can differ in many aspects, such as heart rate, movement, speed, or PM concentrations, which would make accurate predictions more difficult. While more detailed activity classification would improve on this point, it would increase the burden on participants.

3.4. The Added Value of the Devices Used

One aim of this research was to determine the respective contributions of the two devices used—the SAT and PPM—to the performance of activity classification. An assessment was conducted in WEKA based on the data collected in Group M, as it showed the best performance between the two groups. The RF and J48 classifiers were used to classify the data from Group M (1) without the data collected with the PPM, i.e., all PM data, temperature, relative humidity, and speed (no PPM), (2) without just the particulate matter data (no PM), and (3) without the data collected with the SAT, i.e., heart rate, M.E.T., and steps (no SAT). The results showed, as evident in Table 9, that in the case of J48, the share of correctly classified instances is not reduced much if the PM or SAT data were removed (by 0.7% and 1.4%, respectively). On the other hand, if the PPM part of the dataset is removed entirely, the share of correctly classified instances falls to 60.1%. The RF models show similar values and trends.

Table 9. Instances correctly classified by the J48 and RF models, based on selectively removing all PPM, only PM, and SAT data from the Group M dataset, respectively.

Classifier	Correctly Classified Instances [%]			
	Baseline	no PPM	no PM	no SAT
J48	76.9	60.1	76.2	75.5
RF	77.2	62.5	76.6	77.2

However, the overall number does not show certain nuances, e.g., the results without the SAT data show a lowered TP rate for sports.out from 0.8 to 0.6 (for J48). Though the difference is not as evident for running and sports.in, there is still a small decline in classification accuracy. This result shows that for specific activities, the SAT data increase accuracy. Similarly, for the dataset without the PM data, accuracy for smoking, running outside, and playing is lower, all of which are activities with increased exposure to PM, though the difference is less pronounced than with the absence of SAT data. On the other hand, removing the entire PPM dataset (for RF) shows worse accuracy for all activities, especially for playing.on.feet (−0.4), smoking (−0.3), play.sedentary (−0.3), and cleaning.wet (−0.2). The latter could be explained by the absence of data on relative humidity.

Collecting data on PM concentrations and other environmental variables is a partial improvement on the 1st and 4th technical challenges listed in Section 1.2. With more specific data on environmental conditions, the characteristics are more distinct and can improve feature extraction. Moreover, as complex activities have several actions associated with them, more data on the overall environment could offset the lack of data on these specific activities.

4. Conclusions

4.1. Summary of Results

Two groups of participants were equipped with devices that measured their exposure to PM and their physical activity, while they logged their activity data with a paper TAD with hourly resolution (group H) and a smartphone app with minute resolution (group M). The primary aims were to evaluate the effectiveness of combining low-cost personal environmental and biometric sensor data for recognising individual activities and to assess the impact of data temporal resolution on the performance of different classifiers. These classifiers were selected for their proven efficacy, understandability, and accessibility, aligning with our objective of empowering lay individuals with HAR technology in participatory urban health studies. Successfully achieving these aims could significantly reduce or eliminate the need for manual activity recording in exposure studies, paving the way for a broader application of machine learning and classification methods in HAR, particularly in participatory-based research settings.

Results showed improved accuracy when (1) the activity time resolution was changed from 1-h to 1-min resolution and (2) more vague activities, e.g., cleaning, cooking, and playing, were divided into more detailed categories. Most misclassified instances belong to activities with vague definitions (resting, playing), while well-defined activities (smoking, cooking, cleaning, running) have fewer misclassified instances. Accuracy increased for all activities, especially playing, smoking, and running, when the environmental sensor data were included.

All the used classifiers for group H showed accuracy above 35%, with RF being the most accurate with 52.9%. As the training data consisted of hourly labelled activities, this meant lower resolution and more errors (some activities do not last an hour, and most do not last exactly a set number of full hours). An improvement in labelling data by the minute was proposed and evaluated with group M, which showed a noticeable improvement in all measures of performance (e.g., the accuracy of ~77% for J48 and RF models). This was an expected outcome, as the sensor data were also recorded in minute intervals and provided a good starting point to achieve activity prediction without resorting to manually recording data, which are prone to errors.

All of the models, for both groups, showed the most misclassified instances with resting. This could be the result of a vague definition of resting in comparison with sleeping, running, and most other activities. An educated guess of how the activities would be ranked from most vague definition to least vague would be: resting, playing, sports, cleaning, cooking, running, smoking, and sleeping. Resting could include naps, sitting behind a computer, reading a book, watching TV, hanging out with friends and family, or going for a short walk. All of these activities could have very different values for the observed variables. This is also true for sports, which is a wide term, and in the case of this dataset, it does not include running. What category should jogging and speed-walking then fall into?

When separated into more specific activities for group M, these activities showed moderate improvement, especially when considering more relevant activities pertaining to exposure to particulate matter. When dividing cooking into two activities, hot and cold, the results show that cooking using a heat source can be identified more easily and produces better results in terms of classification. A similar trend is present for cleaning, though the results are not as clear as for cooking.

On the other hand, sleeping or smoking are quite well-defined activities where there is little room for subjectivity. Even if smoking indoors includes using vaporizers, hookahs, pipes, or other gadgets, the observed variables would still presumably show similar results (elevated levels of PM, sedentary activity in enclosed space, not moving, relatively high heart rate, during the day, etc.), as would sleeping (in a chair, on a bed, on transport, taking a nap, etc.).

4.2. Limitations and Future Work

The shallow algorithms used in this research could be replaced by deep learning algorithms that could provide more accurate data. Additional steps could be taken to improve accuracy, e.g., noise removal, scaling, feature extraction, segmentation, and hyperparameter tuning. On the other hand, these approaches would be less understandable to non-experts. It could also limit accessibility in terms of the available software. In participatory-based research, frequently led by non-experts in the field of ML, the selection of algorithms and approaches should be considered based on several factors, apart from accuracy—computation requirements, visualisation options, understandable and explainable architectures/principles, etc. Similarly, more complex approaches used in participatory research could be exclusionary, as they would be more difficult to understand for lay individuals, even more so for individuals with less technical skills and knowledge.

Moreover, more ambiguous or subjectively defined activities should be separated into better-defined activities, as listed above. Although this would impose greater challenges when collecting data for the participants, it could provide more detailed final results and improve the classification. The consequences of these challenges can be seen in Table 1, with fewer recorded instances of the more specific activities in Group M, leading to, in some cases, an unbalanced dataset. A necessary focus would be to evaluate which of these activities are more relevant to the specific study or research and only use the classification models to predict those.

Overall, several additional suggestions and possible improvements are proposed for future research:

- Use of direct movement sensor (accelerometer, gyroscope, magnetometer) data from the SAT.
- Addition of possible other variables to be measured with SAT, e.g., skin temperature and conductivity.
- Utilisation of the data from smartphones (light, movement, location, indoor/outdoor, crowd density, barometer, accelerometer, gyroscope, magnetometer, etc.).

- Fusion of data with government monitoring station data to improve correlations of the measured temperature and humidity.
- Use of static sensor data at home, at the workplace, or in the car to improve or correct measurements made by wearable sensors.
- improvement of an app for logging activity data by providing the participant with (a) a warning when the devices detect a possible change of activity due to changes in parameters and (b) providing suggestions for possible activities ranked from most likely to least likely based on this research.

Half of the challenges listed in Section 1.2 remain unaddressed within the scope of this research, i.e., challenges 2, 3, and 6. While the demonstrated approach in this work does not require detailed personal data (as described in challenge 6), arguably the predictive performance could potentially be improved by including personal characteristics and GPS tracking. Moreover, the inclusion of ambient environmental data do not provide any tangible solutions to increase the quantity of annotated data or reduce sensory data heterogeneity. An argument could be made, to a degree, that collecting data on more variables could require fewer instances of annotated data.

An important improvement to participatory studies would be to reduce the burden of participants filling out time activity diaries while simultaneously reducing the chance of human error. This research shows that machine learning, informed by low-cost personal environmental monitors, can improve the process of recording activity data by reducing or potentially, in the future, completely freeing study participants from recording their activities. Combining the results of this research with environmental stressors measured with portable, low-cost sensors will provide a more detailed picture of exposure and intake dose on an individual scale. Further research is needed to test, validate, and improve these approaches.

As low-cost sensors become more widely used and individuals are able to gain access to more information about their living environment, researchers must provide adequate tools to assess and improve accuracy. A promising step forward would be to reduce the input of individuals and increase the role of machine learning. This research shows that a novel approach to using classification methods with data from low-cost portable environmental and activity sensors can be used to recognise specific activities without direct manual human input.

Supplementary Materials: The following supporting information can be downloaded at: https://www.mdpi.com/article/10.3390/s23249890/s1, Figure S1: Average values for all variables, per activity, for group H, Figure S2: Average values for all variables, per activity for group M, Table S1: IBk confusion matrix—group H, Table S2: IBk confusion matrix—group M, Table S3: J48 confusion matrix—group H, Table S4: J48 confusion matrix—group M, Table S5: Random Forest confusion matrix—group H, Table S6: Random Forest confusion matrix—group M, Table S7: Feature merits (importance scores) and ranks for group M.

Author Contributions: R.N., S.D. and D.K. conceptualised the work. R.N. handled the software, validation, visualisations, and wrote the main manuscript text. R.N. and S.D. formed the methodology. Data collection was done by R.N., D.K., J.A.R. and T.K. The process was supervised by D.K. and S.D., project administration handled by D.K. and D.S., and funding acquired by D.S. All authors approved the content of the manuscript. All authors have read and agreed to the published version of the manuscript.

Funding: This work has received funding from the European Union's Horizon 2020 Programme for Research, Technological Development, and Demonstration under grant agreement No. 690105 (Integrated Climate forcing and Air pollution Reduction in Urban Systems (ICARUS)). This work reflects only the authors' views, and the European Commission is not responsible for any use that may be made of the information it contains. Funding was received from the Young Researchers Programme, P1-0143 programme "Cycling of substances in the environment, mass balances, modelling of environmental processes, and risk assessment", and P2-0103 programme "Knowledge technologies", all funded by the Slovenian Research Agency.

Institutional Review Board Statement: The study was conducted in accordance with the Declaration of Helsinki. Ethical approval for the ICARUS project in Slovenia was obtained from the National Medical Ethics Committee of the Republic of Slovenia (approval nr. 0120-388/2018/6 on 22 August 2018). The data in this paper were selected only from participants in Slovenia. All methods were performed in accordance with the relevant guidelines and regulations.

Informed Consent Statement: Informed consent was obtained from all individual participants involved in the study.

Data Availability Statement: The datasets generated and/or analysed during the current study are not publicly available due to privacy issues, but are available from the corresponding author on reasonable request.

Acknowledgments: The authors acknowledge and thank all the participants who volunteered to contribute to the sampling campaign in Ljubljana, Slovenia.

Conflicts of Interest: The authors have no relevant financial or non-financial interests to disclose.

References

1. Bauman, A.; Bittman, M.; Gershuny, J. A Short History of Time Use Research; Implications for Public Health. *BMC Public Health* **2019**, *19*, 607. [CrossRef] [PubMed]
2. Chatzitheochari, S.; Fisher, K.; Gilbert, E.; Calderwood, L.; Huskinson, T.; Cleary, A.; Gershuny, J. Using New Technologies for Time Diary Data Collection: Instrument Design and Data Quality Findings from a Mixed-Mode Pilot Survey. *Soc. Indic. Res.* **2018**, *137*, 379–390. [CrossRef] [PubMed]
3. Elevelt, A.; Lugtig, P.; Toepoel, V. Doing a Time Use Survey on Smartphones Only: What Factors Predict Nonresponse at Different Stages of the Survey Process? *Surv. Res. Methods* **2019**, *13*, 195–213. [CrossRef]
4. Ferro, A.R.; Kopperud, R.J.; Hildemann, L.M. Elevated Personal Exposure to Particulate Matter from Human Activities in a Residence. *J. Expo. Sci. Environ. Epidemiol.* **2004**, *14*, S34–S40. [CrossRef] [PubMed]
5. Semple, S.; Apsley, A.; Azmina Ibrahim, T.; Turner, S.W.; Cherrie, J.W. Fine Particulate Matter Concentrations in Smoking Households: Just How Much Secondhand Smoke Do You Breathe in If You Live with a Smoker Who Smokes Indoors? *Tob. Control* **2015**, *24*, e205–e211. [CrossRef] [PubMed]
6. Chen, R.; Aherrera, A.; Isichei, C.; Olmedo, P.; Jarmul, S.; Cohen, J.E.; Navas-Acien, A.; Rule, A.M. Assessment of Indoor Air Quality at an Electronic Cigarette (Vaping) Convention. *J. Expo. Sci. Environ. Epidemiol.* **2018**, *28*, 522–529. [CrossRef] [PubMed]
7. Rosati, J.A.; Thornburg, J.; Rodes, C. Resuspension of Particulate Matter from Carpet Due to Human Activity. *Aerosol Sci. Technol.* **2008**, *42*, 472–482. [CrossRef]
8. Corsi, R.L.; Siegel, J.A.; Chiang, C. Particle Resuspension during the Use of Vacuum Cleaners on Residential Carpet. *J. Occup. Environ. Hyg.* **2008**, *5*, 232–238. [CrossRef]
9. Ma, J.; Li, C.; Kwan, M.-P.; Kou, L.; Chai, Y. Assessing Personal Noise Exposure and Its Relationship with Mental Health in Beijing Based on Individuals' Space-Time Behavior. *Environ. Int.* **2020**, *139*, 105737. [CrossRef]
10. Chatzidiakou, L.; Krause, A.; Kellaway, M.; Han, Y.; Martin, E.; Kelly, F.J.; Zhu, T.; Barratt, B.; Jones, R.L. Automated Classification of Time-Activity-Location Patterns for Improved Estimation of Personal Exposure to Air Pollution. Available online: https://www.researchsquare.com (accessed on 27 September 2022).
11. Freeman, N.C.G.; Saenz de Tejada, S. Methods for Collecting Time/Activity Pattern Information Related to Exposure to Combustion Products. *Chemosphere* **2002**, *49*, 979–992. [CrossRef]
12. Steinle, S.; Reis, S.; Sabel, C.E. Quantifying Human Exposure to Air Pollution—Moving from Static Monitoring to Spatio-Temporally Resolved Personal Exposure Assessment. *Sci. Total Environ.* **2013**, *443*, 184–193. [CrossRef] [PubMed]
13. Crosbie, T. Using Activity Diaries: Some Methodological Lessons. *J. Res. Pract.* **2006**, *2*, 1–13.
14. Robinson, J.A.; Novak, R.; Kanduč, T.; Maggos, T.; Pardali, D.; Stamatelopoulou, A.; Saraga, D.; Vienneau, D.; Flückiger, B.; Mikeš, O.; et al. User-Centred Design of a Final Results Report for Participants in Multi-Sensor Personal Air Pollution Exposure Monitoring Campaigns. *Int. J. Environ. Res. Public Health* **2021**, *18*, 12544. [CrossRef] [PubMed]
15. Wu, X.; Bennett, D.H.; Lee, K.; Cassady, D.L.; Ritz, B.; Hertz-Picciotto, I. Feasibility of Using Web Surveys to Collect Time–Activity Data. *J. Expo. Sci. Environ. Epidemiol.* **2012**, *22*, 116–125. [CrossRef] [PubMed]
16. Wu, J.; Jiang, C.; Houston, D.; Baker, D.; Delfino, R. Automated Time Activity Classification Based on Global Positioning System (GPS) Tracking Data. *Environ. Health* **2011**, *10*, 101. [CrossRef] [PubMed]
17. Dewulf, B.; Neutens, T.; Van Dyck, D.; de Bourdeaudhuij, I.; Int Panis, L.; Beckx, C.; Van de Weghe, N. Dynamic Assessment of Inhaled Air Pollution Using GPS and Accelerometer Data. *J. Transp. Health* **2016**, *3*, 114–123. [CrossRef]
18. Morawska, L.; Thai, P.K.; Liu, X.; Asumadu-Sakyi, A.; Ayoko, G.; Bartonova, A.; Bedini, A.; Chai, F.; Christensen, B.; Dunbabin, M.; et al. Applications of Low-Cost Sensing Technologies for Air Quality Monitoring and Exposure Assessment: How Far Have They Gone? *Environ. Int.* **2018**, *116*, 286–299. [CrossRef] [PubMed]
19. Thompson, J.E. Crowd-Sourced Air Quality Studies: A Review of the Literature & Portable Sensors. *Trends Environ. Anal. Chem.* **2016**, *11*, 23–34. [CrossRef]

20. Škultéty, E.; Pivarčiová, E.; Karrach, L. The Comparing of the Selected Temperature Sensors Compatible with the Arduino Platform. *Manag. Syst. Prod. Eng.* **2018**, *26*, 168–171. [CrossRef]
21. Sá, J.P.; Alvim-Ferraz, M.C.M.; Martins, F.G.; Sousa, S.I.V. Application of the Low-Cost Sensing Technology for Indoor Air Quality Monitoring: A Review. *Environ. Technol. Innov.* **2022**, *28*, 102551. [CrossRef]
22. Demrozi, F.; Pravadelli, G.; Bihorac, A.; Rashidi, P. Human Activity Recognition Using Inertial, Physiological and Environmental Sensors: A Comprehensive Survey. *IEEE Access* **2020**, *8*, 210816–210836. [CrossRef] [PubMed]
23. Majidzadeh Gorjani, O.; Proto, A.; Vanus, J.; Bilik, P. Indirect Recognition of Predefined Human Activities. *Sensors* **2020**, *20*, 4829. [CrossRef] [PubMed]
24. Hussein, D.; Bhat, G. SensorGAN: A Novel Data Recovery Approach for Wearable Human Activity Recognition. *ACM Trans. Embed. Comput. Syst.* **2023**. [CrossRef]
25. Saeedi, S.; El-Sheimy, N. Activity Recognition Using Fusion of Low-Cost Sensors on a Smartphone for Mobile Navigation Application. *Micromachines* **2015**, *6*, 1100–1134. [CrossRef]
26. Kozina, S.; Gjoreski, H.; Gams, M.; Luštrek, M. Efficient Activity Recognition and Fall Detection Using Accelerometers. In *Evaluating AAL Systems Through Competitive Benchmarking*; Botía, J.A., Álvarez-García, J.A., Fujinami, K., Barsocchi, P., Riedel, T., Eds.; Springer: Berlin/Heidelberg, Germany, 2013; pp. 13–23.
27. Su, X.; Tong, H.; Ji, P. Activity Recognition with Smartphone Sensors. *Tsinghua Sci. Technol.* **2014**, *19*, 235–249. [CrossRef]
28. Shoaib, M.; Bosch, S.; Incel, O.; Scholten, H.; Havinga, P. Complex Human Activity Recognition Using Smartphone and Wrist-Worn Motion Sensors. *Sensors* **2016**, *16*, 426. [CrossRef] [PubMed]
29. Tedesco, S.; Sica, M.; Ancillao, A.; Timmons, S.; Barton, J.; O'Flynn, B. Accuracy of Consumer-Level and Research-Grade Activity Trackers in Ambulatory Settings in Older Adults. *PLoS ONE* **2019**, *14*, e0216891. [CrossRef]
30. Oniani, S.; Woolley, S.I.; Pires, I.M.; Garcia, N.M.; Collins, T.; Ledger, S.; Pandyan, A. Reliability Assessment of New and Updated Consumer-Grade Activity and Heart Rate Monitors. In Proceedings of the SENSORDEVICES 2018: The Ninth International Conference on Sensor Device Technologies and Applications, IARIA, Venice, Italy, 16–20 September 2018; p. 6.
31. Añazco, E.V.; Lopez, P.R.; Lee, S.; Byun, K.; Kim, T.-S. Smoking Activity Recognition Using a Single Wrist IMU and Deep Learning Light. In Proceedings of the 2nd International Conference on Digital Signal Processing, Tokyo, Japan, 25–27 February 2018; Association for Computing Machinery: New York, NY, USA, 2018; pp. 48–51.
32. Chen, K.; Zhang, D.; Yao, L.; Guo, B.; Yu, Z.; Liu, Y. Deep Learning for Sensor-Based Human Activity Recognition: Overview, Challenges and Opportunities. *ACM Comput. Surveys* **2021**, *54*, 1–40. [CrossRef]
33. Liu, H.; Hartmann, Y.; Schultz, T. A Practical Wearable Sensor-Based Human Activity Recognition Research Pipeline. In Proceedings of the 15th International Joint Conference on Biomedical Engineering Systems and Technologies—WHC, Virtual, 9–11 February 2022; SciTePress: Setúbal, Portugal, 2022; pp. 847–856.
34. Chen, H.; Gouin-Vallerand, C.; Bouchard, K.; Gaboury, S.; Couture, M.; Bier, N.; Giroux, S. Leveraging Self-Supervised Learning for Human Activity Recognition with Ambient Sensors. In Proceedings of the 2023 ACM Conference on Information Technology for Social Good, Lisbon, Portugal, 6–8 September 2023; Association for Computing Machinery: New York, NY, USA, 2023; pp. 324–332.
35. Lee, T.-H.; Kim, H.; Lee, D. Transformer Based Early Classification for Real-Time Human Activity Recognition in Smart Homes. In Proceedings of the 38th ACM/SIGAPP Symposium on Applied Computing, Tallinn, Estonia, 7 June 2023; Association for Computing Machinery: New York, NY, USA, 2023; pp. 410–417.
36. Niu, H.; Ung, H.Q.; Wada, S. Source Domain Selection for Cross-House Human Activity Recognition with Ambient Sensors. In Proceedings of the 2022 21st IEEE International Conference on Machine Learning and Applications (ICMLA), Nassau, Bahamas, 12–14 December 2022; pp. 754–759.
37. Apicella, G.; D'Aniello, G.; Fortino, G.; Gaeta, M.; Gravina, R.; Tramuto, L.G. A Situation-Aware Wearable Computing System for Human Activity Recognition. In Proceedings of the 2022 IEEE International Conference on Dependable, Autonomic and Secure Computing, International Conference on Pervasive Intelligence and Computing, International Conference on Cloud and Big Data Computing, International Conference on Cyber Science and Technology Congress (DASC/PiCom/CBDCom/CyberSciTech), Physical Conference, Calabria, Italy, 12–15 September 2022; pp. 1–7.
38. Zhang, S.; Li, Y.; Zhang, S.; Shahabi, F.; Xia, S.; Deng, Y.; Alshurafa, N. Deep Learning in Human Activity Recognition with Wearable Sensors: A Review on Advances. *Sensors* **2022**, *22*, 1476. [CrossRef]
39. Aha, D.; Kibler, D. Instance-Based Learning Algorithms. *Mach. Learn.* **1991**, *6*, 37–66. [CrossRef]
40. Quinlan, J.R. *C4.5: Programs for Machine Learning*; Elsevier: Amsterdam, The Netherlands, 2014; ISBN 978-0-08-050058-4.
41. John, G.H.; Langley, P. Estimating Continuous Distributions in Bayesian Classifiers. In Proceedings of the Eleventh Conference on Uncertainty in Artificial Intelligence, Montreal, Canada, 18–20 August 1995; Morgan Kaufmann: San Mateo, CA, USA, 1995; pp. 338–345.
42. Breiman, L. Random Forests. *Mach. Learn.* **2001**, *45*, 5–32. [CrossRef]
43. Friedman, J.H. *Stochastic Gradient Boosting*; Stanford University: Stanford, CA, USA, 1999.
44. Chang, C.-C.; Lin, C.-J. LIBSVM—A Library for Support Vector Machines. *ACM Transact. Intell. Syst. Technol.* **2011**, *2*, 1–27. [CrossRef]
45. Dohnálek, P.; Gajdoš, P.; Peterek, T. Human Activity Recognition: Classifier Performance Evaluation on Multiple Datasets. Available online: https://www.jvejournals.com/article/15013 (accessed on 18 June 2020).

46. Rodrigues, L.M.; Mestria, M. Classification Methods Based on Bayes and Neural Networks for Human Activity Recognition. In Proceedings of the 2016 12th International Conference on Natural Computation, Fuzzy Systems and Knowledge Discovery (ICNC-FSKD), Changsha, China, 13–15 August 2016; pp. 1141–1146.
47. Bustoni, I.A.; Hidayatulloh, I.; Ningtyas, A.M.; Purwaningsih, A.; Azhari, S.N. Classification Methods Performance on Human Activity Recognition. *J. Phys. Conf. Ser.* **2020**, *1456*, 012027. [CrossRef]
48. Chau, N.H. Estimation of Air Temperature Using Smartphones in Different Contexts. *J. Inf. Telecommun.* **2019**, *3*, 494–507. [CrossRef]
49. Barna, A.; Masum, A.K.M.; Hossain, M.E.; Bahadur, E.H.; Alam, M.S. A Study on Human Activity Recognition Using Gyroscope, Accelerometer, Temperature and Humidity Data. In Proceedings of the 2019 International Conference on Electrical, Computer and Communication Engineering (ECCE), Cox's Bazar, Bangladesh, 7–9 February 2019; pp. 1–6.
50. Shelke, S.; Aksanli, B. Static and Dynamic Activity Detection with Ambient Sensors in Smart Spaces. *Sensors* **2019**, *19*, 804. [CrossRef] [PubMed]
51. Nia, N.G.; Kaplanoglu, E.; Nasab, A.; Qin, H. Human Activity Recognition Using Machine Learning Algorithms Based on IMU Data. In Proceedings of the 2023 5th International Conference on Bio-engineering for Smart Technologies (BioSMART), Paris, France, 7–9 June 2023; pp. 1–8.
52. Noorani, S.H.; Raheel, A.; Khan, S.; Arsalan, A.; Ehatisham-Ul-Haq, M. Identification of Human Activity and Associated Context Using Smartphone Inertial Sensors in Unrestricted Environment. In Proceedings of the 2023 International Conference on Communication, Computing and Digital Systems (C-CODE), Islamabad, Pakistan, 17–18 May 2023; pp. 1–6.
53. Dernbach, S.; Das, B.; Krishnan, N.C.; Thomas, B.L.; Cook, D.J. Simple and Complex Activity Recognition through Smart Phones. In Proceedings of the 2012 Eighth International Conference on Intelligent Environments, Washington, DC, USA, 26–29 June 2012; pp. 214–221.
54. Sousa Lima, W.; Souto, E.; El-Khatib, K.; Jalali, R.; Gama, J. Human Activity Recognition Using Inertial Sensors in a Smartphone: An Overview. *Sensors* **2019**, *19*, 3213. [CrossRef] [PubMed]
55. Frank, E.; Hall, M.A.; Witten, I.H. *The WEKA Workbench. Online Appendix for "Data Mining: Practical Machine Learning Tools and Techniques"*, 4th ed.; Morgan Kaufmann: Burlington, MA, USA, 2016.
56. Kocman, D.; Kanduč, T.; Novak, R.; Robinson, J.A.; Mikeš, O.; Degrendele, C.; Sáňka, O.; Vinkler, J.; Prokeš, R.; Vienneau, D.; et al. Multi-Sensor Data Collection for Personal Exposure Monitoring: ICARUS Experience. *Fresenius Environ. Bull.* **2022**, *31*, 8349–8354.
57. Sarigiannis, D.; Karakitsios, S.; Chapizanis, D.; Hiscock, R. D4.2 Methodology for Properly Accounting for SES in Exposure Assessment; ICARUS: Zeebrugge, Belgium, 2018.
58. Robinson, J.A. User Experience Evaluation of Novel Air Quality Sensing Technologies for Citizen Engagement in Environmental Health Studies. Ph.D. Thesis, Jožef Stefan Institute, Ljubljana, Slovenia, 2022.
59. Castaneda, D.; Esparza, A.; Ghamari, M.; Soltanpur, C.; Nazeran, H. A Review on Wearable Photoplethysmography Sensors and Their Potential Future Applications in Health Care. *Int. J. Biosens. Bioelectron.* **2018**, *4*, 195–202. [CrossRef]
60. Garmin; Subsidiaries, G.L. or its Garmin Vívosmart® 3 | Fitness Activity Tracker | Specs. Available online: https://www.garmin.com/en-US/p/567813 (accessed on 15 November 2023).
61. Dorn, D.; Gorzelitz, J.; Gangnon, R.; Bell, D.; Koltyn, K.; Cadmus-Bertram, L. Automatic Identification of Physical Activity Type and Duration by Wearable Activity Trackers: A Validation Study. *JMIR mHealth uHealth* **2019**, *7*, e13547. [CrossRef]
62. Montes, J.; Tandy, R.; Young, J.; Lee, S.-P.; Navalta, J.W. Step Count Reliability and Validity of Five Wearable Technology Devices While Walking and Jogging in Both a Free Motion Setting and on a Treadmill. *Int. J. Exerc. Sci.* **2020**, *13*, 410–426.
63. Chow, H.-W.; Yang, C.-C. Accuracy of Optical Heart Rate Sensing Technology in Wearable Fitness Trackers for Young and Older Adults: Validation and Comparison Study. *JMIR mHealth uHealth* **2020**, *8*, e14707. [CrossRef] [PubMed]
64. Passler, S.; Bohrer, J.; Blöchinger, L.; Senner, V. Validity of Wrist Worn Activity Trackers for Estimating VO2max and Energy Expenditure. *Int. J. Environ. Res. Public. Health* **2019**, *16*, 3037. [CrossRef] [PubMed]
65. Reddy, R.K.; Pooni, R.; Zaharieva, D.P.; Senf, B.; Youssef, J.E.; Dassau, E.; Iii, F.J.D.; Clements, M.A.; Rickels, M.R.; Patton, S.R.; et al. Accuracy of Wrist-Worn Activity Monitors During Common Daily Physical Activities and Types of Structured Exercise: Evaluation Study. *JMIR mHealth uHealth* **2018**, *6*, e10338. [CrossRef] [PubMed]
66. Bulot, F.M.J.; Johnston, S.J.; Basford, P.J.; Easton, N.H.C.; Apetroaie-Cristea, M.; Foster, G.L.; Morris, A.K.R.; Cox, S.J.; Loxham, M. Long-Term Field Comparison of Multiple Low-Cost Particulate Matter Sensors in an Outdoor Urban Environment. *Sci. Rep.* **2019**, *9*, 1–13. [CrossRef] [PubMed]
67. Cowell, N.; Chapman, L.; Bloss, W.; Pope, F. Field Calibration and Evaluation of an Internet-of-Things-Based Particulate Matter Sensor. *Front. Environ. Sci.* **2022**, *9*, 798485. [CrossRef]
68. Masic, A.; Bibic, D.; Pikula, B.; Blazevic, A.; Huremovic, J.; Zero, S. Evaluation of Optical Particulate Matter Sensors under Realistic Conditions of Strong and Mild Urban Pollution. *Atmos. Meas. Tech.* **2020**, *13*, 6427–6443. [CrossRef]
69. Connolly, R.E.; Yu, Q.; Wang, Z.; Chen, Y.-H.; Liu, J.Z.; Collier-Oxandale, A.; Papapostolou, V.; Polidori, A.; Zhu, Y. Long-Term Evaluation of a Low-Cost Air Sensor Network for Monitoring Indoor and Outdoor Air Quality at the Community Scale. *Sci. Total Environ.* **2022**, *807*, 150797. [CrossRef]
70. Novak, R.; Kocman, D.; Robinson, J.A.; Kanduč, T.; Sarigiannis, D.; Horvat, M. Comparing Airborne Particulate Matter Intake Dose Assessment Models Using Low-Cost Portable Sensor Data. *Sensors* **2020**, *20*, 1406. [CrossRef]

71. Novak, R.; Petridis, I.; Kocman, D.; Robinson, J.A.; Kanduč, T.; Chapizanis, D.; Karakitsios, S.; Flückiger, B.; Vienneau, D.; Mikeš, O.; et al. Harmonization and Visualization of Data from a Transnational Multi-Sensor Personal Exposure Campaign. *Int. J. Environ. Res. Public. Health* **2021**, *18*, 11614. [CrossRef]
72. COING Inc. Clockify—The #1 Time Tracker for Teams. Available online: https://clockify.me/apps (accessed on 1 June 2021).
73. R Core Team. *R: A Language and Environment for Statistical Computing*; R Foundation for Statistical Computing: Vienna, Austria, 2020.
74. Meier, R.; Eeftens, M.; Phuleria, H.C.; Ineichen, A.; Corradi, E.; Davey, M.; Fierz, M.; Ducret-Stich, R.E.; Aguilera, I.; Schindler, C.; et al. Differences in Indoor versus Outdoor Concentrations of Ultrafine Particles, PM2.5, PMabsorbance and NO2 in Swiss Homes. *J. Expo. Sci. Environ. Epidemiol.* **2015**, *25*, 499–505. [CrossRef]
75. Nadali, A.; Arfaeinia, H.; Asadgol, Z.; Fahiminia, M. Indoor and Outdoor Concentration of PM10, PM2.5 and PM1 in Residential Building and Evaluation of Negative Air Ions (NAIs) in Indoor PM Removal. *Environ. Pollut. Bioavailab.* **2020**, *32*, 47–55. [CrossRef]
76. Zhang, C.; Cao, K.; Lu, L.; Deng, T. A Multi-Scale Feature Extraction Fusion Model for Human Activity Recognition. *Sci. Rep.* **2022**, *12*, 20620. [CrossRef] [PubMed]
77. Herm, L.-V.; Heinrich, K.; Wanner, J.; Janiesch, C. Stop Ordering Machine Learning Algorithms by Their Explainability! A User-Centered Investigation of Performance and Explainability. *Int. J. Inf. Manag.* **2023**, *69*, 102538. [CrossRef]
78. Kim, M.; Kim, D.; Jin, D.; Kim, G. Application of Explainable Artificial Intelligence (XAI) in Urban Growth Modeling: A Case Study of Seoul Metropolitan Area, Korea. *Land* **2023**, *12*, 420. [CrossRef]
79. Wanner, J.; Herm, L.-V.; Heinrich, K.; Janiesch, C. The Effect of Transparency and Trust on Intelligent System Acceptance: Evidence from a User-Based Study. *Electron. Mark.* **2022**, *32*, 2079–2102. [CrossRef]
80. Kamakshi, V.; Krishnan, N.C. Explainable Image Classification: The Journey So Far and the Road Ahead. *AI* **2023**, *4*, 620–651. [CrossRef]
81. Wu, X.; Kumar, V.; Ross Quinlan, J.; Ghosh, J.; Yang, Q.; Motoda, H.; McLachlan, G.J.; Ng, A.; Liu, B.; Yu, P.S.; et al. Top 10 Algorithms in Data Mining. *Knowl. Inf. Syst.* **2008**, *14*, 1–37. [CrossRef]
82. Robnik-Sikonja, M.; Kononenko, I. An Adaptation of Relief for Attribute Estimation in Regression. In Proceedings of the Fourteenth International Conference on Machine Learning, Nashville, TN, USA, 8–12 July 1997; Fisher, D.H., Ed.; Morgan Kaufmann: San Francisco, CA, USA, 1997; pp. 296–304.
83. Feinstein, A.R.; Cicchetti, D.V. High Agreement but Low Kappa: I. The Problems of Two Paradoxes. *J. Clin. Epidemiol.* **1990**, *43*, 543–549. [CrossRef]
84. Zec, S.; Soriani, N.; Comoretto, R.; Baldi, I. High Agreement and High Prevalence: The Paradox of Cohen's Kappa. *Open Nurs. J.* **2017**, *11*, 211–218. [CrossRef]
85. McHugh, M.L. Interrater Reliability: The Kappa Statistic. *Biochem. Med.* **2012**, *22*, 276–282. [CrossRef]

Disclaimer/Publisher's Note: The statements, opinions and data contained in all publications are solely those of the individual author(s) and contributor(s) and not of MDPI and/or the editor(s). MDPI and/or the editor(s) disclaim responsibility for any injury to people or property resulting from any ideas, methods, instructions or products referred to in the content.

Article

Exploring Regularization Methods for Domain Generalization in Accelerometer-Based Human Activity Recognition

Nuno Bento [1,*], Joana Rebelo [1], André V. Carreiro [1], François Ravache [2] and Marília Barandas [1]

[1] Associação Fraunhofer Portugal Research, Rua Alfredo Allen 455/461, 4200-135 Porto, Portugal
[2] ICOM France, 1 Rue Brindejonc des Moulinais, 31500 Toulouse, France
* Correspondence: nuno.bento@fraunhofer.pt

Abstract: The study of Domain Generalization (DG) has gained considerable momentum in the Machine Learning (ML) field. Human Activity Recognition (HAR) inherently encompasses diverse domains (e.g., users, devices, or datasets), rendering it an ideal testbed for exploring Domain Generalization. Building upon recent work, this paper investigates the application of regularization methods to bridge the generalization gap between traditional models based on handcrafted features and deep neural networks. We apply various regularizers, including sparse training, Mixup, Distributionally Robust Optimization (DRO), and Sharpness-Aware Minimization (SAM), to deep learning models and assess their performance in Out-of-Distribution (OOD) settings across multiple domains using homogenized public datasets. Our results show that Mixup and SAM are the best-performing regularizers. However, they are unable to match the performance of models based on handcrafted features. This suggests that while regularization techniques can improve OOD robustness to some extent, handcrafted features remain superior for domain generalization in HAR tasks.

Keywords: Human Activity Recognition; deep learning; Domain Generalization; regularization; accelerometer

1. Introduction

Human Activity Recognition (HAR) addresses the problem of identifying specific kinds of physical activities or movements performed by a person based on data that can be collected by several types of sensors [1]. It is a critical technology that supports several applications, including remote patient monitoring, locomotor rehabilitation, security, and pedestrian navigation [2]. This work focuses on HAR relying on inertial sensors, such as accelerometers, which measure the acceleration of a body, or gyroscopes, which measure angular velocity. These sensors are usually combined in Inertial Measurement Units (IMUs), which are present in most smartphones and smartwatches, and nearly ubiquitous in our daily life [3]. This translates into an increasing availability of sensor data, which, along with its importance in several fields, has motivated the growth of HAR in the past years [1].

Despite being a widely studied field, there are still challenges to be faced in HAR, one of which is the difficulty in developing models that generalize effectively across different domains [4]. This results in HAR models that perform well when tested on a randomly selected portion of a meticulously acquired dataset, but exhibit a performance decline when tested in realistic Out-of-Distribution (OOD) settings. These settings are characterized by a domain shift (or distribution shift) between the source and target domains [5]. In the HAR context, this can occur when the models are tested across different users, devices, sensor positions, or data acquisition setups [6–8].

The problem of distribution shift can be found in most data-related fields. A straightforward solution involves collecting data from the target domain and adapting the model, which was initially trained on source data, using the target data. This approach, known as domain adaptation, has been extensively explored [5]. However, it presupposes the availability of target data, a condition that may not always be met in real-world scenarios. In order to simultaneously tackle the domain shift and the absence of target data, the problem of domain generalization originated. Domain Generalization (DG) focuses on leveraging only source data to develop models that generalize to OOD target domains [5].

In traditional HAR approaches, features are extracted manually through signal processing techniques before being used as input to a machine learning model [9]. More recently, deep learning has attracted attention as a potential tool for HAR tasks [10]. In this modern approach, features are automatically extracted during the training process [10]. Given the large number of learnable parameters associated with deep learning models, they should be able to learn more complex and discriminative features [11]. This capability is expected to help deep learning thrive in DG scenarios. Nevertheless, several limitations have been identified upon deploying deep learning models, such as the convergence to solutions that rely on spurious correlations [12]. In our previous work, Bento et al. [13] compared the effectiveness of Handcrafted (HC) features versus deep neural representations for DG in HAR. Our findings revealed that while deep learning models initially outperformed those based on HC features, this trend was reversed as the distance from the training distribution increased, creating a gap between these methods in the OOD regime.

Our work attempts to bridge this gap by using regularization, which primarily focuses on mitigating overfitting, consequently leading to improved generalization performance [14,15]. For that purpose, several regularization methods are compared by following a methodology introduced in Bento et al. [13], leveraging five public datasets that are homogenized, so that they can be arranged in different combinations, creating multiple OOD settings.

The research questions addressed by this work are the following:

1. How do different regularization methods impact the Domain Generalization performance of human activity recognition models?
2. Can regularization methods bridge the OOD performance gap between deep neural networks and models based on HC features?

2. Related Work

Concerning classical machine learning approaches using HC features, several algorithms have been proposed for the recognition of human activities. Despite the considerable progress made by these algorithms in HAR, they may not capture more complex signal patterns, which can hinder their generalization performance [10]. To overcome this limitation, research has turned to Deep Learning (DL) models, which can automatically extract high-level features from raw data [10].

With that in mind, recent work [16–20] has compared traditional Machine Learning (ML) with DL approaches for HAR. Their findings consistently demonstrate that deep learning outperforms traditional methods. However, it should be noted that, in these experiments, the data splits were created by randomly shuffling the datasets. As a result, the training and test sets contain samples from distinct domains, therefore mitigating the distribution shift in their evaluations. As such, models optimized for these data splits may achieve suboptimal results in a real-world environment [21].

In studies where data splits took into account the distribution shift caused by different domains [13,20,22,23], the ability of DL methods to generalize has been put into question, as traditional ML models achieved similar or even better results, in some cases.

One of the reasons why DL models may not generalize well is that they are known to suffer from overfitting since they possess many parameters and their optimization process is not perfect [14]. One of the ways to prevent overfitting is by using regularization methods, which can be seen as applying constraints to the training process or the models in the form of penalties applied to parameter norms (e.g., L_2 regularization), elimination of parameters (e.g., dropout), early stopping, among other techniques [14]. As well as these popular regularization techniques, recent work has yielded progressively superior methods [24–27]. Mixup regularization works by performing a linear interpolation between input/target pairs and has been shown to outperform previous methods such as dropout and weight decay [24,25]. The optimization algorithms used for training can also be considered a form of regularization [28]. Methods that attempt to regularize stochastic gradient descent include averaging weights over various iterations [29] or actively searching for flat minima [26]. Sparsity is another form of regularization that can improve both generalization performance and model efficiency in deep learning by promoting the use of fewer non-zero parameters, leading to simpler models [30]. Sparse training is an efficient and effective way to add this type of regularization to a neural network [27,30,31].

Distributionally Robust Optimization (DRO) is a promising approach for addressing the need for optimizing models for Domain Generalization [21,32]. These methods usually regularize the training process by considering the distribution shift between the existing domains. Invariant Risk Minimization (IRM) [21] and Variance—Risk Extrapolation (V-REx) [33] introduce penalties to the loss function with the objective of learning representations that are invariant across multiple domains. Ahuja et al. [34] showed that adding a penalty based on the information bottleneck principle to IRM improves generalization—IB-IRM.

Some of the aforementioned regularization methods have been investigated as a potential solution to the OOD generalization problem in HAR. Gagnon et al. [35] included a HAR dataset in their Domain Generalization benchmark. Their results indicate a 9.07% drop in accuracy from 93.35% In-Distribution (ID) to 84.28% OOD on a dataset where different devices worn in different positions characterize the possible domains. IB-IRM [34] was the best-performing method. However, results did not improve significantly over empirical risk minimization (ERM), which is still a strong baseline [36]. Lu et al. [37] introduced a semantic-aware version of Mixup, which outperformed several Domain Generalization methods in HAR tasks. They presented results across different users, datasets, and positions. However, handcrafted features were not addressed in their work. Trabelsi et al. [20] compared three deep learning approaches and a random forest classifier with handcrafted features as input. Similarly to the experiments in our work, the datasets were homogenized by including only shared activities and separating the test sets by user. They concluded that only one of the deep learning approaches outperformed the baseline model with handcrafted features. Regularization methods were not studied in their work.

Our previous work, Bento et al. [13] showed that while DL methods outperformed traditional ML approaches when the training and test sets were split randomly, as the distance between the distributions grows, the tendency inverts, with methods based on DL usually performing worse in OOD settings. This paper builds on that work, adding different regularization methods to the models in order to assess if and by how much the OOD performance gap between HC features and deep representations is reduced. Our experiments include four Domain Generalization settings with different distances between training and test sets. To the best of our knowledge, this is the first attempt at comparing regularization methods for Domain Generalization in HAR.

3. Methodology

3.1. Datasets

The data employed in this study are the same as that used in Bento et al. [13]. Therefore, the datasets are composed of human activity data collected with smartphones and wearable IMUs. All the datasets are publicly available, and a comprehensive description of each is presented in Table 1.

The datasets were selected according to three criteria: (a) a sampling frequency superior or equal to 50 Hz; (b) most of the main human activities in the literature (walking, sitting, standing, running, upstairs, and downstairs); and (c) at least one common sensor position with another of the chosen datasets.

For this study, only the accelerometer data were used. In addition to the three accelerometer channels (x, y, and z) produced directly by the sensors, a fourth channel comprising the accelerometer magnitude was computed and utilized in the classification process. Five-second windows without overlap were then extracted from those four channels.

Table 1. Description of the datasets, including number of subjects, activities, devices, sample rate, positions, and sources.

Dataset	Subjects	Activities	Devices	Sample Rate	Positions	Source
PAMAP2	9	Sitting, lying, standing, walking, ascending stairs, descending stairs, running.	3 IMUs	100 Hz	Wrist, chest, and ankle.	[38,39]
SAD	10	Sitting, standing, walking, ascending stairs, descending stairs, running and biking.	5 smartphones	50 Hz	Jeans pocket, arm, wrist, and belt.	[40]
DaLiAc	19	Sitting, lying, standing, walking outside, ascending stairs, descending stairs, treadmill running.	4 IMUs	200 Hz	Hip, chest, and ankles.	[41]
MHEALTH	10	Sitting, lying, standing, walking, climbing/descending stairs, jogging, running.	3 IMUs	50 Hz	Chest, wrist, and ankle.	[42,43]
RealWorld	15	Sitting, lying, standing, walking, ascending stairs, descending stairs, running/jogging.	6 IMUs	50 Hz	Chest, forearm, head, shin, thigh, upper arm, and waist.	[44]

Data homogenization consisted of resampling all the datasets to 50 Hz and mapping activity labels of all datasets to a shared naming convention: "walking", "running", "sitting", "standing", and "stairs". Note that the class "stairs" does not differentiate between ascending and descending stairs. Given the discrepancy between the number of windows generated by each dataset, only one-third of the windows from the RealWorld dataset were randomly sampled and used in the experiments. Table 2 displays the final distribution of windows and activities for each dataset.

For further details regarding the datasets used and decisions concerning data preprocessing, please refer to Bento et al. [13].

Table 2. Distribution of samples and activity labels per dataset. The # symbol represents the number of samples. Retrieved from [13].

	Activity	Datasets (%)					Total	
		PAMAP2	SAD	DaLiAc	MHEALTH	Real World	%	#
	Run	10.5	16.9	20.0	33.3	19.1	18.3	7975
	Sit	19.8	16.9	10.6	16.7	17.0	16.3	7102
	Stairs	23.6	32.2	12.3	16.7	30.0	26.3	11,460
	Stand	20.4	16.9	10.6	16.7	16.4	16.2	7047
	Walk	25.7	16.9	46.5	16.7	17.5	22.8	9927
Total	%	12.7	24.4	15.3	4.96	42.6	-	-
	#	5541	10,620	6644	2160	18,546	-	43,511

3.2. Handcrafted Features

The TSFEL library [45] was used to extract features from the windows produced from each public dataset. Features including individual coefficients and audio-related features were excluded to reduce the computation time. This resulted in a total of 192 features per window.

The following steps were used to split the samples based on their task (see Section 4) and perform Z-score normalization with statistical information regarding the training set. The classification algorithms used were a Logistic Regression (LR) and a Multilayer Perceptron (MLP).

Additional details regarding feature extraction and preprocessing can be found in Bento et al. [13].

3.3. Deep Learning

The architectures used in our experiments were different variations of convolutional neural networks. We chose the two best-performing architectures from [13], which were CNN-base and ResNet. Refer to the original paper for a detailed explanation of the used architectures and training process.

For the hybrid models, the HC features are concatenated with the flattened representations of each model and fed to a fusion layer before entering the final classification layer. An illustration of the hybrid version of CNN-base is shown in Figure 1.

For all these models, the input windows were scaled by Z-score normalization, with mean and standard deviation computed across all the windows of the train set.

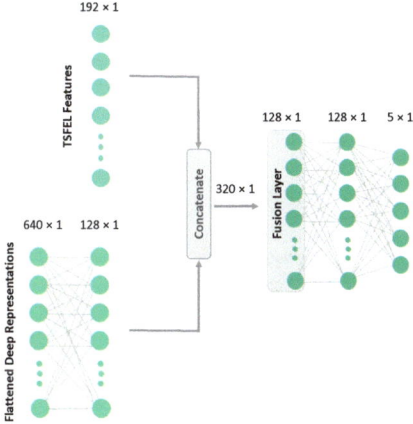

Figure 1. Simplified illustration of the hybrid version of CNN-base (excluding the CNN backbone for ease of visualization) [13].

3.4. Regularization

In this study, several regularization methods are compared:

- Mixup regularization [24,25]: It linearly interpolates input/target pairs to create new examples, which are used to make decision boundaries smoother and avoid overfitting.
- Sharpness-Aware Minimization (SAM) [26]: Optimization method that actively seeks flat minima. This type of minima was shown to be less prone to overfitting [46].
- GraNet [27]: It is a state-of-the-art method for sparse training that gradually reduces the number of non-zero weights during training.
- IRM [21]: It attempts to learn invariant representations by minimizing the sum of the squared norms of the gradients across multiple environments.
- V-REx [33]: It has the same purpose of IRM, but instead it minimizes the gradient variance across environments.
- IB-IRM [34]: It introduces a term to the IRM loss corresponding to the variance in the model parameters, following the information bottleneck principle.

3.5. Evaluation

Various metrics are used in research literature to assess model performance. These include accuracy, sensitivity, specificity, precision, recall, and f1-score [2]. However, due to the frequent occurrence of class imbalance in many public HAR datasets (as indicated in Table 2), f1-score was chosen as the primary performance metric, as it proved to be more resilient than accuracy in these types of situations [47]. For the sake of comparing deep learning models and traditional models using HC features, f1-scores were used as the comparison metric. This comparison was carried out across multiple OOD scenarios and took into consideration five public HAR datasets.

4. Experiments and Results

The goal of this study is to assess the improvement brought by using different regularization techniques on Domain Generalization tasks involving models based on HC features and deep neural networks. To that end, various combinations of model architectures and regularization methods were implemented and evaluated. A scheme of the full pipeline used for the experiments is presented in Figure 2.

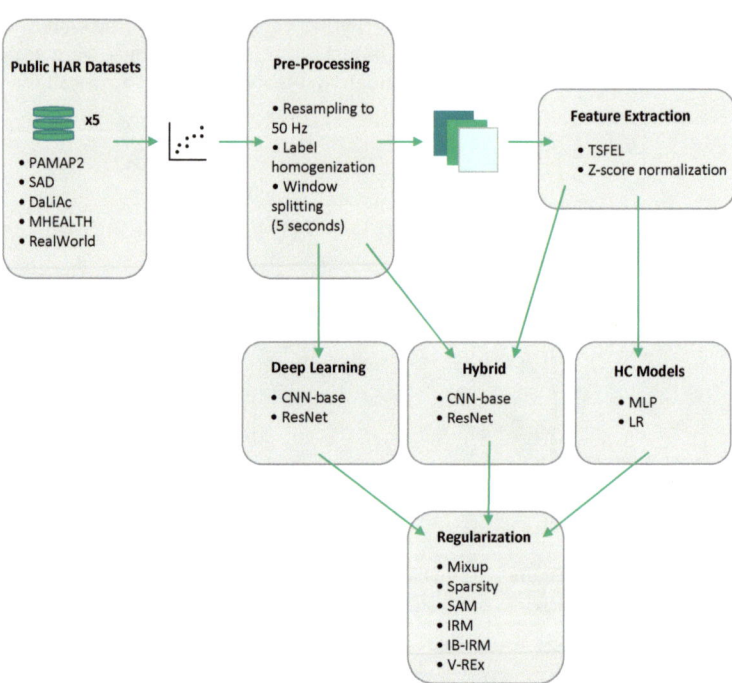

Figure 2. Scheme of the experimental pipeline.

HAR is a task where different domains naturally occur, becoming a Domain Generalization task if these domains are preserved when splitting the data into training and test sets. To measure how different a test set is from the training set, or how OOD it is, Bento et al. [13] computed Wasserstein distance ratios. Following that study, our experiments were conducted over the same four domain generalization settings, comprised of a baseline ID setting and three OOD settings [13]: (a) splitting by user within the same dataset (OOD-U); (b) leaving a dataset out for testing (OOD-MD); and (c) training on a dataset and leaving another for testing (OOD-SD). Test sets in the OOD-U setting were closer to the training distribution, being further away in the OOD-MD and OOD-SD settings. Non-exhaustive hyperparameter optimization was applied to the regularization methods on a small private HAR dataset. The chosen hyperparameters for each method are specified in Table 3.

Table 3. Chosen hyperparameters for each regularization method.

Method	Hyperparameter	Value
Mixup	α	0.1
GraNet	prune rate	0.5
	initial density	0.5
	final density	0.1
SAM	base optimizer	Adam
	ρ	0.05
IRM	λ	100
V-REx	β	10
IB-IRM	λ	100
	γ	10

Table 4 presents the results for the first experiment, which combines neural network architectures and regularizers. The DRO models (IRM, V-REx, and IB-IRM) used in this experiment require the formulation of different environments (i.e., domains) in the training set. For the ID and OOD-U settings, the environments are split by the user. For OOD-MD, each environment can be trivially devised as a dataset. However, for the OOD-SD, only a single dataset is present in the training set, so there is no trivial way to simulate the distribution shift that occurs between the training and test sets. Consequently, this setting was removed from the experiment. In Table 4, it can be verified that ResNet is the best-performing deep learning architecture, as it consistently shows higher f1-scores compared to CNN-base. For CNN-base, only SAM improved over the baseline model without regularization. Still, this improvement was not significant, and the performance was far from its hybrid version (CNN-base hybrid). For the ResNet architecture, Mixup, SAM, IRM, and IB-IRM improved over the baseline. Mixup achieved the best results (76.44%), marginally surpassing the score of the hybrid version (76.29%). Overall, Mixup and SAM can be considered the best-performing regularizers since the scores either improved or remained approximately the same in both architectures. The larger improvement verified on the ResNet may be due to the increased effectiveness of regularization methods in overparameterized regimes [12,48].

DRO methods are known to heavily depend on the chosen hyperparameters [36]. This may have hindered their performance. For these methods, hyperparameter optimization was performed over only three different values (10, 100, and 1000) for their regularization weights since the regularization methods should require as little computational overhead as possible.

As a second experiment, the best architecture (ResNet) and the two best regularizers (Mixup and SAM) were chosen, so that it could be assessed whether a combination of the best regularizers could further improve the generalization of deep learning models. The results are presented in Table 5. Since none of the DRO methods were considered, the OOD-SD setting could be recovered for this experiment, as none of the remaining methods require information about the environments. After adding the OOD-SD setting, the ResNet hybrid (66.77%) slightly outperformed the regularized ResNet models (66.42% and 66.48%). However, the difference is minimal, so we can consider that their performance is approximately the same. In the OOD-MD setting, ResNet with Mixup regularization (71.18%) outperformed some classical models. Despite that, this improvement loses significance when assessing the average scores.

The best-performing model across both experiments was TSFEL + LR, followed by the remaining HC feature-based models, which includes TSFEL + MLP and the hybrid models. Despite the effectiveness of regulariztion, it was insufficient for deep learning models to reach the desired OOD performance levels. Still, as regularizers did not improve the overall results of any of the models that make use of HC features (i.e., TSFEL + LR,

TSFEL + MLP and hybrid models), it can be observed that these regularizers can help bridge the generalization gap between deep learning models and models based on HC features.

Table 4. Average f1-score in percentage (%) over all the tasks in all the settings except OOD-SD. Values in bold indicate the best performance for each setting. * indicates the experiments without regularization performed in [13].

Model	Setting			Avg. OOD
	ID	OOD-U	OOD-MD	
CNN-base *	92.10 ± 5.06	80.79 ± 9.68	66.94 ± 5.19	73.87 ± 5.49
CNN-base Mixup	91.18 ± 4.56	80.66 ± 9.33	67.05 ± 3.88	73.86 ± 5.05
CNN-base Sparse	92.01 ± 5.02	80.75 ± 10.04	66.11 ± 5.48	73.43 ± 5.72
CNN-base SAM	91.44 ± 4.86	81.46 ± 10.56	66.68 ± 4.49	74.07 ± 5.74
CNN-base IRM	90.70 ± 5.41	80.24 ± 9.89	66.86 ± 4.25	73.55 ± 5.38
CNN-base V-REx	88.69 ± 7.29	80.31 ± 9.71	67.05 ± 5.68	73.68 ± 5.63
CNN-base IB-IRM	84.33 ± 14.98	80.55 ± 10.60	65.59 ± 4.90	73.07 ± 5.84
CNN-base hybrid *	93.48 ± 4.35	85.28 ± 6.64	67.74 ± 3.37	76.51 ± 3.72
ResNet *	92.46 ± 4.73	81.16 ± 9.60	67.22 ± 4.89	74.19 ± 5.39
ResNet Mixup	92.10 ± 4.59	81.71 ± 9.91	71.18 ± 4.18	76.44 ± 5.38
ResNet Sparse	92.48 ± 4.61	80.86 ± 10.80	67.10 ± 3.32	73.98 ± 5.65
ResNet SAM	92.01 ± 4.46	81.36 ± 10.56	68.82 ± 3.57	75.09 ± 5.57
ResNet IRM	91.40 ± 4.68	80.66 ± 9.80	69.00 ± 5.53	74.83 ± 5.63
ResNet V-REx	90.64 ± 5.54	80.74 ± 9.61	67.00 ± 5.66	73.87 ± 5.58
ResNet IB-IRM	85.14 ± 15.26	80.63 ± 10.00	68.69 ± 5.52	74.66 ± 5.71
ResNet hybrid *	**93.79 ± 4.21**	84.71 ± 7.72	67.87 ± 3.40	76.29 ± 4.22
TSFEL + MLP *	92.87 ± 4.70	**87.09 ± 5.35**	70.11 ± 3.57	78.60 ± 3.22
TSFEL + LR *	90.54 ± 5.15	87.08 ± 5.55	**71.94 ± 3.19**	**79.51 ± 3.20**

Table 5. Average f1-score in percentage over all the tasks in a given setting. Values in bold indicate the best performance for each setting. * indicates the experiments without regularization performed in [13].

Model	Setting				Avg. OOD
	ID	OOD-U	OOD-MD	OOD-SD	
ResNet *	92.46 ± 4.73	81.16 ± 9.60	67.22 ± 4.89	46.57 ± 4.84	64.98 ± 3.94
ResNet Mixup	92.10 ± 4.59	81.71 ± 9.91	71.18 ± 4.18	46.56 ± 6.27	66.48 ± 4.15
ResNet Mixup SAM	92.03 ± 4.27	82.22 ± 10.32	70.04 ± 4.00	46.99 ± 6.25	66.42 ± 4.24
ResNet hybrid *	**93.79 ± 4.21**	84.71 ± 7.72	67.87 ± 3.40	47.73 ± 2.11	66.77 ± 2.90
ResNet hybrid Mixup SAM	93.30 ± 3.54	83.83 ± 9.00	69.87 ± 2.47	46.60 ± 4.22	66.77 ± 3.41
TSFEL + MLP *	92.87 ± 4.70	87.09 ± 5.35	70.11 ± 3.57	**51.45 ± 5.31**	69.55 ± 2.78
TSFEL + LR *	90.54 ± 5.15	87.08 ± 5.55	**71.94 ± 3.19**	50.97 ± 3.29	**70.00 ± 2.40**
TSFEL + LR Mixup SAM	90.38 ± 5.03	87.05 ± 5.23	71.67 ± 3.97	50.41 ± 3.58	69.71 ± 2.49
TSFEL + MLP Mixup SAM	93.03 ± 4.58	**87.46 ± 5.97**	70.39 ± 2.71	51.26 ± 4.00	69.71 ± 2.56

5. Discussion

The work conducted in this study evaluated the differences in the performance of different regularization techniques on Domain Generalization tasks applied to HAR classification.

In the first experiment, a comparison between various combinations of neural network architectures and regularizers was carried out. ResNet outperformed CNN-base consistently across different ID and OOD settings. This result aligns with the general perception about the superiority of ResNet in handling a broad range of tasks due to its deeper architecture and residual connections, enabling it to learn more complex representations [49]. Regarding regularization methods, Mixup and SAM were considered to be the best-performing methods in both architectures. This agrees with the original premises behind these techniques, as Mixup attempts to improve generalization by enforcing a smoother decision boundary [24], while SAM adds robustness to label noise [26].

In the second experiment, a combination of the two best regularizers (Mixup and SAM) and the best architecture (ResNet) was performed to assess if it further improves the generalization of deep learning models. Combining the two methods did not yield better results than using Mixup alone. This experiment also showed that regularization only improved the performance of deep learning models that did not include HC features. The fact that the average OOD score of the ResNet hybrid did not change with the use of regularizers may indicate that the use of these features as an auxiliary input can already have a regularizing effect.

Overall, despite none of the deep learning models being able to surpass the performance of models based on HC features, regularization improved the generalization ability of deep Llearning models and was as effective as the auxiliary features for the ResNet architecture. Moreover, as regularizers did not improve the results for the classical methods, they were clearly able to reduce the generalization gap.

This study shows that HC features still have their place in modern machine learning, as they can be more robust to distribution shift and allow simple classifiers to achieve better results.

We also showed that merging all the datasets for training (OOD-MD) resulted in a performance gain of 20% when compared to using a single dataset for training (OOD-SD). This indicates that each dataset has limited information. In practice, if a larger space of possible scenarios can be covered during the acquisition process, it will result in a more diverse dataset and, consequently, in better generalization. This means that an ML-based HAR system can improve if data are recorded using a wider range of devices, users, sensor positions, and physical environments, among other possible factors of variation.

Our work has some limitations, as the choice of datasets was limited to the field of HAR and only a few regularization techniques were tested. Additional research should explore datasets from different fields and a wider range of increasingly novel regularization methods to comprehensively understand their effects on domain generalization. Future work could also investigate the use of different neural network architectures, such as transformers, or even neural architecture search [50] since it has been shown that, in some cases, the choice of model architecture may have more impact than the loss function [51]. Domain-specific regularization methods [37] were also demonstrated to have the potential to vastly improve the generalization of deep learning models. Despite that, these methods suffer from the caveat of not being directly applicable to other tasks.

6. Conclusions

This study has addressed the impact of different regularization methods on the domain generalization performance of HAR models and whether these methods can bridge the OOD performance gap between deep neural networks and HC feature-based models. Our experimental results indicate that state-of-the-art regularization methods, such as Mixup and SAM, can improve OOD generalization and reach the results of hybrid models. However, it was not enough to be on par with classical ML approaches. We conclude that the use of HC features, regularizers, and diverse training data may enable more robust HAR systems.

Overall, this study contributes to the understanding of how regularization methods can impact the Domain Generalization performance of HAR models and their potential to narrow the OOD performance gap between deep neural networks and traditional approaches. These insights will be valuable for researchers and practitioners working on HAR and related fields, helping them build more robust and generalizable models for real-world applications.

Author Contributions: Conceptualization: N.B., J.R., A.V.C. and M.B.; data curation, N.B., J.R. and M.B.; methodology, N.B., J.R. and M.B.; software, N.B. and J.R.; validation, N.B., J.R., A.V.C., F.R. and M.B.; formal analysis, N.B. and J.R.; writing—original draft preparation, N.B. and J.R.; writing—review and editing, N.B., J.R., A.V.C., F.R. and M.B.; visualization, N.B., J.R. and M.B.; supervision, A.V.C. and M.B. All authors have read and agreed to the published version of the manuscript.

Funding: This work is financed by national funds through FCT, Fundação para a Ciência e a Tecnologia, I.P., within the scope of SAIFFER project under the Eureka Eurostars program (E!114310).

Institutional Review Board Statement: Not applicable.

Informed Consent Statement: Not applicable.

Data Availability Statement: Not applicable.

Conflicts of Interest: The authors declare no conflict of interest.

References

1. Hussain, Z.; Sheng, M.; Zhang, W.E. Different approaches for human activity recognition: A survey. *arXiv* **2019**, arXiv:1906.05074.
2. Sousa Lima, W.; Souto, E.; El-Khatib, K.; Jalali, R.; Gama, J. Human activity recognition using inertial sensors in a smartphone: An overview. *Sensors* **2019**, *19*, 3213. [CrossRef] [PubMed]
3. Zhang, S.; Li, Y.; Zhang, S.; Shahabi, F.; Xia, S.; Deng, Y.; Alshurafa, N. Deep learning in human activity recognition with wearable sensors: A review on advances. *Sensors* **2022**, *22*, 1476. [CrossRef] [PubMed]
4. Qin, X.; Wang, J.; Chen, Y.; Lu, W.; Jiang, X. Domain Generalization for Activity Recognition via Adaptive Feature Fusion. *ACM Trans. Intell. Syst. Technol.* **2022**, *14*, 1–21. [CrossRef]
5. Zhou, K.; Liu, Z.; Qiao, Y.; Xiang, T.; Loy, C.C. Domain Generalization: A Survey. *IEEE Trans. Pattern Anal. Mach. Intell.* **2023**, *45*, 4396–4415. [CrossRef]
6. Soleimani, E.; Nazerfard, E. Cross-subject transfer learning in human activity recognition systems using generative adversarial networks. *Neurocomputing* **2021**, *426*, 26–34. [CrossRef]
7. Wang, J.; Zheng, V.W.; Chen, Y.; Huang, M. Deep transfer learning for cross-domain activity recognition. In Proceedings of the 3rd International Conference on Crowd Science and Engineering, Singapore, 28–31 July 2018; pp. 1–8.
8. Hoelzemann, A.; Van Laerhoven, K. Digging deeper: Towards a better understanding of transfer learning for human activity recognition. In Proceedings of the 2020 International Symposium on Wearable Computers, Virtual Event, 12–17 September 2020; pp. 50–54.
9. Ariza-Colpas, P.P.; Vicario, E.; Oviedo-Carrascal, A.I.; Butt Aziz, S.; Piñeres-Melo, M.A.; Quintero-Linero, A.; Patara, F. Human Activity Recognition Data Analysis: History, Evolutions, and New Trends. *Sensors* **2022**, *22*, 3401. [CrossRef]
10. Wang, J.; Chen, Y.; Hao, S.; Peng, X.; Hu, L. Deep learning for sensor-based activity recognition: A survey. *Pattern Recognit. Lett.* **2019**, *119*, 3–11. [CrossRef]
11. Nafea, O.; Abdul, W.; Muhammad, G.; Alsulaiman, M. Sensor-based human activity recognition with spatio-temporal deep learning. *Sensors* **2021**, *21*, 2141. [CrossRef]
12. Sagawa, S.; Raghunathan, A.; Koh, P.W.; Liang, P. An investigation of why overparameterization exacerbates spurious correlations. In Proceedings of the International Conference on Machine Learning, PMLR, Virtual Event, 13–18 July 2020; pp. 8346–8356.
13. Bento, N.; Rebelo, J.; Barandas, M.; Carreiro, A.V.; Campagner, A.; Cabitza, F.; Gamboa, H. Comparing Handcrafted Features and Deep Neural Representations for Domain Generalization in Human Activity Recognition. *Sensors* **2022**, *22*, 7324. [CrossRef]
14. Goodfellow, I.; Bengio, Y.; Courville, A. *Deep Learning*; MIT Press: Cambridge, MA, USA, 2016. Available online: http://www.deeplearningbook.org (accessed on 16 July 2023).
15. Kukačka, J.; Golkov, V.; Cremers, D. Regularization for deep learning: A taxonomy. *arXiv* **2017**, arXiv:1710.10686.
16. Chen, Y.; Xue, Y. A deep learning approach to human activity recognition based on single accelerometer. In Proceedings of the 2015 IEEE International Conference on Systems, Man, and Cybernetics, Hong Kong, China, 9–12 October 2015; pp. 1488–1492.
17. Zebin, T.; Scully, P.J.; Ozanyan, K.B. Human activity recognition with inertial sensors using a deep learning approach. In Proceedings of the 2016 IEEE Sensors, Orlando, FL, USA, 30 October–3 November 2016; pp. 1–3.
18. Lee, S.M.; Yoon, S.M.; Cho, H. Human activity recognition from accelerometer data using Convolutional Neural Network. In Proceedings of the 2017 IEEE International Conference on Big Data and Smart Computing (Bigcomp), Jeju Island, Republic of Korea, 13–16 February 2017; pp. 131–134.
19. Ferrari, A.; Micucci, D.; Mobilio, M.; Napoletano, P. Hand-crafted features vs residual networks for human activities recognition using accelerometer. In Proceedings of the 2019 IEEE 23rd International Symposium on Consumer Technologies (ISCT), Ancona, Italy, 19–21 June 2019; pp. 153–156.
20. Trabelsi, I.; Françoise, J.; Bellik, Y. Sensor-based Activity Recognition using Deep Learning: A Comparative Study. In Proceedings of the 8th International Conference on Movement and Computing, Chicago, IL, USA, 22–25 June 2022; pp. 1–8.
21. Arjovsky, M.; Bottou, L.; Gulrajani, I.; Lopez-Paz, D. Invariant risk minimization. *arXiv* **2019**, arXiv:1907.02893.
22. Baldominos, A.; Cervantes, A.; Saez, Y.; Isasi, P. A comparison of machine learning and deep learning techniques for activity recognition using mobile devices. *Sensors* **2019**, *19*, 521. [CrossRef]
23. Boyer, P.; Burns, D.; Whyne, C. Out-of-distribution detection of human activity recognition with smartwatch inertial sensors. *Sensors* **2021**, *21*, 1669. [CrossRef] [PubMed]
24. Zhang, H.; Cisse, M.; Dauphin, Y.N.; Lopez-Paz, D. mixup: Beyond empirical risk minimization. *arXiv* **2017**, arXiv:1710.09412.

25. Verma, V.; Lamb, A.; Beckham, C.; Najafi, A.; Mitliagkas, I.; Lopez-Paz, D.; Bengio, Y. Manifold mixup: Better representations by interpolating hidden states. In Proceedings of the International Conference on Machine Learning, PMLR, Long Beach, CA, USA, 10–15 June 2019; pp. 6438–6447.
26. Foret, P.; Kleiner, A.; Mobahi, H.; Neyshabur, B. Sharpness-aware minimization for efficiently improving generalization. *arXiv* **2020**, arXiv:2010.01412.
27. Liu, S.; Chen, T.; Chen, X.; Atashgahi, Z.; Yin, L.; Kou, H.; Shen, L.; Pechenizkiy, M.; Wang, Z.; Mocanu, D.C. Sparse training via boosting pruning plasticity with neuroregeneration. *Adv. Neural Inf. Process. Syst.* **2021**, *34*, 9908–9922.
28. Neyshabur, B. Implicit regularization in deep learning. *arXiv* **2017**, arXiv:1709.01953.
29. Neu, G.; Rosasco, L. Iterate averaging as regularization for stochastic gradient descent. In Proceedings of the Conference On Learning Theory, PMLR, Stockholm, Sweden, 6–9 July 2018; pp. 3222–3242.
30. Louizos, C.; Welling, M.; Kingma, D.P. Learning sparse neural networks through L_0 regularization. *arXiv* **2017**, arXiv:1712.01312.
31. Evci, U.; Gale, T.; Menick, J.; Castro, P.S.; Elsen, E. Rigging the lottery: Making all tickets winners. In Proceedings of the International Conference on Machine Learning, PMLR, Virtual, 13–18 July 2020; pp. 2943–2952.
32. Sagawa, S.; Koh, P.W.; Hashimoto, T.B.; Liang, P. Distributionally robust neural networks for group shifts: On the importance of regularization for worst-case generalization. *arXiv* **2019**, arXiv:1911.08731.
33. Krueger, D.; Caballero, E.; Jacobsen, J.H.; Zhang, A.; Binas, J.; Zhang, D.; Le Priol, R.; Courville, A. Out-of-distribution generalization via risk extrapolation (rex). In Proceedings of the International Conference on Machine Learning, PMLR, Virtual, 18–24 July 2021; pp. 5815–5826.
34. Ahuja, K.; Caballero, E.; Zhang, D.; Gagnon-Audet, J.C.; Bengio, Y.; Mitliagkas, I.; Rish, I. Invariance principle meets information bottleneck for out-of-distribution generalization. *Adv. Neural Inf. Process. Syst.* **2021**, *34*, 3438–3450.
35. Gagnon-Audet, J.C.; Ahuja, K.; Darvishi-Bayazi, M.J.; Dumas, G.; Rish, I. WOODS: Benchmarks for Out-of-Distribution Generalization in Time Series Tasks. *arXiv* **2022**, arXiv:2203.09978.
36. Rosenfeld, E.; Ravikumar, P.; Risteski, A. The risks of invariant risk minimization. *arXiv* **2020**, arXiv:2010.05761.
37. Lu, W.; Wang, J.; Chen, Y.; Pan, S.J.; Hu, C.; Qin, X. Semantic-discriminative mixup for generalizable sensor-based cross-domain activity recognition. *Proc. ACM Interact. Mob. Wearable Ubiquitous Technol.* **2022**, *6*, 1–19. [CrossRef]
38. Reiss, A.; Stricker, D. Introducing a New Benchmarked Dataset for Activity Monitoring. In *ISWC '12: Proceedings of the 2012 16th Annual International Symposium on Wearable Computers (ISWC)*; IEEE Computer Society: Washington, DC, USA, 2012; pp. 108–109.
39. Reiss, A.; Stricker, D. Creating and Benchmarking a New Dataset for Physical Activity Monitoring. In *PETRA '12, Proceedings of the 5th International Conference on PErvasive Technologies Related to Assistive Environments, Heraklion, Greece, 6–8 June 2012*; Association for Computing Machinery: New York, NY, USA, 2012. [CrossRef]
40. Shoaib, M.; Bosch, S.; Incel, O.D.; Scholten, H.; Havinga, P.J. Fusion of smartphone motion sensors for physical activity recognition. *Sensors* **2014**, *14*, 10146–10176. [CrossRef] [PubMed]
41. Leutheuser, H.; Schuldhaus, D.; Eskofier, B.M. Hierarchical, Multi-Sensor Based Classification of Daily Life Activities: Comparison with State-of-the-Art Algorithms Using a Benchmark Dataset. *PLoS ONE* **2013**, *8*, e75196. [CrossRef]
42. Banos, O.; Garcia, R.; Holgado-Terriza, J.A.; Damas, M.; Pomares, H.; Rojas, I.; Saez, A.; Villalonga, C. mHealthDroid: A Novel Framework for Agile Development of Mobile Health Applications. In Proceedings of the Ambient Assisted Living and Daily Activities, Belfast, UK, 2–5 December 2014; Pecchia, L., Chen, L.L., Nugent, C., Bravo, J., Eds.; Springer: Cham, Switzerland, 2014; pp. 91–98. [CrossRef]
43. Banos, O.; Villalonga, C.; Garcia, R.; Saez, A.; Damas, M.; Holgado-Terriza, J.A.; Lee, S.; Pomares, H.; Rojas, I. Design, implementation and validation of a novel open framework for agile development of mobile health applications. *Biomed. Eng. Online* **2015**, *14*, S6. [CrossRef]
44. Sztyler, T.; Stuckenschmidt, H. On-body Localization of Wearable Devices: An Investigation of Position-Aware Activity Recognition. In Proceedings of the 2016 IEEE International Conference on Pervasive Computing and Communications (PerCom), Sydney, NSW, Australia, 14–18 March 2016; pp. 1–9. [CrossRef]
45. Barandas, M.; Folgado, D.; Fernandes, L.; Santos, S.; Abreu, M.; Bota, P.; Liu, H.; Schultz, T.; Gamboa, H. TSFEL: Time series feature extraction library. *SoftwareX* **2020**, *11*, 100456. [CrossRef]
46. Keskar, N.S.; Mudigere, D.; Nocedal, J.; Smelyanskiy, M.; Tang, P.T.P. On large-batch training for deep learning: Generalization gap and sharp minima. *arXiv* **2016**, arXiv:1609.04836.
47. Logacjov, A.; Bach, K.; Kongsvold, A.; Bårdstu, H.B.; Mork, P.J. HARTH: A Human Activity Recognition Dataset for Machine Learning. *Sensors* **2021**, *21*, 7853. [CrossRef]
48. Hu, T.; Wang, W.; Lin, C.; Cheng, G. Regularization matters: A nonparametric perspective on overparametrized neural network. In Proceedings of the International Conference on Artificial Intelligence and Statistics, PMLR, Virtual, 18–24 July 2021; pp. 829–837.
49. He, K.; Zhang, X.; Ren, S.; Sun, J. Deep residual learning for image recognition. In Proceedings of the IEEE Conference on Computer Vision and Pattern Recognition, Las Vegas, NV, USA, 27–30 June 2016; pp. 770–778.

50. Zoph, B.; Le, Q.V. Neural architecture search with reinforcement learning. *arXiv* **2016**, arXiv:1611.01578.
51. Izmailov, P.; Kirichenko, P.; Gruver, N.; Wilson, A.G. On feature learning in the presence of spurious correlations. *Adv. Neural Inf. Process. Syst.* **2022**, *35*, 38516–38532.

Disclaimer/Publisher's Note: The statements, opinions and data contained in all publications are solely those of the individual author(s) and contributor(s) and not of MDPI and/or the editor(s). MDPI and/or the editor(s) disclaim responsibility for any injury to people or property resulting from any ideas, methods, instructions or products referred to in the content.

Article

MeshID: Few-Shot Finger Gesture Based User Identification Using Orthogonal Signal Interference

Weiling Zheng [1], Yu Zhang [2], Landu Jiang [3], Dian Zhang [4] and Tao Gu [2],*

[1] School of Computing Technologies, RMIT University, 124 La Trobe Street, Melbourne, VIC 3000, Australia; zwlcle@hotmail.com
[2] School of Computing, Macquarie University, 4 Research Park Drive, North Ryde, NSW 2109, Australia; y.zhang@mq.edu.au
[3] Base of Red Bird MPhil, HKUST(GZ) University, No.1 Du Xue Rd., Guangzhou 511458, China; landu.jiang@mail.mcgill.ca
[4] College of Computer Science and Software Engineering, Shenzhen University, 3688 Nanhai Blvd, Shenzhen 518060, China; zhangd@szu.edu.cn
* Correspondence: tao.gu@mq.edu.au

Abstract: Radio frequency (RF) technology has been applied to enable advanced behavioral sensing in human-computer interaction. Due to its device-free sensing capability and wide availability on Internet of Things devices. Enabling finger gesture-based identification with high accuracy can be challenging due to low RF signal resolution and user heterogeneity. In this paper, we propose MeshID, a novel RF-based user identification scheme that enables identification through finger gestures with high accuracy. MeshID significantly improves the sensing sensitivity on RF signal interference, and hence is able to extract subtle individual biometrics through velocity distribution profiling (VDP) features from less-distinct finger motions such as drawing digits in the air. We design an efficient few-shot model retraining framework based on first component reverse module, achieving high model robustness and performance in a complex environment. We conduct comprehensive real-world experiments and the results show that MeshID achieves a user identification accuracy of 95.17% on average in three indoor environments. The results indicate that MeshID outperforms the state-of-the-art in identification performance with less cost.

Keywords: device-free behavioral sensing; orthogonal signal interference; user identification

1. Introduction

Human Activity Recognition (HAR) and Human Behavior Recognition (HBR) technologies are integral components of Human-Computer Interaction (HCI) systems. They enable computers to interpret and respond to human actions and behaviors, enhancing the overall user experience. HAR and HBR systems [1] utilize various sensors and algorithms to analyze data such as movement patterns, gestures and physiological signals, facilitating seamless interaction between humans and computers. However, they raise security concerns about potential misuse or unauthorized access to users' data. Through robust user identification methods such as biometrics, passwords, or behavioral analysis, HCI systems can mitigate the risk of unauthorized access. Vision technology [2,3] can identify different users through physical activity characteristics captured from image frames using high resolution cameras, but it is susceptible to failure in the presence of luminous changes and obstacles within the line-of-sight (LoS) [4], thus exacerbating significant concerns regarding user privacy. In stark contrast, RF sensors (such as WiFi, RFID, and Zigbee) offer numerous advantages, including freedom from illumination constraints, reduced privacy apprehensions, equitable penetration and widespread availability on IoT devices. As a result, they are widely proposed to enable advanced device-free behavioral sensing [5,6]. Towards RF-based behavioral sensing, existing systems propose a variety of behavioral characteristics including daily activities [7–10], vital signs [11,12] and gestures [13–15]. Although these systems have demonstrated their effectiveness with fair accuracy in laboratory settings, they may still encounter real-world constraints. Extracting biometric features from daily activities usually require users consistently performing a set of pre-defined activities for long-term tracking [5]. The motion of vital signs (e.g., heartbeats [16] and respiration rate [17]) remains fragile, and identification may be prone to failure due to body movement artifacts and ambient noise. In contrast, performing gestures (e.g., finger gesture) for identification can practically mitigate the impact of motion noise and offer considerable user-friendliness [18–20]. Gesture-based interaction stands as the most common and efficient method of HCI [21–23]. Gesture recognition technology is mature and capable of achieving high accuracy. When users write in the air, they commonly use their forefinger naturally. In our experiment, we adhere to this habit by employing forefinger gestures for user identification. Enabling user identification through finger gestures with high accuracy is a non-trivial challenge, which requires addressing two key challenges.

Firstly, as the movement of the finger motion is quite small, the amplitude change in the RF reflected signal caused by finger motion could be very faint, hence it is difficult to identify minor variance among users' biometrics extracted from the limited signal variance. According to the theory of the Fresnel model [24,25], when a user moves his/her hand across the boundaries of Fresnel zones, the CSI of the signal will form a series of peaks and valleys. We regard this variance pattern as a kind of RF signal resolution. If we have more peak values in an unit area we can say that the RF signal resolution is higher. Once the motion becomes smaller, like finger motion, the number of peak values is reduced and the amplitude change in CSI diminishes much smaller, resulting in low accuracy for identification. The key question we have is how to fundamentally improve the RF signal resolution (i.e., CSI variance). Inspired by Young's double-slit interference experiment [26], we use a pair of transmitters (double-source) in the same frequency to induce the signal interference. The double-source interference produces numerous dense narrow beams. In other word, it greatly increases the number of boundaries in the same unit area compared with traditional methods. By setting up two orthogonal double-source pairs, the sensing area will be covered by a dense signal mesh, hence the signal resolution is enormously improved.

Secondly, due to user heterogeneity (e.g., different users, preferences, and surroundings), in reality, the data distribution of users' biometrics may become complex and unpredictable to fail user identification. The performance of the traditional deep learning (DL) technique relies heavily on collecting a large amount of user data as a prerequisite, especially assuming that the represented data distribution is relatively stationary without dynamic changes. Towards robust and efficient model retraining, we utilize a one-shot learning approach based on the Siamese network [27–29], with two core techniques: first component reverse (FCR) extraction and convolution block attention module (CBAM), achieving high model robustness and performance in heterogeneous scenarios (e.g., identifying unseen users). A unique velocity distribution profiling (VDP) is calculated from a double-source interference pattern, reflecting the personal motion features.

When users perform finger gestures in a complex environment, the input feature space of extracted biometrics contains both non-related features (i.e., common features shared by the same gestures of users and ambient noise) and user-specific features (i.e., personal features), but the issue is that the non-related features may strongly affect the performance of identifying users. To improve it, we design a first component reverse (FCR) extraction, inspired by principal component analysis (FCR) extraction, inspired by principal component analysis (PCA), hat removes the non-related features (i.e., first component in PCA) and helps extract user-specific features from the input feature space, boosting our CBAM-based Siamese network with a superior identification capability.

To address the above issues, we propose MeshID, a novel RF-based user identification approach leveraging signal interference for accurate finger gesture tracking. MeshID is able to significantly improve the sensing sensitivity by leveraging double-source signal interference and extracting subtle individual biometrics from less distinct finger motions. Due to the effect of enhancing CSI variance, our mesh approach can mitigate the multipath effects and contribute to resisting the interference from ambient environments. By applying an efficient CBAM-based few-shot deep learning framework with FCR extraction, MeshID achieves high model robustness and can be easily adapted to new users with complex surroundings.

The main contribution are summarized as follows:

- We present MeshID, an RF-based user identification approach based on a beam structure caused by double-source orthogonal signal interference. The system is able to detect user in-air finger motion with high accuracy, which could support for a variety of smart home/building applications especially for user identification.
- To the best of our knowledge, MeshID is the first solution that derives the unique velocity distribution profiling (VDP) features from signal patterns leveraging the novel mesh beam structure. It fully enables user identification by essentially enhancing the signal sensing sensitivity compared to traditional RF-based approaches
- We design a first component reverse (FCR) extraction method to emphasize user-specific features and remove non-related features, hence improving the identification accuracy and model capability. We propose a one-shot learning framework with CBAM in a Siamese network for model retraining robustness and efficiency.
- We evaluate MeshID in comprehensive real-world experiments. The results demonstrate that MeshID is able to achieve an average identification accuracy of 95.17% on average by performing single finger gestures in three indoor environments.

The rest of the paper is organized as follows. Section 2 discusses the preliminary knowledge of the method. Section 3 presents the design of MeshID. Section 4 demonstrates the comprehensive evaluation results. Section 5 shows the key discussions. Section 6 reviews the related works. Section 7 concludes the paper and discusses future works.

2. Preliminary

In this section, we introduce the fundamental concept of Channel State Information (CSI) and then explore the double-source interference phenomenon for finger gesture-based user identification.

CSI is a fine-grained physical layer that depicts how RF signals propagate between a transmitter and a receiver [30,31]. It captures the slight change in the surrounding objects in both the time domain and spatial domains. The CSI channel H is modeled by $\mathcal{Y}(f,t) = H(f,t)\mathcal{X}(f,t) + \mathbb{N}$, where $\mathcal{Y}$ is the received signal, $\mathcal{X}$ is the pre-defined modulated signal and $\mathbb{N}$ is the noise vector.

Empirical Study of double-source interference. In tradition, for a pair of transmitters and receivers, as shown in Figure 1a, the signal variance pattern caused by the reflection of target activity is usually identified as Fresnel zones. Small finger motions in the same zone usually have a small impact on the signal variance, while user activity across different zones causes a large signal variance. We may regard these zones as a kind of "sensing sensitivity". The zone number can be a measure of sensitivity to sensing. For example, in a 1×1 m area (without losing generality, the frequency is set as 5.76 GHz), roughly we may have only seven Fresnel zones.

To increase the boundaries of sensing, our intuition is to apply RF signal interference. Inspired by Young's double-slit interference experiment [26], we use a pair of transmitters with same frequency to induce the signal interference, resulting in a stripe pattern in a parallel way. The simulation results of signal reflection are depicted in Figure 1b. The reason why we can see such a fringes pattern is that multiple sources signals can interfere with each other constructively, where the amplitude of combined signal is greater than the individual one, while interference destructively where the amplitude of the combined signal is smaller than the individual one or same with the original. We call the above phenomenon double-source interference. Double-source interference obeys the Huygens-Fresnel Principle [32], hence the Fresnel zones still exist under this double-source setting, as shown in Figure 1d.

The double-source fringes pattern and the Fresnel zones are overlapping with each other, the Fresnel zones will be divided into several "mesh" cells. Within this dense mesh, if use activity is crosses the cell borders, it also causes a larger signal variance. That is, in the same 1×1 m area, we may roughly have more than 83 separated areas in total. The boundaries of "mesh" with fringes are notably increased by interference compared to the Fresnel zones, which indicates that the sensing sensitivity on RF signal interference increases. The interference of multiple waves are generally regarded as a not-so-trivial negative effect leading to unpredictable signal patterns and poor signal quality in wireless sensing, many studies aim to mitigate interference for better accuracy. However, our objective is to utilize such negative effect in user identification and change the effect from "negative" to "positive".

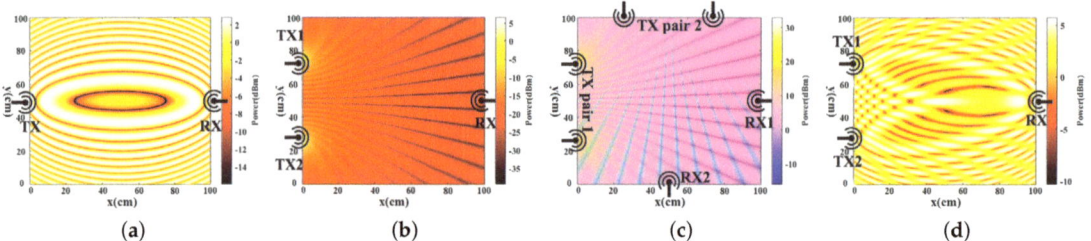

Figure 1. The simulation heatmaps: (**a**) Fresnel zones from single source; (**b**) Interference patterns from double sources; (**c**) Mesh of the formed by orthogonal signal interference; (**d**) Fusion of interference and Fresnel zone from double sources.

User identification based on orthogonal double-source interference: Since that a finger gesture may lead to motions in different directions (e.g., drawing a digit "0"), one transmitter pair may only be sensitive to the motions in the perpendicular direction to the transmitters, hence it may lose useful motion features. To address this, we propose a new setup of orthogonal antenna pairs to enable a dense interference pattern (i.e., dense mesh) from both vertical and horizontal directions. Our basic idea is that if we deploy another pair of transmitters with the same distance, but in another direction (e.g., parallel to the x-axis) and with a slight frequency difference, such transmitters also cause a stripe result but in a vertical way. The amplitude of the received signal appears like a dense mesh with two orthogonal interference transmitter pairs, as shown in Figure 1c. We name such pairs of transmitters as orthogonal antenna pairs. To observe the mesh pattern of orthogonal double-source interference clearly, the Fresnel zone is omitted from this figure. Consequently, we are able to achieve a higher sensing sensitivity on RF signal compared to traditional methods.

The received CSI comprises a mixture of signals, including the line-of-sight component, double-source component, and others. We utilize the Complementary Ensemble Empirical Mode Decomposition (CEEMD) to separate the double-source interference component from the CSI data. With this feature, we propose a novel VDP feature to capture the fine-grained finger motions of users within the sensing area. The VDP encapsulates the motion change pattern of the user and the corresponding potential biometric features such as the motion speed. Since users have diversity motion habits and behavioral habits (e.g., drawing digit "0" clockwise or anticlockwise, different pauses and speeds when performing gesture), even when executing the same gesture, the moving finger interacts with RF signals differently, resulting in distinct patterns. As shown in Figure 2, Figure 2a–c are the CSI variances of three different users drawing the same digit "0" based on their own writing habits. Figure 2d–f are the corresponding VDPs that have different velocity patterns. Therefore, we can depict users' behavioral characteristics by leveraging biometric feature extraction methods and bring the opportunity for user authentication.

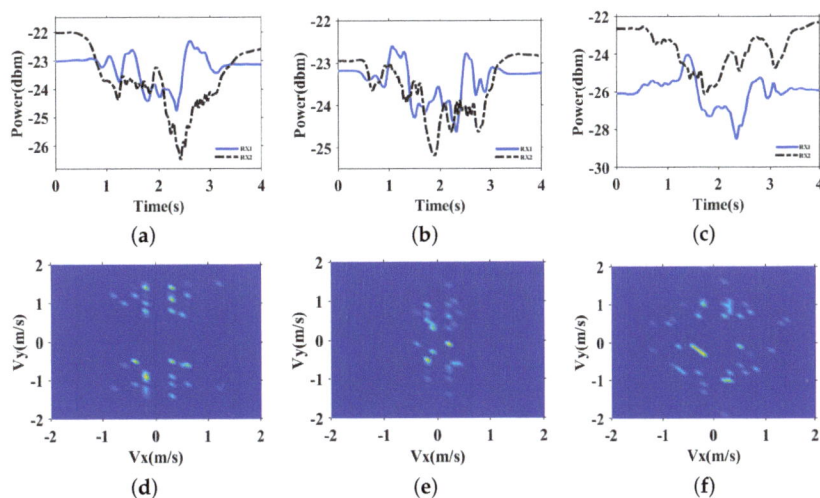

Figure 2. The VDP of different users when writing digit 0. (**a**) CSI of user 1. (**b**) CSI of user 2. (**c**) CSI of user 3. (**d**) VDP of user 1. (**e**) VDP of user 2. (**f**) VDP of user 3.

3. Methodology

3.1. Framework

As shown in Figure 3, our design consists of four components: (1) data collection from orthogonal double-source interference, (2) noise removal and gesture recognition, (3) FCR extraction analysis to extract user-specified features, (4) data transform and (5) CBAM-based few-shot learning for identification.

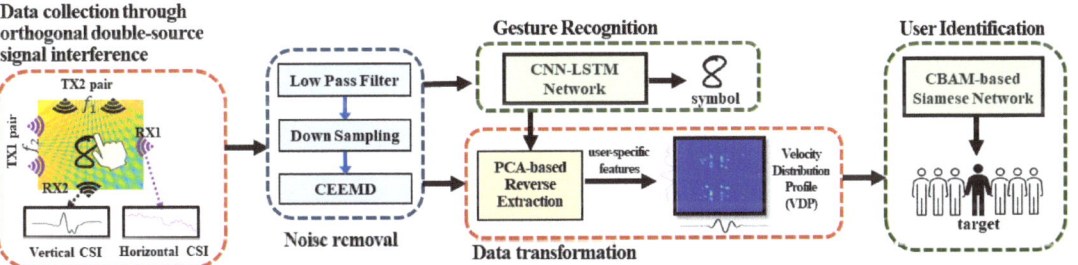

Figure 3. MeshID design overview.

Firstly, we introduce our design of the orthogonal signal interference. The signal interference pattern is like a dense mesh. The user is able to perform finger gestures within this area. Under this setting, CSI information is collected from both horizontal and vertical transmitters. Secondly, a low pass filter algorithm is used to remove the high-frequency noise and a down sampling algorithm is used to reduce the data size in order to effectively process the data. Then, we use Complementary Ensemble Empirical Mode Decomposition (CEEMD) [33] on the data to obtain ensemble Intrinsic Mode Function (IMF) which is related to the interference pattern. Thirdly, with the IMF data from the CSI series, we utilize a CNN-based LSTM method to recognize the user's gesture. Then, we leverage FCR extraction to remove the effect of the shared component, so that it can remove the correlated non-related features among different users and leave the user-specified features of the user. Fourthly, we estimate the user instantaneous moving speed according to the fringe-based variance pattern, then we generate the VDP for following user identification. Finally, we can identify users by leveraging the CNN-based Siamese Network.

3.2. Double-Source Interference

In free-space scenarios, where two RF waves with the same frequency f travel along two different paths to converge at the same destination point, the difference of two path lengths is equivalent to a multiple of wavelength λ, which is known as constructive interference [34] Similarly, if the two waves are out of phase by half of the wavelength, the result is destructive interference. In this case, the combined signals produce several fringes, where the high points of fringes are called the crests, and the low points of the fringes are called the troughs.

Given this phenomenon, we may deliberately utilize the aforementioned double-source signals, each possessing the same radio frequency, to create a combined signal H_{sum}, which can be defined as follow:

$$H_{sum}(f,t) = H_1(f,t,r_1) + H_2(f,t,r_2) \tag{1}$$

where H_1 is the received signal travel with distance r_1 from the first antenna, and H_2 is from the second antenna and travels with distance r_2.

The width of the neighbouring fringes can be calculated by three primary major parameters: the distance between the transmitter node pairs, the wavelength of the radio wave, and the distance from the transmitter pairs. The node distance between nodes and the wavelength of the radio wave are the known values. The impact range of finger motion usually is small compared to the sensing area, hence the distance from the transmitter pair can be regarded as the middle of the sensing area. The fringes are symmetrical starting from the central line (thick red line in the figure) of the receiver antenna, we refer to the upper fringes as $fringe_m^u$ and bottom fringes as $fringe_m^b$, m is the number of a specific fringe number, as shown in Figure 4. The width $\Delta dist$ between fringe $fringe_m^u$ and $fringe_{m+1}^u$ (or between fringe $fringe_m^b$ and $fringe_{m+1}^b$) at a position whose distance to the y axis is l, can be calculated by

$$\Delta dist = m\lambda \sqrt{4 + \frac{l^2}{d^2 - m^2\lambda^2}} \tag{2}$$

where d is the node distance between the transmitter antennas pairs, λ is the wavelength.

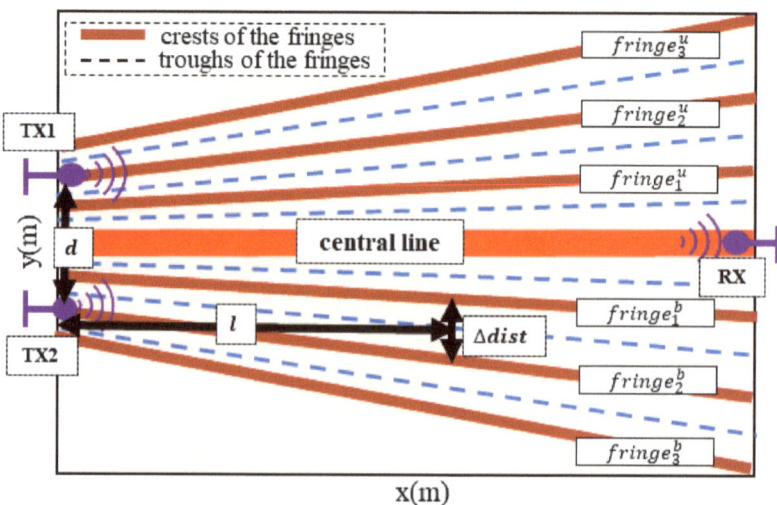

Figure 4. Illustration of double-source interference.

3.3. Orthogonal Signal Interference

In previous instances of interference fringes, employing a pair of antennas in transmission resulted in fringes appearing predominantly along one direction (e.g., horizontal direction in Figure 1b). Introducing another pair of antennas orthogonally (e.g., vertically, as shown in Figure 1c), operating at slightly different frequencies, produces interference fringes along a different axis (vertical direction). Therefore, the combined interference signal fringes appear like a mesh. We can effectively capture the fine-grained finger motion from any direction, thereby enhancing signal resolution.

We may have CSI data from both vertical and horizontal antenna pairs at the same time. The orthogonal RF signals can be represented as below:

$$\mathcal{M}(RX_1, RX_2) = \begin{bmatrix} H_{RX_1}(f,t) \\ H_{RX_2}(f,t) \end{bmatrix} \quad (3)$$

where H_{RX_1} is the received signal at receiver RX1, and H_{RX_2} is the received signal at receiver RX2.

3.4. Noise Removal and Gesture Recognition

The raw wireless signals are inherently noisy due to the multipath effect. In order to effectively identify the human, we should remove noise first, then perform the data transformation for enhanced analysis.

3.4.1. Low Pass Filtering

The CSI variance caused by human motion has a relatively low frequency specified as f_l, while the high-frequency data usually contain environment noise. Therefore, we utilize a a low-pass FIR filter to remove the environmental noise. The cut-off angular frequency used in the low pass filter is calculated as $\omega_c = 2\pi \frac{f_l}{f_s}$.

3.4.2. Down Sampling

To expedite calculations and enhance processing speed, we initially employ a downsampling (interpolation) method to stretch the input CSI matrix. Assuming the original sampling rate is f_s, and the data length of the original data is m. After the re-sampling, the data length is n with the new sampling rate $\tilde{f}_s = f_s * \frac{m}{n}$. The CSI matrix after re-sampling can be rewritten as $\widetilde{\mathcal{M}}(\tilde{f}_s) = F(\mathcal{M}, f_s)$ where $\mathcal{M}$ is the input orthogonal RF signal.

3.4.3. Interference Pattern Extraction

Given that interference theory remains constrained by the principles of the Fresnel Zone, the interference pattern can be covered by the Huygens-Fresnel Principle. Figure 5 shows an example of how both the Fresnel zone and environment noise can detrimentally impact the interference pattern. We place the receiver antenna on the right side of the drawing area. The transmitter is placed on the left side. Horizontal zones/fringes will exist with a single sourcesetting/double source setting, as shown in Figure 1a,b. While a user draws a straight line from center top to center bottom and just crosses the Line-of-Sight (LoS), the amplitude variance with a single source setup is shown in Figure 5a, we identify the start point and the endpoints with a red dashed line. The figure clearly shows that only one peak value is shown within this range. For purposes of a fair comparison, we repeat the same finger motion under a double-source interference setup. We also identify the start point and end point with red dashed lines. The amplitude variance is shown in Figure 5b. The result indicates that three peak values are distributed in the same range compared with a single source. We utilize a green line to separate the interference fringes based on the peaks. On the basis of Equation (2), we could estimate the width of the fringes. Theoretically, motion features based on personal velocity can be derived if the boundaries of the fringes are accurately identified. Therefore, there is a pressing need for an efficient method to detect these fringe boundaries with precision.

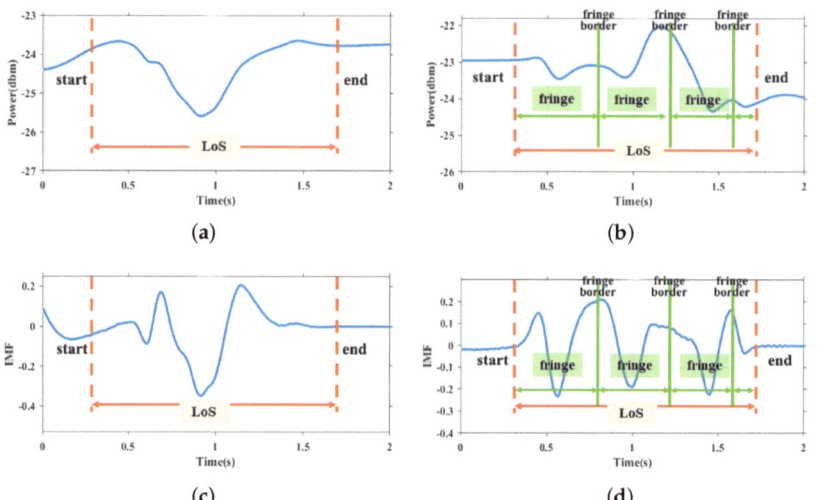

Figure 5. Comparison of Fresnel Zones patterns and interference patterns when user's finger moves perpendicular to the LoS. (**a**) CSI of 1st Fresnel zone with single source. (**b**) CSI of Interference fringes with double sources. (**c**) IMF of 1st Fresnel zones with single source. (**d**) IMF of Interference fringes with double sources.

We apply Complementary Ensemble Empirical Mode Decomposition (CEEMD) on the CSI matrix $\widetilde{M}$ to detect the fringe boundaries.

$$\widetilde{M}(t) = \sum_{i=1}^{N} f(IMF_i(t), \varepsilon_i) + res_N(t) \quad (4)$$

where $f(IMF_i)$ is the ith Intrinsic Mode Function (IMF) component, ε_i is the residue of added white noises, and $res_N(t)$ is the residue.

To simplify the pattern extraction, we can roughly divide the received CSI signals into four layers: noises, interference variances, Fresnel zone variances and trends related to with the distance of transmitter and target objects. Hence, $N = 3$ in our system, and the residue of the decomposition represented the trend of the CSI series. We will use IMF_2 as the input of the following calculation. The results of CEEMD processing are shown in Figure 5c and 5d respectively.

3.5. Finger Gesture Recognition

In this subsection, we describe the gesture recognition methods to recognize users' finger gestures. Unlike some existing work, they require the collection of lots of gesture samples for training the gesture recognition network to meet the requirement for real-world applications that may have few gesture samples for first-time usage. Also, the recognition system is required to offer a short response for real-time applications. We employ Convolutional Neural Network (CNN) based Long Short-term Memory (LSTM) for finger gesture recognition. In our system, only a few gesture samples of a user are needed for re-training the network. The gesture recognition network contains four 1d CNN layers to suppress the data size to shorten the training time. Next, the features are fed into a two-layer LSTM. Based on the known gesture, we use a general Principal Component Analysis (PCA) algorithm to analyze the CSI value for removing the common features of the same gesture.

3.6. First Component Reverse Extraction

Principal Component Analysis (PCA) has been widely used for signal denoising to enhance classification performance [35,36]. Most of the existing works usually keep the top three principal components (especially the first component) of the data and ignore the rest, as PCA ranks the principal components in descending order in terms of their variance. The design of a first component reverse (FCR) extraction method is shown in Figure 6.

In our user identification scenario, the first component derived from the received CSI data of the same finger motion may majorly contain the common features shared among different users when they are performing the same gestures. As shown in Figure 7a–c, when three users perform the same gesture digit "8", the major CSI variance from the same gesture is quite similar. The major CSI variance is caused by the finger motion. We recognize the first component of PCA from these similar variances as the common feature. It is a non-related feature for user identification. Instead, our aim is to extract user-specified features. After remove the non-related features, the rest variances are influenced by diverse body structures, environmental factors, and other sources of noise. Although non-related features may aid in recognize the defined finger gestures across different people, they can adversely impact user identification. We aspire to incorporate more user-specific features for accurate identification. As shown in Figure 7, the CSI result of three different users are Figure 7a, 7b and 7c, respectively. All three users are writing the digit "8" freely. Figure 7d–f are the corresponding CSI by applying FCR extraction. The first component of different users is almost similar to each other. Upon removing these similar components, user-specific features become more pronounced. Therefore, in our design, we remove the principal component to effectively reduce the impact of such non-related features, and extract the characteristic of personal information (i.e., user-specified features) accordingly. This methodology aids in identifying different users by filtering out interference from non-related features without sacrificing personalized information.

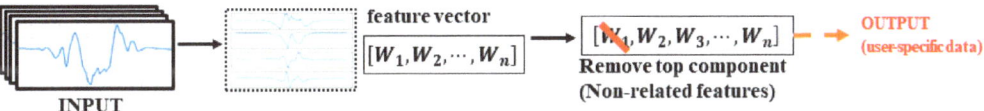

Figure 6. Architecture of FCR extraction.

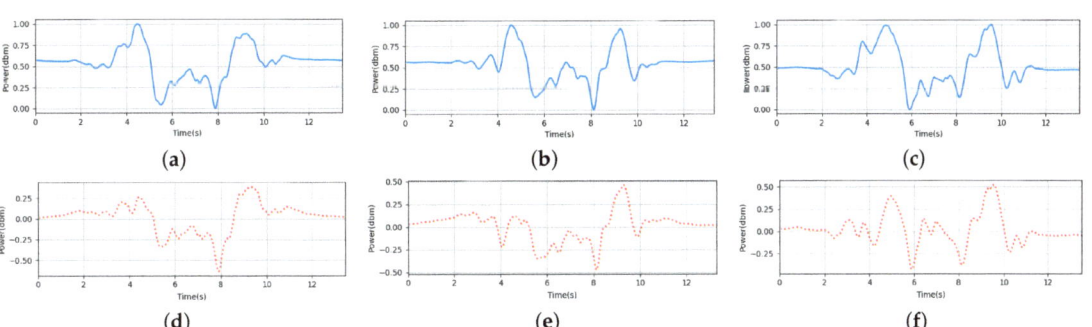

Figure 7. When three different users perform same gesture (digit "8"), the comparison of CSI amplitude before and after FCR extraction. (**a**) User 1 (before FCR extraction). (**b**) User 2 (before FCR extraction). (**c**) User 3 (before FCR extraction). (**d**) User 1 (after FCR extraction). (**e**) User 2 (after FCR extraction). (**f**) User 3 (after FCR extraction).

Applying the FCR extraction algorithm, we will have a coefficient matrix $[\mathcal{W}_1, \mathcal{W}_2, \ldots, \mathcal{W}_n]$. We remove the first component to enhance the performance of one-shot learning for user identification

$$\mathcal{X} = \mathcal{X}_{IMF} - \mathcal{X}_{IMF} \times \mathcal{W}_1 \tag{5}$$

where $\mathcal{W}_1$ is the coefficient of first component and $\mathcal{X}_{IMF}$ is the IMF component of CSI.

3.7. Interference-Based Velocity Distribution Profiling

In observational experiments, we notice that users exhibit brief pause intervals during gesture execution, causing fluctuations in instance motion velocity. Each user possesses a unique instance velocity profile due to his/her inherent writing behavior. We calculate the user velocity profile based on the x_{IMF_n} by analyzing the interference pattern. It's been observed that when a target user moves his/her finger across from the boundary of an interference fringe to the middle part of the neighbour interference fringe, a maximum value A_i^{max} and a minimum value A_i^{min} of the CSI amplitude occur in corresponding CSI time series, as shown in Figure 5d. For two neighboring extreme values, we derive the width of the fringe $\Delta dist$ which is calculated by Equation (2). Hence, the instantaneous velocity is

$$v_{(inst)}(t) = \frac{\Delta dist}{\Delta t} \tag{6}$$

where Δt is decided by the sampling rate.

For each receiver, we segment the signal with a small time window win_l, and calculate the velocity profile by the interference fringes. The two velocity profiles from two receivers RX1 and RX2 can be identified as the horizontal velocity profile $V(v_x)$ and vertical velocity profile $V(v_y)$, because in the MeshID system, receivers are perpendicular to each other. For each time window, we search for the extreme values of the one-dimensional IMF time series to identify the corresponding fringe boundaries, and use them to derive the instantaneous velocity $V(v_x, v_y)$. A two-dimensional VDP matrix $VDP[M \times N]$ with size $[M \times N]$ by quantizing the discrete instantaneous velocity from horizontal and vertical directions. The VDP combines the velocity distribution of the two velocity profiles. VDP can effectively extracts the real-world features of the users since it reflects real-world movements. On one hand, when a user performs the same gestures at different movements with fixed antennas but different movements, the corresponding VDPs are similar to each other. On the other hand, as Figure 2 shows, VDP is different from user to user.

3.8. CBAM-Based Siamese Network

In order to facilitate efficient model retraining for addressing user heterogeneity, we leverage the Siamese neural network for few-shot learning in user identification. It aims to train only few data from unseen users, requiring less model retraining preparation to achieve satisfactory performance. Additionally, the few-shot learning takes advantage from features of previously learned VDP samples, showcasing its capability to identify new individuals with reduced effort. As shown in Figure 8, our proposed model comprises two twin networks that share same parameters and weights.

The purpose of the Siamese network is to minimize the pairwise distances between personal drawing features from the same people and maximize the distances of features from different people. The process can be illustrated as follows:

$$\delta(\mathcal{X}_{(i)}, \mathcal{X}_{(j)}) = \begin{cases} \min \|\mathcal{F}(\mathcal{X}_{(i)}) - \mathcal{F}(\mathcal{X}_{(j)})\|, & U_{(i)} = U_{(j)} \\ \max \|\mathcal{F}(\mathcal{X}_{(i)}) - \mathcal{F}(\mathcal{X}_{(j)})\|, & U_{(i)} \neq U_{(j)} \end{cases} \tag{7}$$

where $\mathcal{F}$ is a non-linear transform based on a twin network.

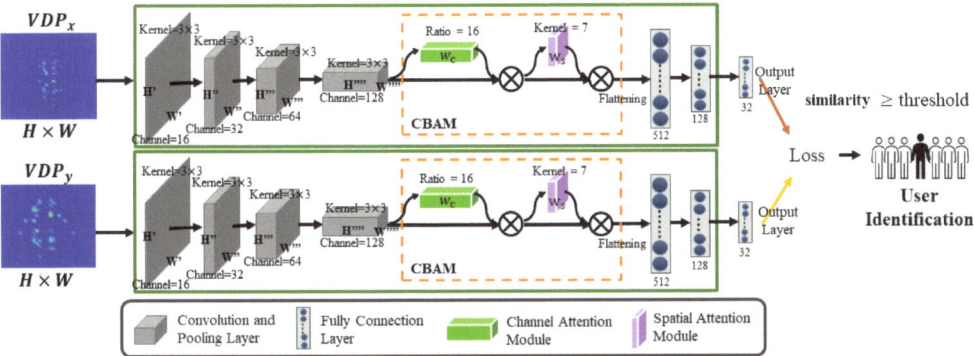

Figure 8. Few-shot learning overview.

Particularly, we adopt the Convolutional Block Attention Module (CBAM) [37] in our Siamese network to emphasize extracting informative features along both the channel and spatial axis. CBAM is an effective attention module for most feed-forward CNN networks. For a given feature map Y_{s1}, CBAM calculates its channel and spatial weight matrix sequentially and then refines the feature map based on these two weight matrices. Specifically, the Siamese network uses the two same networks to extract features, examine the similarity of the input anchor VDP image and store the VDP image for user identification. Therefore, though the CBAM-based Siamese network with a learned VDP feature of an individual, MeshID can authenticate users with a high identification accuracy.

In the training phase, the VDP matrix/spectrogram with size $[M \times N]$ triplets (anchor VDP image, negative VDP image, positive VDP image) are fed into the CBAM-based Siamese model. The basic idea is that the distance between drawing patterns of the same person on the same character should be smaller than that between the drawing patterns of different people. We take combined VDP (i.e., horizontal antenna pair RX1 and vertical antenna pair RX2) as inputs to each stream (sub-network). The architecture of each sub-network is mainly divided into three modules.

More specifically, the input VDP matrix/spectrogram will first be processed by a batch normalization $\mathcal{B}$. Then, the model learns the features from four convolutional layers $\mathcal{C}(n_c, k_s)$. Here n_c is the number of feature maps and k_s is the kernel size. We have used a 3×3 kernel size in the convolutional layer with a stride of 1. The stride determines how many pixels the filter shifts. The depth (number of channels) of the features in each convolutional layer is shown in Figure 8. The pooling layer is used to reduce the size of the features. We have employed max-pooling which retains only the maximum value within a pool. Afterward, the flattening layer is applied, which involves transforming the two-dimensional matrix into a column matrix (vector). This column vector is then passed to the fully connected layer. We will use a non-linear ReLU function for activation.

The CNN architecture of sub-network is abbreviated as $Y_{s1} = \mathcal{F}_1(\mathcal{X})$: $\mathcal{B} \to \mathcal{C}(16) \to \mathcal{P} \to \mathcal{B} \to \mathcal{C}(32) \to \mathcal{P} \to \mathcal{B} \to \mathcal{C}(64) \to \mathcal{P} \to \mathcal{B} \to \mathcal{C}(128)$, where $\mathcal{P}$ is the max-pooling layer. We deploy CBAM in the second module $\mathcal{F}_2$, which adaptively refines the feature map, defined as $Y_{s2} = \mathcal{F}_2(Y_{s1})$. Finally, three fully connected layers $\mathcal{F}_3$ are applied to encode the output of CBAM as the feature vector $\mathcal{X}$. A person can be identified by calculating the pairwise distance with the template in the database.

4. Evaluation

In this section, we begin by outlining our experimental equipment, setup and system workflow. Subsequently, we present our experimental results and comparison with other algorithms. Finally, we conduct an assessment of each component of MeshID.

4.1. Experiment Setup

In principle, our approach is fundamentally applicable to a wide range of RF-based devices, e.g., WiFi, Universal Software Radio Peripheral (USRP), Bluetooth and RFID. To flexibly design carrier signals and baseband waveforms, we utilize USRP devices (Ettus Research, Austin, United States) for system implementation. Specifically, we have two USRP N210 devices (RX1 and RX2), each equipped with an omnidirectional receiver antenna. At the same time, we have two NI USRP 2953R devices with two omnidirectional antenna transmitter pairs. The transmitter sends a simple sinusoidal waveform at a fixed frequency. The transmitters are connected to the PXI Chassis(Ettus Research, Austin, United States). All of the devices are synchronized to the CDA-2990 Clock Distribution Device(Ettus Research, Austin, United States).

To capture two-dimensional CSI variance, we set the two transmitter pairs orthogonally. When performing gestures, the drawing area typically aligns parallel to the user's orientation. Therefore, the optimal orientation is that placing two transmitter pairs and two receivers orthogonally in front of the user so that the finger could cross more double-source interference fringes. Utilizing only one transmitter pair would render the system sensitive primarily to motions perpendicular to the transmitter pair.

Specifically, the transmitter pair 1 (TX_1) operates at 5.76 GHz is placed horizontally positioned with a 50 cm apart, while the transmitter pair 2 (TX2) operates at 5.72 GHz is placed vertically positioned, also with a 50 cm apart, as shown in Figure 9a. The optimal node distance of a transmitter pair is 50 cm, which will be discussed in later Section 4.2. The distances between TX1 and RX1, TX2 and RX2 are both 1 m. Since the frequency of the TX1 pair and TX2 pair have a slight difference, the transmitted signals only interfere with each other inside each transmitter pair. The devices are shown in Figure 9b. Both receivers and transmitters are deployed on a customized shelve with an orthogonal setting. Our algorithms are performed in a DELL server with an i7-6850K 3.6 GHz Processor and 64 G RAM. The operation system of the server is Windows 10 with 64-bit.

We thoroughly evaluate our prototype across three different indoor environments: a standard office (3.4 m × 3.8 m), a meeting room (5 m × 7 m), and a hallway (2.8 m × 35 m), as shown in Figure 10. During data collection, ambient individuals within these environments were not required to vacate the premises.

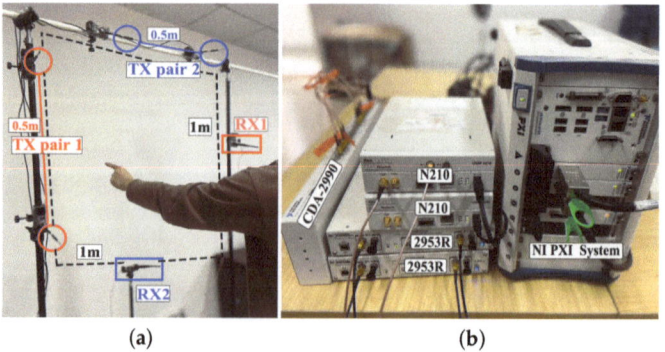

Figure 9. Experimental setup. (**a**) Setting. (**b**) Devices.

Our experiments involve the collection of two datasets. During data collection, users are instructed to freely perform in-air gestures at their desired speed and size. The first dataset contains 2268 gesture samples (6 users × 3 scenes × 6 gestures × 21 samples). We select 6 in-air finger gestures from three categories: digits, letters, and symbols for user identification, as they are the most commonly used in passwords. Specifically, we use "3" and "6" for digits, "d" and "M" for letters, and "@" and "&" for symbols. The second dataset serves to evaluate the system. The data is collected from another 18 users. This dataset comprises a total of 3600 gestures samples (18 users × 10 gestures × 20 samples). Users are instructed to perform 10 Arabic numerals "0–9" to further analyze user-to-user feature variations. Digits are a basic and familiar form of characters for most people. They are universally understood and accepted across different languages and cultures, making them accessible to a wide range of users without language barriers. In total, our dataset comprises 24 users, including 18 males and 6 females with heights ranging from 155 cm to 185 cm and weights ranging from 45 kg to 80 kg. Standard cross-validation techniques are employed in our evaluation process.

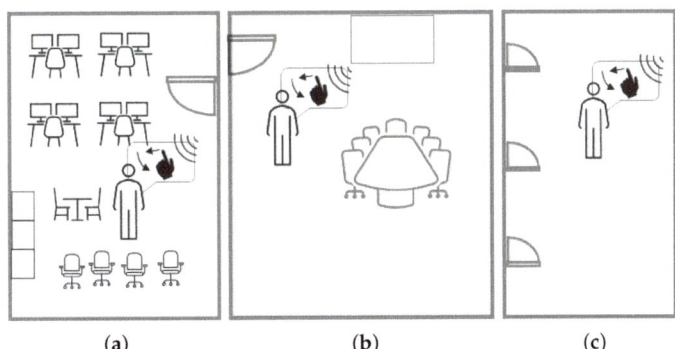

Figure 10. Three indoor environments. (**a**) Office. (**b**) Meeting room. (**c**) Hallway.

4.2. Impact of Node Distance between Transmitters

In MeshID, the resolution of the signal is determined by the density of the mesh, which in turn relies on the node distance between each transmitter pair. This subsection explores the impact of such node distance on the mesh pattern. Since the interference setting mandates the node distance to be a multiple of the wavelength (approximately 5 cm in our setup), within the 100 cm × 100 cm area, we varied the node distance from $1 \times \lambda$ to $10 \times \lambda$, incremented by 5λ. The theoretical interference patterns are illustrated in Figure 11a–c. The findings indicate a direct proportional relationship between the node distance of each transmission pair and its density when the radio frequency remains constant. Opting for a denser mesh pattern necessitates a larger node distance, and vice versa.

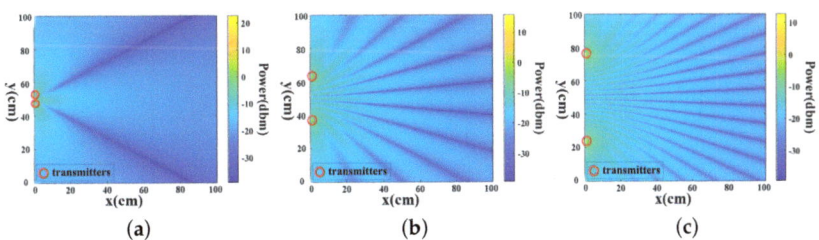

Figure 11. Double–source interference pattern at three different distance of a TX pair. (**a**) Distance of a TX pair is λ. (**b**) Distance of a TX pair is $5 \times \lambda$. (**c**) Distance of a TX pair is $10 \times \lambda$.

We employ the first dateset for the subsequent two subsections, where six users are tasked with performing six gestures across three categories. To be specific, "3" and "6" for digits, "d" and "M" for digits, and "@" and "&" for symbols. Performance evaluation for gesture recognition and user identification is conducted using the standard cross-validation method. Few samples (4 in our model) of the one user are used for retraining and the rest of the samples from this users are used for testing. The data from another five users are only used for pretraining).

The experiments results for three different transmitter node distances (e.g., 30 cm, 50 cm and 70 cm) are shown in Figures 12 and 13. The results demonstrate that using 5 cm node distance setting achieve the highest average accuracy for both gesture recognition accuracy and user identification. In Figure 12, with the node distance set to 50 cm, the gesture recognition accuracy of three types of symbols are 93.75%, 95.83% and 93.23%, respectively. The average accuracy of gesture recognition across the three different transmitter node distances is 88.19%, 94.27% and 88.72%, respectively. Figure 13 illustrates the impact of employing three different node distances on user identification. The average accuracy for user identification in different settings is 91.12%, 95% and 95.13%, respectively. The system may fail in gesture recognition, but it still possible to identify user successfully. This is attributed to the fact that although the FCR extraction is trained based on the result of gesture recognition, but we transformed CSI time series to several components, only few top components (the first component in our FCR design) is removed. Consequently, user-specified features for user identification may still be retained. Hence, the user identification could has better performance than gesture recognition.

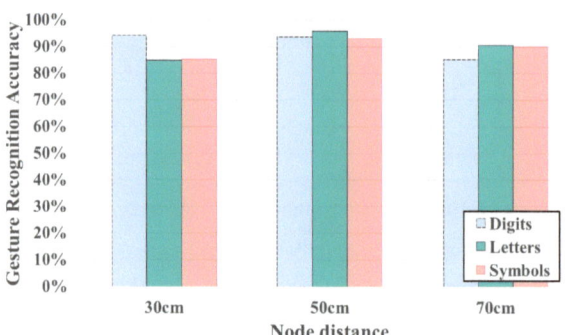

Figure 12. Gesture recognition accuracy under different distance of a transmitter pair.

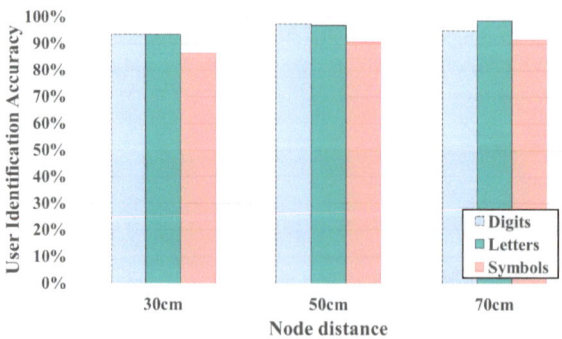

Figure 13. User identification accuracy under different distance of a transmitter pair.

Additionally, the results indicate minor differences between different node distances in MeshID. Hence, the antennas placement in MeshID is flexible to accommodate various real-world scenarios with practicability. Considering our scenario is to identify the users through finger gestures, we need to make sure that the mesh cell size is able to match the finger width (approximately 2–3 cm) of an average person. Therefore, we default to setting node distance as $10 \times \lambda$ ($\lambda = 5$ cm) by default in our subsequent experiments.

4.3. Performance of MeshID

Table 1 presents the results of gesture recognition and user identification across three different environments, spanning from a compact office space to an expansive hallway. These assessments utilize the first dataset and adhere to the standard cross-validation procedure. Results reveal that the gesture recognition accuracy of MeshID achieves 94.27% in an office, 94.1% in a meeting room and 93.4% in a hallway. Notably, since the differences of the results across these environments are minor, MeshID is able to achieve a fair robustness in adapting to diverse indoor settings. MeshID attains identification accuracy of 94.3% in an office, 95.03% in a meeting room and 97.1% in a hallway, respectively. Notably, the highest identification accuracy is recorded in the hallway environment, attributed to the minimal presence of multipath effects. On average, MeshID system achieves an identification accuracy of 95.48%. According to the results, the identification accuracy on average of three gesture categories is 95.25% for digits, 95.52% for letters, and 95.67% for symbols. It can be observed that all types of gestures perform well in three indoor environments. Delving into specific gestures, the average accuracy of gesture "3" is 93.3%, gesture "6" is 97.2%, gesture "d" is 97.5%, gesture "M" is 93.5%, gesture "@" is 99.1% and gesture "&" is 92.2%. Digits are universally understood and accepted across different languages and cultures, making them accessible to a wide range of users without language barriers. Therefore, to make the evaluation more general, we choose digits as the identification gestures for the following experiments.

Table 1. MeshID performance in different environments.

Recognition	Gesture	Office	Meeting Room	Hallway
Gesture Recognition	Digits	93.8%	95.8%	93.2%
	Letters	96.9%	92.2%	93.2%
	Symbols	93.2%	94.8%	92.2%
User Identification	Digits	92.9%	97.4%	95.5%
	Letters	90.9%	97%	98.8%
	Symbols	99.2%	90.8%	97.1%

4.4. Performance of User Identification

To further investigate the performance of our user identification system, we employ a larger second dataset for subsequent evaluations. 18 users are asked to perform the same finger gestures 20 times. Data from the remaining 6 users are only used for intrude detection in later. Other 12 users are evaluated using the standard cross-validation method. Without loss of generality, we test digital numbers ("0" to "9") 20 times for each user. Consistency was maintained in the stroke order for each digit, with users instructed to write the digits in a clockwise manner (e.g., writing digit "0" clockwise). The user identification results are shown in Figure 14. The average identification accuracy across the 10 gestures stands at 93.19%. With the exception of digit "1", the accuracy of all other digits exceeds 84%. Specifically, four digits demonstrate outstanding accuracy: 98.9% for digit "0", 96.3% for digit "4", 96.8% for digit "5", and 98.4% for digit "8". Conversely, the identification accuracy for finger gesture "1" and "9" is comparatively lower than others. This discrepancy could be attributed to the simplicity of the strokes for "1" and "9", where different users might not exhibit significant variations in their finger gestures. Conversely, for the remaining finger gestures, we observe a notably high identification accuracy.

To further evaluate the security level of the system, we conducted an intrusion detection scenario where we enlist 6 unseen users to act as spoofers. These 6 spoofers are replicate the gestures of target users in an attempt to bypass the user identification process. We employ the true negative value to measures the probability that MeshID correctly identifies an unauthorized user. The results, presented in Table 2, reveal a detection accuracy exceeding 80% for all 6 spoofers, with four of the six users achieving approximately 90%. While our primary focus lies on user identification rather than binary classification, the framework of MeshID could still achieves an overall detection accuracy of 91.7% with no prior information of the testing environments.

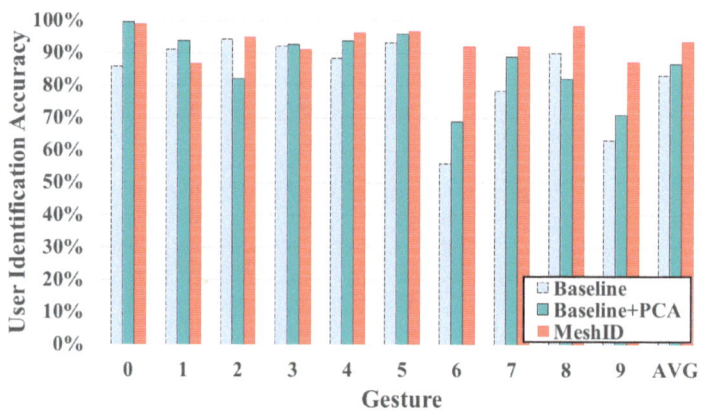

Figure 14. Comparison of key components.

Table 2. Detection accuracy of spoofer.

Spoofer	s1	s2	s3	s4	s5	s6	Avg
Detection accuracy	95%	95%	85%	100%	80%	95%	91.7%

4.5. Impact of Signal Interference

As mentioned in Section 3, we have learned in theory why signal interference can improve the sensing sensitivity. To further understand how the signal interference affects the performance, we conducted a comparative analysis of experimental results using spectrograms based on our double-source interference setting and the traditional single-source setup. We transform the CSI signals obtained from both settings to Continuous Wavelet Transform (CWT) spectrograms. Since the VDPs serve as a unique feature based on interference patterns, we opted for the more commonly used CWT for the comparison.

In single-source experiments, only one ominidirectional antenna is connected to a transmitter, resulting in no signal interference within the finger movement area, in contrast to the double-source setup. In double-source interference scenario, two ominidirectional antennas, referred to as an antenna pair, are connected to a transmitter. Both antennas operate under the same transmission settings, including frequency. Data from 10 users of second dataset are used in this experiment. The other setting remains consistent with those previously introduced.

Figure 15 illustrates the Cumulative Distribution Function (CDF) of the identification error rate with interference and without interference. We can see that the average accuracy of double-source interference setting significantly surpasses that of the single-source setup. Specifically, the inclusion of interference in the signal enhances the system's identification accuracy by 28.3% compared to scenarios without interference. Therefore, double-source interference setup creates a fine-grained signal mesh within the designated area, leading to high user identification accuracy.

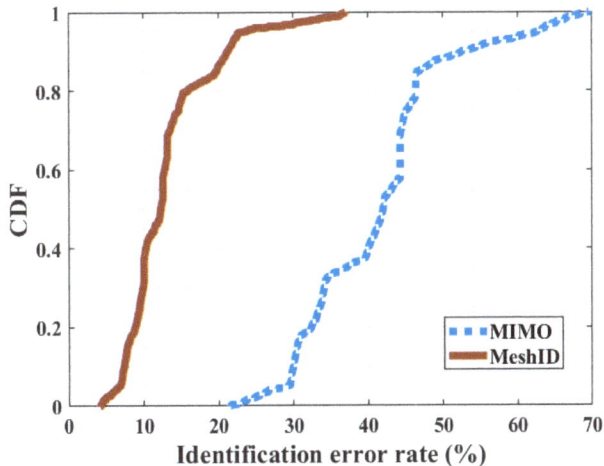

Figure 15. CDFs of MIMO and MeshID.

4.6. Impact of Ambient People Moving

In a typical public room, it is a common condition that other people may dynamically move around when the user is performing finger gestures. Since the wireless signal is sensitive to ambient environment changes, ambient people moving may easily affect the CSI variances, which may have a side effect on user identification. We study the performance of MeshID under the impact of ambient people moving. Figure 16 showcases identification results when a single gesture is performed while an individual moves within a 3-m range. Similarly, Figure 17 depicts results when two consecutive gestures are executed amidst ambient movement. The presence of ambient movement causes a slight decrease in identification error rates, which remain within acceptable bounds for most scenarios. Furthermore, increasing the number of finger gestures performed by the user enhances identification accuracy. This result demonstrates that MeshID is performed as robust to the impact of ambient people moving. Theoretically, the proposed interference wave, comprising a superposition of two waveforms with same frequency, results in stronger CSI variances, facilitating more resilient feature extraction for user identification compared to traditional single-wave setups. Consequently, MeshID effectively mitigates the adverse effects of ambient movement, enhancing overall robustness.

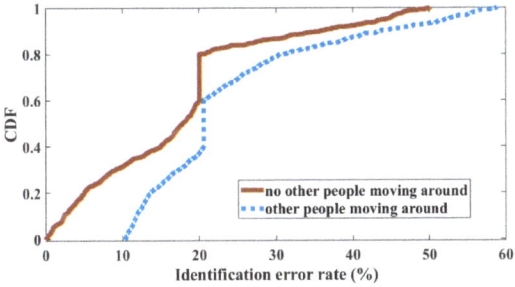

Figure 16. User identification error rate of user performing one gesture when ambient people walk around.

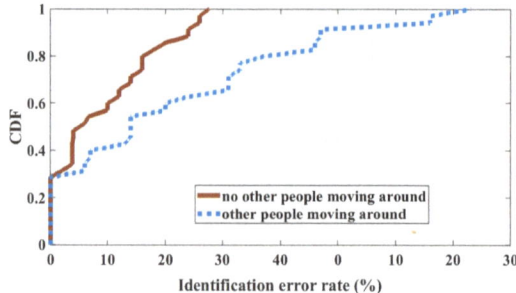

Figure 17. User identification error rate of user performing two sequential gestures when ambient people walk around.

4.7. Performance of FCR Extraction and CBAM

FCR extraction and the CBAM Module are two key components of MeshID. The former removes the non-related features from the input feature space, while the latter extracts features from both channel and spatial axes and focuses on the places with more important information. In order to investigate how these two components affect the performance, we compare our algorithm with a traditional CNN and test the impact of each component.

As shown in Figure 14, employing only the basic CNN yields an average accuracy of 83.2%. Introducing the PCA reverse extraction algorithm enhances the average accuracy to 86.8%, surpassing the basic CNN by 3.6%. Furthermore, integrating both the PCA reverse extraction algorithm and the CBAM Module elevates the average accuracy to 93.5%, marking a 6.7% improvement. Therefore, both components significantly contribute to enhancing user identification accuracy.

4.8. Comparison with Baseline Approaches

MeshID could be a more robust and flexible extended authentication component in existing recognition systems. We evaluated our system on user identification by comparing it with two alternative state-of-the-art approaches, FingerPass [38] and FreeAuth [18]. Both of them are leveraging wireless information for user authentication. Specifically, FingerPass utilizes segmented CSI time series as learning features and adopts the LSTM-based DNN model. FreeAuth proposes a CNN-based method for extracting CSI features. FreeAuth applies a Recurrent Neural Network (RNN) model to extract users' gesture characteristics and maximize the individual uniqueness characterized by a Gaussian Mixture Model (GMM). To control the variable, the training and evaluation process for the baselines follow the same rules of MeshID (e.g., the number of training epoch). We utilize the same dataset, which is our first dataset, for training and evaluation purposes. The comparison results of user identification are summarized in Table 3. Both MeshID and FreeAuth exhibit superior identification results, achieving over 90% accuracy, compared to FingerPass when tested with seven users.

Table 3. Method Comparison with Different User Number.

User Number	FingerPass	FreeAuth	MeshID
7 users	83.6%	93.93%	97.4%
8 users	78%	87.24%	96.5%
9 users	76.8%	70.76%	95.3%
10 users	71.1%	68.69%	94.7%
11 users	62.6%	67.0%	94%
12 users	60.9%	66.3%	93.4%

To ascertain the relationship between system performance and the number of users, we evaluated the three systems with varying numbers of users, ranging from 7 to 12. FingerPass and FreeAuth experience a significant degradation in identification accuracy as the number of users increases. However, our system maintains an average identification accuracy of over 90% for up to 12 users. MeshID achieves better overall performance than the other two approaches, specifically, outperforming FingerPass by about 6% and FreeAuth by nearly 30% with up to 12 users. We noticed that emphasizing user-specified features can mitigate the side effect of the same gesture, and improve the robustness of the system with a larger user number.

It satisfies the demands of most families and some small groups. The performance of MeshID keeps consistency as the number of users increases from 7 to 12. In reality, the challenge of user heterogeneity increases significantly with a large number of users (e.g., thousands of users). However, collecting and labeling finger gestures from such a massive user pool for evaluation purposes poses considerable'difficulties.

5. Discussion

Impact of user height. Different user height may impact the identification accuracy. To investigate the relationship between identification accuracy and user height, we analyzed the height distribution of users. As the statistical result shows in Figure 18, shorter users tend to achieve higher identification accuracy. The height distribution, ranging from 155 cm to 185 cm, is represented on the left axis. In our experiments, we fixed antenna pairs at relatively low positions to accommodate users of different heights. Consequently, taller users may experience more reflected signals from lower parts of their body, such as the chest. Despite this challenge, MeshID maintains a high capability of identification, as gesture pattern retain their uniqueness with interference settings. However, it's worth noting that antenna height adjustments can be made to suit different scenarios.

Impact of user weight. The weight of the users varies from 45 kg to 80 kg. The right axis in Figure 18 represents the weight of the user. The figure shows that the identification decrease is not caused primarily by user weight. Although some of the statistical values show that identification accuracy is better when user has lower weight. That is because user who is shorter usually has lower weight. To delve deeper into this relationship, we conducted a focused analysis on data from users within the same height range. Surprisingly, our findings reveal no discernible pattern in the distribution of identification accuracy, indicating that weight alone does not dictate accuracy levels.

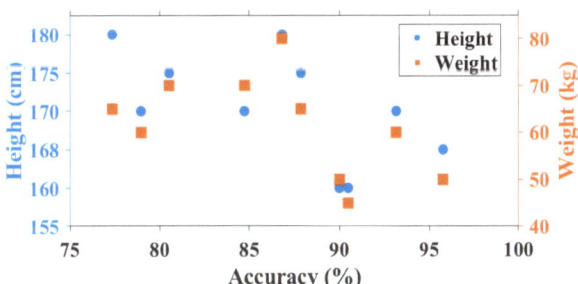

Figure 18. Statistics of users.

Impact of motion speed. In our experiments, users are free to perform gestures de-pending on their habits. Different motion speeds may result in different sample lengths.However, RF devices usually have a related high sampling rate (e.g., more than 250 Hz), hence the sampling interval is less than 0.004 s. It is short enough to capture the normal motion finger. The sampling rate of USRP is much higher; it could be up to 200 MHz. In our setup, the signal sampling rate is 651 Hz, It is adequate for different motion speed samples. There-fore, the motion speed is not the significant factor that affects the identification accuracy.

Impact of different environments. The multipath effect is a phenomenon prevalent in radio frequency applications where signals travel multiple paths from transmitter, encountering reflections, diffraction, and scattering off objects and surfaces in the environment. Consequently, the receiver detects various versions of the transmitted signal, each arriving at slightly different times and exhibiting diverse amplitudes and phases. In principle, RF systems in outdoor environments usually perform better than indoor. Since in indoor environments, with numerous reflective surfaces and obstacles, the multipath effect becomes particularly pronounced, leading to signal degradation and complicating signal processing. We conducts MeshID in three environments: an office, a meeting room and a hallway. According to the identification results, which are 94.3% in the office, 95.03% in the meeting room and 97.1% in the hallway. These outcomes suggest that reduced multipath effects likely contribute to the system's enhanced performance.

Latency. It takes about 1 s to 30 s to perform a gesture in our evaluation. This timeframe is sufficient to satisfy most users, including those with physical limitations who can only draw gestures slowly. We also implement MeshID on a desktop with 12th Gen Intel(R) Core(TM) i3-12100F 3.30 GHz. The time consumption of MeshID mainly comes from the noise removal module and the data transformation module. It demands, on average, 1.743 s to identify the user when eight users are in the database after segmenting the gesture. We believe it is adequate for the majority of identification application scenarios for users.

From the results, it is convincing that the proposed mesh and one-shot DL approach can address the challenge of user heterogeneity, enabling MeshID with high robustness.

6. Related Work

6.1. Behavioral Identification

Behavioral identification as a subset of biometric authentication has been well proposed using a range of sensing technologies. Vision technology [39,40] can identify different users through physical activity characteristics captured from image frames using high-resolution cameras, but it may easily fail in the conditions of luminous changes and obstacles placed in line-of-sight (LoS) [4], in particular raising severe user privacy concerns. Bioelectrical technology [41–44] can utilize bioelectrical sensors, e.g., electrocardiogram (ECG), electromyogram (EMG) and electroencephalogram (EEG), to precisely extract unique biomedical information through body's electrical activities. Ashraf et al. [45] propose a fusion system that uses biometric features of the iris and foot. It achieves a very high accuracy of 98%. Ref. [46] utilizes the phase transform method and Fourier decomposition method to identify individual ECG features. Since these sensors are required to be attached carefully to the user's body, such wearable requirements may compromise user experience, leading to inconvenience in reality [47]. In contrast, since our identification system is essentially developed by RF technology, MeshID can enable user-friendly device-free identification with the advantages of being illumination-free and having fewer privacy concerns.

6.2. RF-Based Behavioral Identification

Existing works focus on different individual behavioral characteristics [10,15], e.g., daily activities, walking gaits, vital signs, and gestures. For gait-based identification, WiWho [48] uses commercial WiFi devices to verify a small group of users using walking gaits. WiFi ID [49] explores the relationship between the feature pattern of subcarrier-amplitude frequency (SAF) based on WiFi CSI and individual walking style, and employs a linear-kernel support vector machine (SVM) to identify users. For daily activity-based identification, E-eye [50] proposes to identify users using the WiFi CSI profiles caused by the activities across the home on a mobile device, while Au-Id [5] uses the reflected RFID signal of users' activities for identification. Since these works usually require a long-term user activity tracking, the real-world applications still remain limited. Besides, a numbers of works propose to enable gait-based identification. Shi et al. [51] extract WiFi features of walking and stationary activities for human authentication. RFree-ID [52] identified human targets based on walking gait by using a phase matrix from tag array. Due to the large range of gait motion, these works may be strongly vulnerable to the impact of ambient environments. Similarly, some works demonstrate vital sign based identification to estimate users' heartbeat rate [16,53,54] and respiratory biometrics [17,55] from RF signals, but the identification may practically fail as the motion of vital signs remains brittle to body movement artifacts and ambient noises. Since performing gestures is user-friendly in a small motion area, gestures-based identification is promising to have much less impact on ambient noises. WiHF [19] is proposed to recognize gestures using WiFi signals for enabling user identification, but the proposed gestures still remain as a large range of arm motions, resulting in much less user-friendliness compared to using in-air finger gestures. FingerPass [38] proposes to identify users through finger gestures with fair accuracy, but the identification performance may still be subject to the issue of low RF signal resolution due to using a traditional RF setup. Unlike these two works, MeshID leverages on the effects of orthogonal signal interference and an attention-based siamese network to fundamentally improve the signal resolution and model retraining, achieving high identification accuracy and robustness.

7. Conclusions and Future Work

This paper presents a novel RF-based user identification scheme that leverages on the proposed mesh and few-shot deep learning approaches to enable highly accurate user identification through finger gestures. MeshID can essentially promote the sensing sensitivity on RF signal to extract sufficient individual biometrics from the movements of finger gestures, accurately identifying different users even in a complex environment. Also, MeshID is able to efficiently retrain the model to ensure high robustness, adapting to an unseen user with little data. In practice, MeshID as an appealing add-on can be easily integrated into existing RF-based gesture recognition systems at low cost. To further investigate the robustness of MeshID, we plan to test our approaches on existing large finger gesture datasets, and develop MeshID for efficient integration with existing RF-based gesture recognition systems. Other methods [56] such as SVM, Logistic Regression and Random Forest will be considered as the improving module in our system. Also, we plan to further evaluate the robustness of our prototype in more real-world scenarios in our future work.

Author Contributions: Conceptualization, W.Z., Y.Z. and L.J.; methodology, W.Z. and L.J.; software, W.Z. and Y.Z.; validation, W.Z.; formal analysis, W.Z.; investigation, W.Z. and D.Z.; resources, W.Z. and D.Z., data curation, W.Z.; writing—original draft preparation, W.Z., Y.Z. and D.Z.; writing—review and editing, W.Z., Y.Z. and L.J.; visualization, W.Z.; supervision, T.G.; project administration, T.G.; funding acquisition, D.Z. All authors have read and agreed to the published version of the manuscript.

Funding: This research was funded by Stable Support Project of Shenzhen (Project No. 20231122145548001), grant number NSFC: 61872247.

Institutional Review Board Statement: The study was conducted in accordance with the Declaration of Helsinki, and approved by the Ethics Committee of RMIT (Project No. 25086).

Informed Consent Statement: Not applicable.

Data Availability Statement: Data will be made available upon request.

Conflicts of Interest: The authors declare no conflicts of interest.

References

1. Liu, H.; Gamboa, H.; Schultz, T. *Sensors for Human Activity Recognition*; MDPI: Basel, Switzerland, 2023. [CrossRef]
2. Zhang, Y.; Wang, C.; Wang, X.; Zeng, W.; Liu, W. Fairmot: On the fairness of detection and re-identification in multiple object tracking. *Int. J. Comput. Vis.* **2021**, *129*, 3069–3087. [CrossRef]

3. Zheng, Z.; Yang, X.; Yu, Z.; Zheng, L.; Yang, Y.; Kautz, J. Joint Discriminative and Generative Learning for Person Re-Identification. In Proceedings of the IEEE/CVF Conference on Computer Vision and Pattern Recognition (CVPR), Long Beach, CA, USA, 15–20 June 2019.
4. Wang, J.; She, M.; Nahavandi, S.; Kouzani, A. A Review of Vision-Based Gait Recognition Methods for Human Identification. In Proceedings of the 2010 International Conference on Digital Image Computing: Techniques and Applications, Sydney, NSW, Australia, 1–3 December 2010.
5. Huang, A.; Wang, D.; Zhao, R.; Zhang, Q. Au-Id: Automatic User Identification and Authentication through the Motions Captured from Sequential Human Activities Using RFID. *Proc. ACM Interact. Mob. Wearable Ubiquitous Technol.* **2019**, *3*, 1–26. [CrossRef]
6. Pokkunuru, A.; Jakkala, K.; Bhuyan, A.; Wang, P.; Sun, Z. NeuralWave: Gait-based user identification through commodity WiFi and deep learning. In Proceedings of the IECON 2018-44th Annual Conference of the IEEE Industrial Electronics Societ, Washington, DC, USA, 21–23 October 2018; pp. 758–765.
7. Moshiri, P.F.; Shahbazian, R.; Nabati, M.; Ghorashi, S.A. A CSI-based human activity recognition using deep learning. *Sensors* **2021**, *21*, 7225. [CrossRef] [PubMed]
8. Damodaran, N.; Haruni, E.; Kokhkharova, M.; Schäfer, J. Device free human activity and fall recognition using WiFi channel state information (CSI). *CCF Trans. Pervasive Comput. Interact.* **2020**, *2*, 1–17. [CrossRef]
9. Cheng, X.; Huang, B. CSI-based human continuous activity recognition using GMM–HMM. *IEEE Sens. J.* **2022**, *22*, 18709–18717. [CrossRef]
10. Gu, T.; Wang, L.; Chen, H.; Tao, X.; Lu, J. Recognizing Multiuser Activities Using Wireless Body Sensor Networks. *IEEE Trans. Mob. Comput.* **2011**, *10*, 1618–1631. [CrossRef]
11. Zhang, D.; Zeng, Y.; Zhang, F.; Xiong, J. Wifi csi-based vital signs monitoring. In *Contactless Vital Signs Monitoring*; Elsevier: Amsterdam, The Netherlands, 2022; pp. 231–255.
12. Khan, M.I.; Jan, M.A.; Muhammad, Y.; Do, D.T.; Rehman, A.U.; Mavromoustakis, C.X.; Pallis, E. Tracking vital signs of a patient using channel state information and machine learning for a smart healthcare system. *Neural Comput. Appl.* **2021**, 1–15. [CrossRef]
13. Ahmed, H.F.T.; Ahmad, H.; Aravind, C. Device free human gesture recognition using Wi-Fi CSI: A survey. *Eng. Appl. Artif. Intell.* **2020**, *87*, 103281. [CrossRef]
14. Tian, Z.; Wang, J.; Yang, X.; Zhou, M. WiCatch: A Wi-Fi based hand gesture recognition system. *IEEE Access* **2018**, *6*, 16911–16923. [CrossRef]
15. Wang, L.; Gu, T.; Chen, H.; Tao, X.; Lu, J. Real-Time Activity Recognition in Wireless Body Sensor Networks: From Simple Gestures to Complex Activities. In Proceedings of the 2010 IEEE 16th International Conference on Embedded and Real-Time Computing Systems and Applications, Macau, China, 23–25 August 2010; pp. 43–52. [CrossRef]
16. Ning, J.; Xie, L.; Wang, C.; Bu, Y.; Xu, F.; Zhou, D.; Lu, S.; Ye, B. RF-Badge: Vital Sign-based Authentication via RFID Tag Array on Badges. *IEEE Trans. Mob. Comput.* **2021**, *22*, 1170–1184. [CrossRef]
17. Liu, J.; Dong, Y.; Chen, Y.; Wang, Y.; Zhao, T. Leveraging Breathing for Continuous User Authentication. In Proceedings of the 24th Annual International Conference on Mobile Computing and Networking, New Delhi, India, 29 October–2 November 2018.
18. Kong, H.; Lu, L.; Yu, J.; Zhu, Y.; Tang, F.; Chen, Y.C.; Kong, L.; Lyu, F. Push the Limit of WiFi-based User Authentication towards Undefined Gestures. In Proceedings of the IEEE INFOCOM 2022—IEEE Conference on Computer Communications, London, UK, 2–5 May 2022; pp. 410–419. [CrossRef]
19. Li, C.; Liu, M.; Cao, Z. WiHF: Enable User Identified Gesture Recognition with WiFi. In Proceedings of the IEEE INFOCOM 2020—IEEE Conference on Computer Communications, Toronto, ON, Canada, 6–9 July 2020.
20. Gu, Y.; Yan, H.; Dong, M.; Wang, M.; Zhang, X.; Liu, Z.; Ren, F. WiONE: One-Shot Learning for Environment-Robust Device-Free User Authentication via Commodity Wi-Fi in Man–Machine System. *IEEE Trans. Comput. Soc. Syst.* **2021**, *8*, 630–642. [CrossRef]
21. Guo, L.; Lu, Z.; Yao, L. Human-machine interaction sensing technology based on hand gesture recognition: A review. *IEEE Trans.-Hum.-Mach. Syst.* **2021**, *51*, 300–309. [CrossRef]
22. Ahmed, S.; Kallu, K.D.; Ahmed, S.; Cho, S.H. Hand gestures recognition using radar sensors for human-computer-interaction: A review. *Remote Sens.* **2021**, *13*, 527. [CrossRef]
23. Xia, H.; Glueck, M.; Annett, M.; Wang, M.; Wigdor, D. Iteratively designing gesture vocabularies: A survey and analysis of best practices in the HCI literature. *ACM Trans.-Comput.-Hum. Interact. (TOCHI)* **2022**, *29*, 1–54. [CrossRef]
24. Rappaport, T. *Wireless Communications: Principles and Practice*; Cambridge University Press: Cambridge, UK, 2001.
25. Zhang, D.; Zhang, F.; Wu, D.; Xiong, J.; Niu, K. Fresnel zone based theories for contactless sensing. In *Contactless Human Activity Analysis*; Springer: Berlin/Heidelberg, Germany, 2021; pp. 145–164.
26. Jenkins, F.A.; White, H.E. Fundamentals of optics. *Indian J. Phys.* **1957**, *25*, 265–266. [CrossRef]
27. Hindy, H.; Tachtatzis, C.; Atkinson, R.; Brosset, D.; Bures, M.; Andonovic, I.; Michie, C.; Bellekens, X. Leveraging siamese networks for one-shot intrusion detection model. *J. Intell. Inf. Syst.* **2023**, *60*, 407–436. [CrossRef]
28. Suljagic, H.; Bayraktar, E.; Celebi, N. Similarity based person re-identification for multi-object tracking using deep Siamese network. *Neural Comput. Appl.* **2022**, *34*, 18171–18182. [CrossRef]
29. Chung, D.; Tahboub, K.; Delp, E.J. A Two Stream Siamese Convolutional Neural Network for Person Re-identification. In Proceedings of the 2017 IEEE International Conference on Computer Vision (ICCV), Venice, Italy, 22–29 October 2017.

30. Caire, G.; Shamai, S. On the capacity of some channels with channel state information. *IEEE Trans. Inf. Theory* **1999**, *45*, 2007–2019. [CrossRef]
31. Yang, Z.; Zhou, Z.; Liu, Y. From RSSI to CSI: Indoor localization via channel response. *ACM Comput. Surv. (CSUR)* **2013**, *46*, 1–32. [CrossRef]
32. Kraus, H.G. Huygens–Fresnel–Kirchhoff wave-front diffraction formulation: Spherical waves. *JOSA A* **1989**, *6*, 1196–1205. [CrossRef]
33. Yeh, J.R.; Shieh, J.S.; Huang, N.E. Complimentary ensemble empirical mode decomposition: A novel noise enhanced data analysis method. *Adv. Adapt. Data Anal.* **2010**, *2*, 135–156. [CrossRef]
34. Coleman, D.D.; Westcott, D.A. *Certified Wireless Network Administrator Official Study Guide*; John Wiley and Sons, Inc.: Hoboken, NJ, USA, 2012.
35. Qi, W.; Zhang, R.; Zhou, Q.; Jing, X. Towards device-free cross-scene gesture recognition from limited samples in integrated sensing and communication. In Proceedings of the IEEE Wireless Communications and Networking Conference, Dubai, United Arab Emirates, 21–24 April 2024; pp. 195–198.
36. Moin, A.; Zhou, A.; Rahimi, A.; Menon, A.; Benatti, S.; Alexandrov, G.; Tamakloe, S.; Ting, J.; Yamamoto, N.; Khan, Y.; et al. A wearable biosensing system with in-sensor adaptive machine learning for hand gesture recognition. *Nat. Electron.* **2021**, *4*, 54–63. [CrossRef]
37. Woo, S.; Park, J.; Lee, J.Y. CBAM: Convolutional Block Attention Module. In Proceedings of the European Conference on Computer Vision (ECCV), Tel Aviv, Israel, 23–27 October 2022.
38. Kong, H.; Lu, L.; Yu, J.; Chen, Y.; Kong, L.; Li, M. FingerPass: Finger Gesture-Based Continuous User Authentication for Smart Homes Using Commodity WiFi. In Proceedings of the Twentieth ACM International Symposium on Mobile Ad Hoc Networking and Computing, Catania, Italy, 2–5 July 2019.
39. Lohr, D.; Komogortsev, O.V. Eye Know You Too: Toward viable end-to-end eye movement biometrics for user authentication. *IEEE Trans. Inf. Forensics Secur.* **2022**, *17*, 3151–3164. [CrossRef]
40. Roth, J.; Liu, X.; Metaxas, D. On continuous user authentication via typing behavior. *IEEE Trans. Image Process.* **2014**, *23*, 4611–4624. [CrossRef] [PubMed]
41. Agrawal, V.; Hazratifard, M.; Elmiligi, H.; Gebali, F. ElectroCardioGram (ECG)-based user authentication using deep learning algorithms. *Diagnostics* **2023**, *13*, 439. [CrossRef] [PubMed]
42. Shams, T.B.; Hossain, M.S.; Mahmud, M.F.; Tehjib, M.S.; Hossain, Z.; Pramanik, M.I. EEG-based Biometric Authentication Using Machine Learning: A Comprehensive Survey. *ECTI Trans. Electr. Eng. Electron. Commun.* **2022**, *20*, 225–241. [CrossRef]
43. Abo-Zahhad, M.; Ahmed, S.M.; Abbas, S.N. A Novel Biometric Approach for Human Identification and Verification Using Eye Blinking Signal. *IEEE Signal Process. Lett.* **2015**, *22*, 876–880. [CrossRef]
44. Belgacem, N.; Fournier, R.; Nait-Ali, A.; Bereksi-Reguig, F. A novel biometric authentication approach using ECG and EMG signals. *J. Med. Eng. Technol.* **2015**, *39*, 226–238. [CrossRef]
45. Ashraf, S.; Ahmed, T. Dual-nature biometric recognition epitome. *Trends Comput. Sci. Inf. Technol.* **2020**, *5*, 8–14.
46. Fatimah, B.; Singh, P.; Singhal, A.; Pachori, R.B. Biometric identification from ECG signals using Fourier decomposition and machine learning. *IEEE Trans. Instrum. Meas.* **2022**, *71*, 1–9. [CrossRef]
47. Barros, A.; Rosário, D.; Resque, P.; Cerqueira, E. Heart of IoT: ECG as biometric sign for authentication and identification. In Proceedings of the 2019 15th International Wireless Communications & Mobile Computing Conference (IWCMC), Tangier, Morocco, 24–28 June 2019; pp. 307–312.
48. Zeng, Y.; Pathak, P.H.; Mohapatra, P. WiWho: WiFi-Based Person Identification in Smart Spaces. In Proceedings of the 2016 15th ACM/IEEE International Conference on Information Processing in Sensor Networks (IPSN), Vienna, Austria, 11–14 April 2016.
49. Zhang, J.; Wei, B.; Hu, W.; Kanhere, S.S. WiFi-ID: Human Identification Using WiFi Signal. In Proceedings of the 2016 International Conference on Distributed Computing in Sensor Systems (DCOSS), Washington, DC, USA, 26–28 May 2016.
50. Wang, Y.; Liu, J.; Chen, Y.; Gruteser, M.; Yang, J.; Liu, H. E-Eyes: Device-Free Location-Oriented Activity Identification Using Fine-Grained WiFi Signatures. In Proceedings of the 20th Annual International Conference on Mobile Computing and Networking, Maui, HI, USA, 7–11 September 2014.
51. Shi, C.; Liu, J.; Liu, H.; Chen, Y. Smart User Authentication through Actuation of Daily Activities Leveraging WiFi-Enabled IoT. In Proceedings of the 18th ACM International Symposium on Mobile Ad Hoc Networking and Computing, Chennai, India, 10–14 July 2017.
52. Zhang, Q.; Li, D.; Zhao, R.; Wang, D.; Deng, Y.; Chen, B. RFree-ID: An Unobtrusive Human Identification System Irrespective of Walking Cofactors Using COTS RFID. In Proceedings of the 2018 IEEE International Conference on Pervasive Computing and Communications (PerCom), Athens, Greece, 19–23 March 2018.
53. Xu, Q.; Wang, B.; Zhang, F.; Regani, D.S.; Wang, F.; Liu, K.J.R. Wireless AI in Smart Car: How Smart a Car Can Be? *IEEE Access* **2020**, *8*, 55091–55112. [CrossRef]
54. Mishra, A.; McDonnell, W.; Wang, J.; Rodriguez, D.; Li, C. Intermodulation-Based Nonlinear Smart Health Sensing of Human Vital Signs and Location. *IEEE Access* **2019**, *7*, 158284–158295. [CrossRef]

55. Wang, Y.; Gu, T.; Luan, T.H.; Yu, Y. Your breath doesn't lie: Multi-user authentication by sensing respiration using mmWave radar. In Proceedings of the 19th Annual IEEE International Conference on Sensing, Communication, and Networking (SECON), Madrid, Spain, 11–14 September 2023; pp. 64–72.
56. Wang, Y.; Cao, J.; Li, W.; Gu, T.; Shi, W. Exploring traffic congestion correlation from multiple data sources. *Pervasive Mob. Comput.* **2017**, *41*, 470–483. [CrossRef]

Disclaimer/Publisher's Note: The statements, opinions and data contained in all publications are solely those of the individual author(s) and contributor(s) and not of MDPI and/or the editor(s). MDPI and/or the editor(s) disclaim responsibility for any injury to people or property resulting from any ideas, methods, instructions or products referred to in the content.

Article

SENS+: A Co-Existing Fabrication System for a Smart DFA Environment Based on Energy Fusion Information

Teng-Wen Chang [1,*], Hsin-Yi Huang [1], Cheng-Chun Hong [1], Sambit Datta [2] and Walaiporn Nakapan [3]

1. College of Design, National Yunlin University of Science and Technology, Douliou 640, Yunlin, Taiwan
2. School of Electrical Engineering, Computing and Mathematical Sciences, Curtin University, Bentley, WA 6102, Australia
3. Parabolab, Bangkok 11000, Thailand
* Correspondence: tengwen@yuntech.edu.tw; Tel.: +886-55342601 (ext. 6510)

Abstract: In factories, energy conservation is a crucial issue. The co-fabrication space is a modern-day equivalent of a new factory type, and it makes use of Internet of Things (IoT) devices, such as sensors, software, and online connectivity, to keep track of various building features, analyze data, and produce reports on usage patterns and trends that can be used to improve building operations and the environment. The co-fabrication user requires dynamic and flexible space, which is different from the conventional user's usage. Because the user composition in a co-fabrication space is dynamic and unstable, we cannot use the conventional approach to assess their usage and rentals. Prototyping necessitates a specifically designed energy-saving strategy. The research uses a "seeing–moving–seeing" design thinking framework, which enables designers to more easily convey their ideas to others through direct observation of the outcomes of their intuitive designs and the representation of their works through design media. The three components of human behavior, physical manufacture, and digital interaction are primarily the focus of this work. The computing system that connects the physical machine is created through communication between the designer and the digital interface, giving the designer control over the physical machine. It is an interactive fabrication process formed by behavior. The Sensible Energy System+ is an interactive fabrication process of virtual and real coexistence created by combining the already-existing technology, the prototype fabrication machine, and SENS. This process analyzes each step of the fabrication process and energy, fits it into the computing system mode to control the prototype fabrication machine, and reduces the problem between virtual and physical fabrication and energy consumption.

Keywords: internet of things; prototyping process; energy-saving; interactive design; user behaviors; ambient agents

---

Citation: Chang, T.-W.; Huang, H.-Y.; Hong, C.-C.; Datta, S.; Nakapan, W. SENS+: A Co-Existing Fabrication System for a Smart DFA Environment Based on Energy Fusion Information. *Sensors* **2023**, *23*, 2890. https://doi.org/10.3390/s23062890

Academic Editors: Hugo Gamboa, Tanja Schultz and Hui Liu

Received: 16 January 2023
Revised: 1 March 2023
Accepted: 1 March 2023
Published: 7 March 2023

Copyright: © 2023 by the authors. Licensee MDPI, Basel, Switzerland. This article is an open access article distributed under the terms and conditions of the Creative Commons Attribution (CC BY) license (https://creativecommons.org/licenses/by/4.0/).

1. Introduction

1.1. Background

With the introduction of technology and software-based services, much research has been done to create smart environments that allow for seamless interaction between users and their immediate surroundings [1]. The focus of smart home/environment architecture is to develop practical collaboration between users and devices to optimize use and improve the quality of experiences and services [2]. The primary goal of a smart environment is to enable users to easily control and regulate appliances with the use of IoT-enabled sensors [3–6], therefore enhancing energy efficiency and reducing wastage.

In factory settings, especially in co-fabrication spaces, energy conservation by human activities is a crucial issue. As identified by the editorial of Volume 1 of this Special Issue [7], how advanced sensors can be integrated from the smart environment with human activity recognition and how one can be further connected to the energy consumed by these human activities provide a way to allow users to be aware of the consequence of their activities and further change these activities for the sake of energy conservation.

Co-fabrication spaces can monitor various building features, such as sensors, software, and online connectivity, analyze data, and generate information on usage patterns and trends that can be used to optimize the environment and building operations [8–12]. It is important to note that co-fabrication users have different needs compared to typical users, as they require dynamic and flexible spaces. Since co-fabrication spaces have a high level of user turnover, the common methods for evaluating usage and rent are not suitable. As a result, prototyping requires a specifically designed energy-saving strategy.

In co-fabrication spaces, rapid prototyping is critical during the user's prototyping process, as it involves high energy consumption stages. During the prototyping process, the user typically works independently and attempts to solve practical problems or refine designs through a trial-and-error process. The emergence of rapid prototyping has lowered the fabrication threshold and increased the production of innovative designs, while also providing feedback on machined parts based on knowledge of the actual fabrication process [13,14].

Presently, many fabricators, designers, and researchers use digital devices, content, teaching guidelines, and even videos to learn, understand, and obtain information about prototyping to reduce the threshold for contact [15–18]. However, there are hidden technical operation issues and professional divisions of labor in the fabrication process that force designers to hand over design drawings to other professionals. This not only increases the design threshold and energy consumption but also causes the designer to spend more time consulting professionals and waiting for fabrication, while preventing the creation of different methods and allowing other experts to intervene in the design.

Collective data fusion approaches are critical in improving the efficiency and effectiveness of design processes that involve multiple disciplines. Such approaches can help in implementing experimental interactive installations and using innovative design methods. Traditional processes can be time-consuming and inefficient and require tailor-made construction processes for fabrication and assembly to test proposed prototypes. One example of a collective data fusion approach is the System of Design–Fabrication–Assembly (ViDA). This approach facilitates co-fabrication, cooperation, and information transmission. The ViDA system provides a platform for designers, engineers, and builders to share data and collaborate on projects, improving the efficiency of the design process. Another approach is the use of the Internet of Things (IoT) to collect data on building operations. Although this approach may not necessarily lead to behavior change, it can help to collect data on the building's energy consumption, air quality, and other environmental factors. These data can be used to identify areas for improvement and optimize building operations. To improve the accuracy of sensing and help users manage the process between virtual data and physical environments, a proposed control mechanism called the human behavior sensor can be used. This mechanism uses gestures, behaviors, and interfaces to enhance the precision of the fabrication process and provide a smooth user experience. The use of a camera on another manipulator can allow for viewpoints to be executed at appropriate times during a task, ensuring that there is always a robust view suitable for monitoring the task. Collective data fusion can enable more efficient and effective design processes by incorporating these approaches. The integration of various design tools, technologies, and disciplines can enable designers to create innovative solutions that are more efficient and effective than traditional approaches.

1.2. Motivation and Approach

The purpose of this study is to integrate the Design–Fabrication–Assembly (DFA) architecture with the real-time interface and smart energy information in the digital twin. The motivation for the project is to improve the usage rate of the facilities of the co-fabrication space, to reduce unnecessary prototyping for users, and to maximize the use of dynamic energy for power consumption. The advantage of the DFA approach is that it can improve the fabrication and design process. DFA is mainly used for new innovative designs, and there is no practical example yet. Many applications in the materials or assembly are unknown, and design and manufacture need to take into account how the finished product will be assembled, and the process requires constant trial and error. The digital twin is a way to connect the virtual model and the physical, finished prototype. If the virtual and physical model can be integrated into a digital interface for designers to interact with, the problem of an digital communication medium in DFA can be solved, and the final prototype can be shown in advance in real time and can be used in a variety of materials.

Through further discussion, the development and problems in the three directions of human behaviors, digital system energy integration, and co-existing fabrication can be understood. The designer interacts with the transmission of digital information, energy consumption forecast, and physical fabrication through behavior, which can be embedded in the process of fabrication systems. Through the interaction of the above three directions and integrating the co-existing fabrication system for a smart DFA environment based on energy fusion information, a tool that supports designers, manufacturers, and assemblers in digital fabrication behaviors through virtual–physical integration technology can be created. In this project, it was hypothesized that every user could have an agent system, and the agent could change the effect of the device according to the user's actions. Users do not have static identities and have multiple uses. Its behavior cannot be calculated using a single ID. Dynamic calculation and analysis are required. In this way, the user can understand not only her consumption and daily behavior, but also her investment in generated energy to improve the user experience. The tool system will provide construction method suggestions for users in the digital fabrication process to modify and manufacture in real time, and perform an interactive fabrication process of coexistence of virtual and real through the interactive interface combined with the real-time information of the system to solve (1) the designer's concern about fabrication differences in understanding, (2) physical objects that can be augmented with virtual information, (3) problems with assemblers during component assembly, and (4) energy consumption alert and representation.

2. Related Research

2.1. Virtual Environment Energy with the Co-Fabrication Space

Smart environments connect computers and other smart devices to everyday settings and tasks, extending pervasive computing and promoting the idea of a world connected to sensors and computers [19]. These sensors and computers are integrated with everyday objects in people's lives and are connected through networks. Human activity recognition (HAR) is the process of interpreting human motion using computer and machine vision technology. Human motion can be interpreted as activities, gestures, or behaviors, which are recorded by sensors. The movement data are then translated into action commands for computers to execute, and human activity recognition code is an important and effective tool [20–22]. The composition of co-fabrication spaces requires smart sensors and apps to create opportunities for more in-depth environmental monitoring compared to formal monitoring networks and to involve the public in environmental monitoring through participatory data collection and monitoring systems [23].

As electronic devices and their applications continue to grow [24], advances in artificial intelligence (AI) have also greatly improved the ability to extract deeply hidden information for accurate detection and interpretation. Additionally, current architectures must accommodate the number of devices dynamically available in the smart environment and the high degree of data heterogeneity to match the complexity of human activity. A key feature expected of backbones connecting smart environments is reliable communication that ensures lossless and low-latency data transmission over the network. Another aspect is to monitor the state of smart devices, handle error cases, and reallocate resources to maintain overall system performance [25,26].

Therefore, HAR proposed in this research contains three major technologies: (1) smart devices [24,26] for human interface and data collectors, (2) an agent-based system (an acting role model for its dynamic role model features [27,28]) as the software framework for communication and coordination among smart devices, and (3) Internet of Things (IoT) technology [29,30] for a hardware framework to collect signals from the physical environment and fabrication devices.

The Sensible Energy System (SENS) [30] configures an agent communication framework, which includes Human Agents, Devices, Local Servers, Consumption Managers, Communication Protocols, Location Servers, Power Switches, and Cloud Servers. Every user is a human agent, and a human agent uses a device in physical space. Once the local server receives the user's information, it interacts with other agents in the system environment through its sensor module, execution module, and communication module. Building upon the existing work in the attribute-based access control model, it is possible to capture the physical context through sensed data (attributes) and perform dynamic reasoning over these attributes and context-driven policies using technology to execute access control decisions. By leveraging the existing structure, it becomes easier to access and capture user data.

Mixed reality (MR) not only occurs in the physical world or virtual world, but it also includes augmented reality technology that combines reality, and VR real-time interaction technology [8,31]. We can use an agent to describe event simulations in combination with physical objects placed in a virtual environment. Users in different roles can manipulate physical objects as physical counterparts to different machines and equipment in virtual space in real time. The system will visualize engineered factories, data, and behaviors and perform further analysis. This method combines dynamic discrete-event simulation seamlessly connected to physical objects placed in the virtual environment to enable co-design. The bridge between simulation and physical objects is made through a digital integration platform. Using physical artifacts as counterparts to the simulation's virtual objects, participants can safely interact with the simulation regardless of their skill or ability. This design provides a framework by which many participants can engage in the process, experiment with the concept (Figure 1), and immediately respond to the adaptation of interactive surfaces.

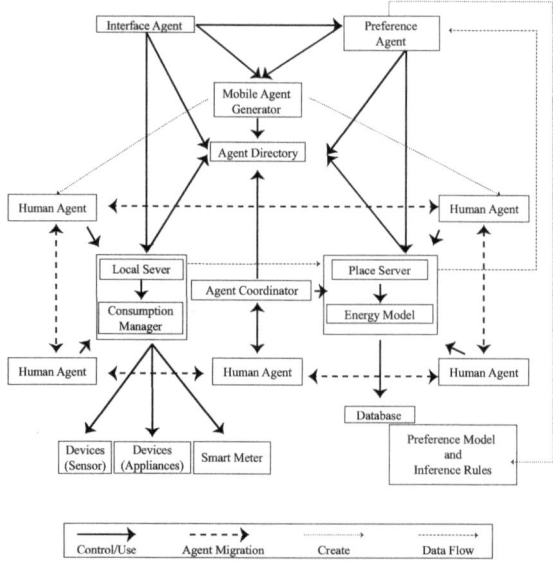

Figure 1. The concept of the user behavior dataflow in the SENS virtual environment [30].

2.2. Sensing the Physical Environment with the Smart Factory

The integration of technology and intelligent services is essential for most ubiquitous intelligent automation systems, especially those that require agile situational management, such as smart homes. Smart environments (SEs) evolve from ubiquitous computing following the idea that a physical world is richly and invisibly interwoven with sensors, actuators, displays, and computational elements, embedded seamlessly in the everyday objects of our lives, which are emotional and connected through a continuous network [19,32]. A smart environment provides a large amount of data because it consists of multiple heterogeneous sensors placed throughout the environment. A scalable and flexible platform integrates such devices and provides applications with the necessary interfaces to interact with information from available resources. Therefore, current architectures must accommodate the number of devices dynamically available on the smart environment and the high degree of data heterogeneity. A key feature expected of backbones connecting smart environments is reliable communication that ensures lossless and low-latency data transmission over the network. Another aspect is to monitor the state of smart devices, handle error cases, and reallocate resources to maintain overall system performance [25,26].

In the co-fabrication space, rapid prototyping is critical in the prototyping process of the user, and it is the stage that consumes the most energy. In the prototyping process, the user mainly works by himself and through practical problem-solving or design experimentation to find solutions. The emergence of rapid prototyping not only decreases the original fabrication threshold but also increases innovative design production. Presently, many manufacturers, designers, and researchers use digital devices, content, teaching guidelines, and even videos to learn, understand, and obtain information for prototyping to decrease the threshold of contact. However, there are some hidden technical operation issues and the professional division of work in the fabrication process that forces the designer to hand over the design drawings to other professionals to complete. When the fabrication process is divided, it not only increases the design threshold and energy consumption but also prevents the creation of different methods and allows other experts to intervene in the design.

Every design is a unique and innovative structure that affects the prototyping process's results. Therefore, the concept of customized small-batch production has become popular. Such development not only allows designers to create many different unique and special designs but also enables designers to undergo continuous trial and error and adjust processes, improving the realization of their ideas. If we can structure the trial–error process and anticipate possible problems in advance, it will reduce the number of attempts by designers to make mistakes. Among others, such a concept can be explored through the proposed DFA framework (Figure 2). DFA is a collective method that systematically maps out the prototyping process of an interactive interface in three stages: (1) the design stage, in which a device prototype is designed, (2) the fabrication stage, in which device components are tested and produced, and (3) the assembly stage, in which the components are assembled and tested [33]. To assist designers in reviewing designs and presenting more specific design concepts, 3D technology has further developed Augmented Reality (AR), Virtual Reality (VR), and Mixed Reality (MR). It is a simulation of 3D sketches on the computer, which allows designers to preview the real size, position, and angle of the virtual model. MR is the integration of both real and virtual worlds to generate new environments and visualizations in which physical and digital objects coexist and interact instantly [34–36].

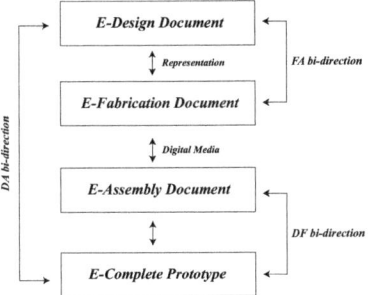

Figure 2. The Design–Fabrication–Assembly (DFA) process.

2.3. Human Interaction and Behaviors with the Co-Fabrication Space

The presence of human beings in various processes introduces a problem as they cause a source of disturbance and unpredictability in the form of performance variation. In the traditional digital fabrication process, designers ignore differences in the fabrication process due to unfamiliarity with the workmanship during the design-to-assembly process. In the process of making prototypes, the effect of different methods on different materials and the differences between various materials will affect the result of the actual work. Despite humans' natural ability to be flexible, their presence still serves as a disturbance within the system. This makes modeling and optimization of these systems considerably more challenging and, in some cases, even impossible. The most common scenario is that the machine is unable to produce the design components, resulting in repeated errors. To enable many ideas in the larger intelligent manufacturing paradigm to be realized, it is crucial to overcome the significant challenge of improving the ability of robotic operators to adapt their behavior to variations in human task performance [37]. ViDA is part of the refinement in the co-fabrication space problem encountered and revolves around the DFA model [38]. The system suggests methods for the model, such as grouping components, wiring groups, and setting sensor positions. It generates several fabrication documents and assigns them to several different machines through different file types to let the user finish this document. Finally, when the fabrication documents are completed and enter the assembly stage, the assembler (Figure 3) comes into play. By the same token, the Co-Existing Fabrication System (CoFabs) [39] aims to enable the cyber-physical integrated system to extend the rapid prototyping application to other fabrication processes and show information on the MR device.

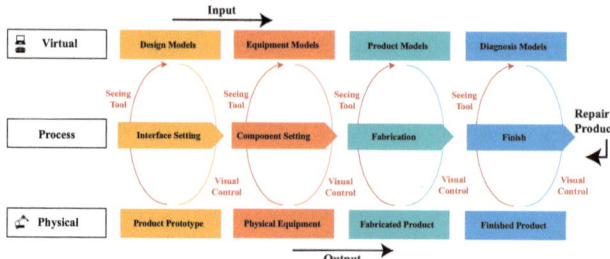

Figure 3. The concept of the visualization system of the DFA (ViDA) process.

AR and MR are rapidly growing in popularity as mainstream technologies that can be used to develop intelligent settings that combine the physical world and virtual items, providing new options for supporting both solitary and group human activities. Augmented reality refers to a medium in which digital information is added or superimposed onto the real world in accordance with the world itself and shown to a user depending on their position and perspective [40]. In an interactive simulation, human behavior is used as the basis for interaction, rather than deliberate interaction, allowing users to further integrate into the experience and enhance the immersion of the interaction through a variety of different behaviors, as well as by performing system functions in a wearable manner [41–45].

To allow the maker to program and operate the fabrication tool more quickly and intuitively, a simple operation is accompanied by complex programming, forming the interaction between the human and the fabrication tool. The MR environment is a way to decrease the designer's fabrication threshold, as the designer can immerse themselves in the MR environment, interact with virtual objects with direct gestures, and even introduce existing physical objects into their designs. The concept of ViDA [38,39] is currently a system operating on the common design software named Rhino [46]. The system can provide dynamic simulation, physical, structural, and environmental analysis, microcontroller programming, and fabrication tool programming, calculate and coordinate component numbers, component attributes, and binding numbers through imported geometric models, and split the prototype structure to generate corresponding mechanisms and component assembly maps. However, it can be hard to imagine the size and effect of the design on a computer screen. Thus, the ViDA system disassembles the design process and generates different stage-specific documents to help different users join the project. When making prototypes, common 3D modeling systems give designers operational control over prototype size and assembly. Fabrication documents, such as assembly sequence, component number, and sensor location maps, provide construction method suggestions in the prototyping process, improving the quality of the prototype. Such technology can make the self-created achieve an immersive effect.

3. Co-Existing Space for the Smart Factory

The aim of this project is to help users manage their activities and identify potential sources of energy consumption based on their choices, activities, and behaviors. Although there are many ways to track user activity and display energy information, they are limited in their effectiveness in changing user behavior and providing visualizations in a timely manner. The co-existing fabrication system for a smart DFA environment based on energy fusion information proposes new avenues for the design and interaction for a smart environment system that optimizes energy recommendations. With this in mind, the following design criteria are proposed for the development of the SENS+ system.

Recognition of human behaviors: The designer interacts with the transmission of digital information, energy consumption forecast, and physical fabrication through behavior, which can be embedded in the process of fabrication systems. Through the interaction of the above three directions and the integration of the co-existing fabrication system for a smart DFA environment based on energy fusion information, a tool that supports designers, manufacturers, and assemblers in digital fabrication behaviors through virtual–physical integration technology can be created.

Recommendation of energy behaviors: The project aims to improve the usage rate of the facilities in the co-fabrication space, reduce unnecessary prototyping for users, and maximize the use of dynamic energy for power consumption. The proposed DFA design framework aims to improve the fabrication and design process. DFA is mainly for new designs, and there is no practical example for many applications in materials or assembly, which requires constant trial and error. The digital twin is a way to connect the virtual model and the physical, finished prototype. If the virtual and physical models can be integrated into a digital interface for designers to interact with, the problem of the analog communication medium in DFA can be solved, and the final prototype can be shown in advance in real time and can be used in a variety of materials.

User interaction and information visualization: It is hypothesized that every user could have an agent system, and the agent could change the effect of the device according to the user's actions. Users do not have static identities and have multiple uses. Their behavior cannot be calculated using a single ID and requires dynamic calculation and analysis. In this way, the user can not only understand their consumption and daily behavior, but also understand their investment in generated energy to improve the user experience. The tool system will provide construction method suggestions for users in the digital fabrication process to modify and manufacture in real time, and perform an interactive fabrication process of coexistence of virtual and physical through the interactive interface combined with real-time information.

3.1. The System

Every design is a unique and innovative structure that affects the process of prototyping results. Therefore, the concept of customized small-batch production has become popular. This development not only allows designers to create many different, unique, and special designs but also enables them to undergo continuous trial-and-error and adjustment processes, improving the realization of their ideas. If we can structure the trial-and-error process and anticipate possible problems in advance, we can reduce the number of attempts made by designers to rectify mistakes. SENS+ provides a framework in which many participants can engage in the process, experiment with the concept, and respond to the adaptation of interactive surfaces. Figure 4 shows the SENS+ system structure, which is different from SENS [30] in terms of human agent and application design. In the SENS+ design process, the user will provide information and documents to the system, which will produce corresponding documents for different role users. SENS+ will focus on construction method suggestions, document readability, and user communication results. During the prototyping process, the user will undergo an iterative try–error process to improve their design. The backward modification of the design is a decentralized fabrication feature, so the prototyping process requires continuous communication to reach a consensus before completion [33]. Finally, SENS+ will (1) record the entire process and use the user behaviors and location data to give them future appointments for energy suggestions, (2) analyze the user-provided file and generate possible construction method suggestions for the user, (3) give the user machine suggestions when they choose a method, and (4) allow the user to set the machine function based on their requirements or suggest settings based on the system's recommendations.

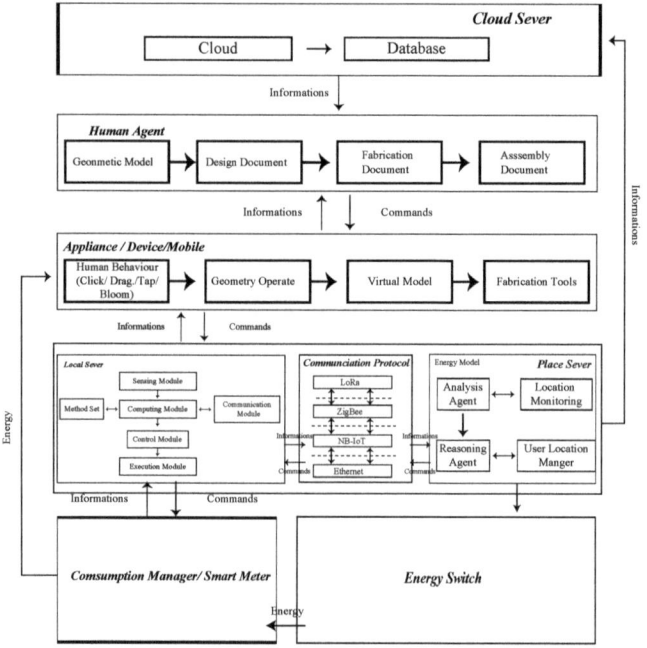

Figure 4. The system structure of this project.

3.2. The Virtual Environment with the Agent and Co-Fabrication Communication Framework

The scenario awareness system is an important component of a smart environment, essential for conducting research. It involves obtaining various signals from the environment and users via home automation [47,48] to provide user services that meet their needs. The acquired situational information is converted into user needs through inference technology, and this process is called a situational awareness service. The activity perception system is an important research field that uses the external context to determine the user's current internal activity context using perception and reasoning, including two main sub-studies of activity recognition and activity prediction. Automation can significantly improve the efficiency of penetration testing [48–51]. The proposed structure used IoT to present a penetration testing methodology, and its automation is based on the belief–desire–

intention (BDI) model, which is one of the classical cognitive architectures of agents to evaluate IoT security. The BDI model provides a method to distinguish between choosing a plan from a library of options or an external planner application and carrying out already-active plans. As a result, BDI agents can balance the time spent thinking through ideas and deciding what to do with the time spent carrying out those plans. A third task, which is left to the system designer and programmer, is making the first blueprints, which are not covered by the model.

Beliefs: Beliefs represent the informational state of the agent, i.e., about the world, including itself and other agents. This stage includes inference rules and allows forward chaining to lead to new beliefs. Beliefs are literals of first-order logic, representing information about the world, updated by the perception of the environment and by the execution of intentions. Using the term belief rather than knowledge recognizes that what an agent believes may not necessarily be true and, in fact, may change in the future. Beliefs are stored in a database, sometimes called a belief base or a belief set, although that is an implementation decision.

Desires: Desires represent the motivational state of the agent and represent objectives or situations that the agent would like to accomplish or bring about. Desires can correspond to the task allocated to the agent and are usually considered logically consistent. Two kinds of desires are usually adopted: To achieve a desire, expressed as a belief formula, and to test a situation formula, which is a belief formula or a disjunction and/or a conjunction of them. Achieving goals involves practical reasoning, test goals involve epistemic reasoning. A goal is a desire that has been adopted for active pursuit by the agent. Usage of the term goals adds the further restriction that the set of active desires must be consistent.

Intentions: Intentions show the agent's deliberate state, or what the agent has decided to do. Intentions are aspirations to which the agent has made a commitment. Plans are sets of instructions, recipes, or related knowledge that an agent can use to achieve one or more of its goals. Plans can merge with other plans, and they are often only partially formed at first, with details being added as the project moves forward. Each plan in the preset library of plans available to BDI agents contains multiple components. The trigger serves as a plan's invocation condition by defining the event that the plan is meant to handle. If the agent detects an event in the event queue, it will consider the plan relevant. The context specifies, as a situation formula, when the plan is applicable, i.e., when the plan should be considered to form an intention. When a plan is filled out, goals are added to the event queue, after which other plans that can deal with similar events are considered. A plan may also include certain "maintenance conditions", which specify the conditions that must endure for the plan to continue to be carried out. For both successful and unsuccessful implementations of the strategy, a set of internal actions is prescribed.

Events: Events serve as the agent's reactionary activity triggers. An event could change goals, set off plans, or change beliefs. Externally produced events may be collected by sensors or other integrated systems. Additionally, internal events may be developed to activate disconnected updates or activity plans. Events in a queue are mapped to perception. There are three different types of events: receiving a message, acquiring a desire, and acquiring a belief. Events are implemented as structures that keep track of past data, such as the responses made to them and their success or failure. The transmission and reception of messages are used to put BDI agents' MAS learning skills into practice.

To integrate the originally separated virtual and physical spaces, some researchers [5,28,52,53] have used field observation, focus groups, and participatory observations to analyze user behaviors. The research used a series of methods to record and analyze their behaviors to design a suitable user interactive structure and usability evaluation. Simulating, predicting, and learning behavior can be done by adopting the Acting Role Model (ARM) [28], agent technologies (SENS) [30], intelligent dynamic interactions with the design system (DARIS), and BDI evaluation and learning [49,54–58] to examine interactive behaviors. The network entity system is generated by combining digital and physical information through the process of the embedded computer, network monitoring, and controlling the entity, calculating its feedback loop through the influence of the entity information [59], forming a co-existing interactive space. The researchers used smart sensors as sensor ports to create a coexistence space for sensory energies. In the environment, each physical object has its own environmental agent. The environmental agents in Figure 5 are divided into human agents, interface agents, preference agents, and sensitive agents. When a user enters the space, their local server starts tracking user behavior and device consumption. Consumption data and user behavior are sent to the Place Server and stored in a database for integration, calculation, and analysis. Finally, the information and details will be shown on the MR and App device.

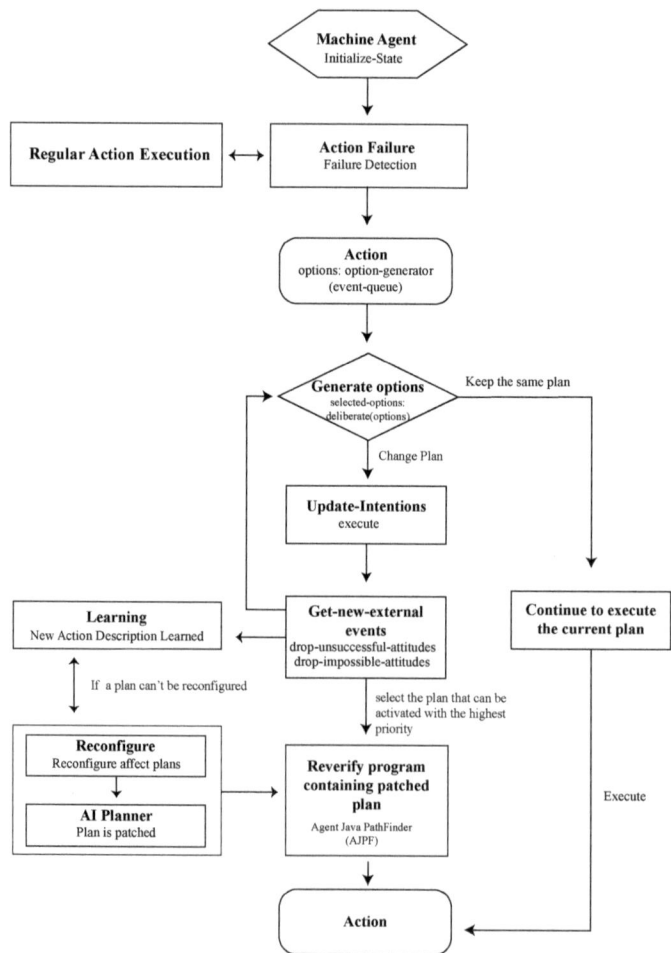

Figure 5. The concept of the communicate process in SENS+ smart sensor agents.

3.3. The Physical Environment with a Smart Factory Architecture

There are many cases in which 3D modeling programs are used in engineering environments for the visualization and evaluation of design elements [60–62]. The information contained in the 3D modeling programs is necessary for presenting the design feasibility and for conducting research analysis. Rapid prototyping and 3D modeling programs allow for more direct and personal interactions with investors, designers, and other professionals involved in the project. The main function of rapid prototyping is to reduce the problems between the designer's ability, the threshold, and operation time. Since the 1990s [63], many fabrication machines have required human control to carry out processes such as metal bending, sewing, and other fabrication tools. However, due to certain dangers in the fabrication process, professional skills are also required to operate these machines. Rapid prototyping aims to automate construction methods through prefab, with some degree of customization and support to avoid problems with usage. The convenience and rapid prototyping brought by digital fabrication have become the development trend of fabrication tools. However, for makers, the technical threshold of these machines is very high, and the manual assembly of parts is as complex as the parts constructed by fabrication tools [64,65]. In the early days, makers needed to communicate with manufacturers through drawings to confirm the physical finished product and subsequently hand it over to the manufacturer to produce the finished product, or even enter the factory to ensure the quality of the physical finished product. Therefore, it is necessary to consider a holistic approach to successfully incorporate fabrication tools, humans, and material mechanisms into the field. Digital fabrication focuses on single-task robots that can be deployed in the field, and designs a system to integrate the physical and virtual environment. The system is divided into a five-layer structure (Figure 6) including a remote terminal, a master control layer, a communication layer, a perception layer, and a driver layer, with the ultimate goal of designing construction sites that work similar to factories.

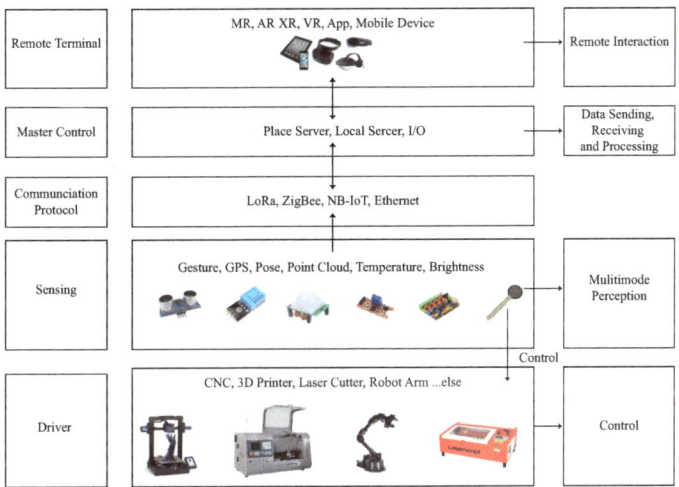

Figure 6. The concept of the architecture design of the SENS+ behaviorsystem.

1. The remote terminal allows users to view information using a variety of tools, such as IoT Explorer, MR, XR, VR, App, and Mobile Devices, making it easy for them to understand the data.
2. The master control layer uses advanced technology such as the NXP RT1062, Raspberry Pi 3B+, and ibeacon to issue control commands and collect data from the entire system. The NXP RT1062 development board connects to Tencent IoT Explorer via an ESP8266 WiFi module to upload sensing data and receive remote control commands using the MQTT protocol stack. Additionally, the Raspberry Pi 3B+ development board's network port communicates with radar to obtain point cloud data, which is then sent to the NXP RT1062 development board after processing.
3. The communication layer serves as the interface between the main control layer and the perception and driver layers. It provides several communication interfaces such as WiFi, LoRa, ZigBee, NB-IoT, Ethernet, and others.
4. The sensing layer contains three types of sensors: user gesture sensors, environmental state sensors, and energy sensors.

5. The driver layer receives commands from the main control layer and completes the movement control of the user, local machine, and device during the prototyping and design phase.

3.4. User Interaction in the MR Device

The fabrication process can pose a challenge for designers during the prototype fabrication phase. The lack of cleanliness in the fabrication process may lead to an inability to accurately evaluate the fabrication results of the prototype. To address this challenge, the Design–Manufacture–Assembly (DFA) design framework has been presented as an integrated prototyping process that allows users to leverage fabrication as part of a collaborative process. Schon and Wiggins [66] proposed a theoretical model called "Seeing–Moving–Seeing", which is primarily used in design because it requires hands-on practice, creative thinking, design thinking, and observation. In this context, designers often allow their ideas to be converted and shown on the design media to better explain the circular thinking related to hands-on implementation and testing. However, as mentioned earlier, the fabrication process can sever the relationship between the designer and the finished product, making it difficult to fully apply the "Seeing–Moving–Seeing" model in digital fabrication.

In the DFA design framework, the designer first designs the virtual model through CAD/CAM, transfers the model to the fabrication machine for fabrication, and then takes the manufactured components back for assembly. During this process, the designer must keep trying and making mistakes to find the best solution. In this system, users may have varying levels of technical knowledge, including tool assembly and machine parameter adjustments. The information is displayed on both the user app and the HoloLens interface (Figure 7). Therefore, the system decomposes each step into different tasks and systematizes and modulates each step through design computing. Cyber-physical technologies are integrated into an interactive fabrication process that controls the physical machine to perform tasks, lowering the technological threshold. All hand icons are converted to gesture input, and all non-hand icons are automated. To perform an action, the user first selects the appropriate icon and then performs a gesture or waits for the system to complete its task. Virtual objects displayed in the user interface (UI) can have three different states. When the gesture icon is not selected, objects that cannot be modified are shown in light gray. When the object can be modified, the object turns yellow, and the object currently being modified is displayed in green. This color display provides information about the current system state. Feedback is provided, especially when grabbing objects, to help users determine whether the system recognizes their engaging gestures and how their movements affect the scene. Through these features, users can perform digital fabrication through simple gestures, integrating the real world with the virtual environment to create new environments and visualization effects. This allows reality and virtualization to coexist and interact immediately.

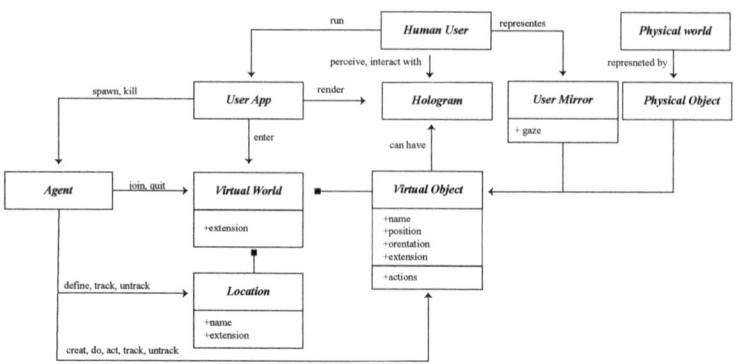

Figure 7. The concept of the co-existing human interaction process of SENS+.

4. Evaluation of User Behavior in Co-Fabrication Space

The co-existing fabrication system for a smart DFA environment integrates real-world and virtual environments, as well as energy usage, based on the agent concept of a system. This system references physical entities and virtual information to each other in a recursive way through a series of physical changes, information analysis, generative fabrication, and energy design suggestions. This approach optimizes the workflow and can be applied to various types of users.

The purpose of the DFA design framework is to improve the fabrication and design process, which is mainly for new designs. However, there is no practical example of DFA applications in materials or assembly, and design and fabrication must consider how the finished product will be assembled, which requires constant trial and error. The agent is a way to connect the virtual model and the physical finished prototype. By integrating the virtual and physical models into a digital interface for designers to interact with, the problem of the analog communication medium in DFA can be solved, and the final prototype can be shown in real time and used in a variety of materials.

To integrate virtual and physical models, the system combines cameras and infrared sensors in a co-existing fabrication system to build an intelligent system of an intelligent DFA environment. The research uses the Seeing–Moving–Seeing–Design–Thinking model to help users refine their ideas. This process of the SENS+ system can be divided into two recursive sequences. One creates a virtual interface layer on top of the physical user perspective, constructing an environment with understandable information for the user. The other captures user gestures and movements to control a robotic arm for handcrafting or fabrication. Users can observe objects in physical space, interact with virtual objects, and recalibrate design properties during fabrication by comparing the appropriate hybrid environment with the virtual model.

To render the model in physical space and generate the model interface, Fologram, a Grasshopper plugin, transfers the geometric information to the SENS+ system and guides the machine end effector along the path. The research designed a UI and computer server system that generate fabrication tool code. The system communicates by scanning a QR code, enabling virtual models to generate data strings and return them to Grasshopper. Additionally, custom C# scripts perform fabrication calculations, such as path calculations that incorporate geometric data, behavioral variables, and remapping parameters between the fabrication tool and UI system. These processes allow for the efficient and accurate translation of virtual models into physical reality.

4.1. Human Behavior Data

The interaction between the user and the space connect devices can modify the user behaviors and change their usage and the consumption at the environment and co-fabrication configuration. A different user can use the devices to understand the operating procedures and energy usage. Table 1 represents an actual co-fabrication space scenario. In this co-fabrication space, the researcher observes and collects information regarding 10 users to analyze their weekly usage in a co-fabrication process in real time.

The research models a variety of human activities using recorded data mining and machine learning approaches, sensor-based activity recognition in the environment, and the developing field of sensor networks. In Figure 8, the X-axis is the record date, the Y-axis is each device's consumption energy in a month, and the lines of different colors represent different devices. The app on the user's mobile phone connects to the environment sensor network for energy consumption predictions, and the environment records the user's real position through the mobile phone's GPS and beacon. Real-time tracking of the user's entry and exit to the building, activity within the co-fabrication space, equipment usage patterns, and energy consumption is carried out. Researchers in the field of sensor-based activity detection believe that by equipping them with powerful computers and sensors that monitor the behavior of agents, these computers can better act on our behalf. HoloLens and other vision sensors that consider color and depth information enable more precise automatic action recognition and integrate a wide range of new applications in smart environments.

Table 1. Co-fabrication space usage.

Space	Machine	A	B	C	D	E	F	G	H	I	J
Laser Cutter Room	Laser Cutter	✓		✓				✓	✓		
	Prototyping	✓	✓	✓			✓	✓	✓		
	Cutting Machine			✓		✓	✓				
	Computer				✓						
	Laser Cutter Teaching				✓					✓	
3D Printer Room	3D Printer Class	✓									
	3D Printer	✓						✓	✓	✓	✓
	CNC Machine	✓						✓			
	Metal Printing							✓			
	Light Curing Printing							✓			
	Resin Printing							✓			
Teaching Space	lecture		✓	✓		✓	✓	✓			
	Administration		✓	✓							
	Conference	✓	✓	✓		✓	✓	✓	✓		✓
	Manager Meeting	✓	✓	✓		✓	✓	✓			
	Teaching Assistant Class	✓	✓	✓							
Muse Space	Project Discuss		✓	✓	✓	✓					
	Idea Thinking										
Scrub Room	Polisher Machine										
	Exhaust Fan				✓						
	Sandblasting Machine				✓		✓				
	Bandsaw Machine				✓				✓		
Metalworking Room	Hand Jointer Machine						✓				
	Angle Slot Machine		✓								
	Vertical Flower Planer		✓	✓	✓						
Model Room	Polisher Machine		✓								
	Hand Jointer Machine				✓				✓		
Robotic Room	Robotic Arm	✓			✓	✓					
	Motor	✓			✓	✓	✓	✓			
	Pressurizer	✓				✓	✓	✓			✓

Figure 8 shows the recorded dates of each device in a month, while Figure 9 shows the daily usage of each user in the space over a month. The graphs indicate that usage still has a different effect, especially if a user stays in the same place each day. The system can use users' demand to change their behaviors and day-to-day conditioning. Therefore, yearly records can help understand users' introductory behaviors and usage. The value increases and decreases allow one to explore daily energy usage differences. In Figure 9, the green line represents co-fabrication space usage, while the blue line represents co-working space usage. It is easy to record the energy used by a single user in a space, but it is difficult to record and distribute energy in a co-fabrication space. This space includes more hidden user behavior, which is different from general usage. For example, one person may lease but divide energy consumption into subdivisions. To address this, the researchers identified users and their locations to help distribute energy usage and make personalized energy suggestions. Overall, the research shows that usage still has a different effect, especially if a user stays in the same place each day. By exploring daily energy usage differences, the system can help change user behaviors and encourage day-to-day conditioning. The yearly records can help understand users' introductory behaviors and usage.

However, with the introduction of distributed storage, it becomes possible to balance the forecast and simulation of daily swings, such as the ones observed in the test research, with consumption patterns. The experiment has shown that human behaviors are complex and usage patterns are influenced by a variety of circumstances. Energy usage fluctuations throughout the day cannot be accurately predicted or specified. However, simulation feedback suggests that providing real-time access to consumption statistics may be a useful strategy for persuading users to alter their usage patterns. Ultimately, the results demonstrate that users can understand consumption data on their devices and that they need access to relevant information at all times.

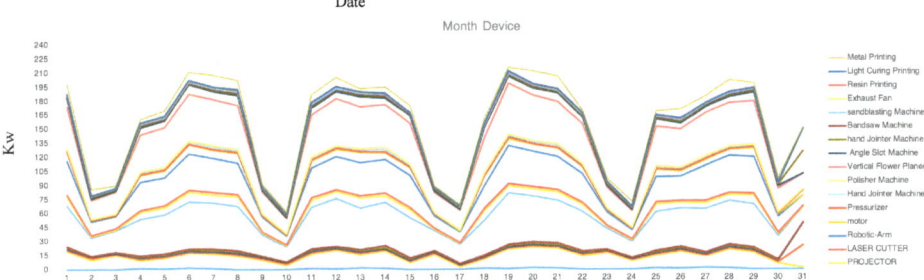

Figure 8. Monthly usage of every device.

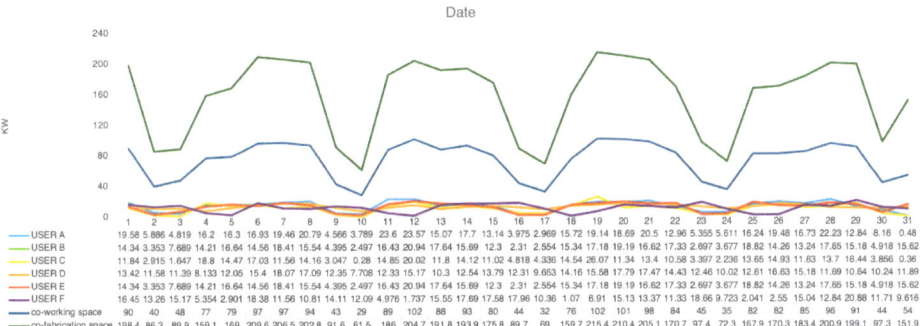

Figure 9. Different space compare data.

4.2. User Interaction with the Scenario and the User Interface

To gain insights into the practical implementation of the design, the researchers conducted an initial experiment using MR technology to superimpose and project complex shapes and shape grammar onto real-life environments. The experiment was divided into two stages. During the first stage (Figure 10a), user employed shape grammars to design 40 × 40 × 40 structures. As user designed and created small models, there were no issues with fabrication, assembly, or differences in construction methods or design comprehension, likely due to the models' small size. In the second stage (Figure 10b), the user plans to scale up the design to a 300 × 300 × 300 funicular weaving structure with a folded plate skin design that has been calculated using shape grammar. However, the fabrication process may encounter some challenges due to the use of a single method, which can affect the assembly of individual components and make the construction process more complex. Consequently, designers will need to devote more time to describing the process and creating images to help others understand it. To address these challenges, users will evaluate and refine the fabrication and assembly methods to optimize the practical implementation of the design. This will involve considering multiple fabrication methods and potential solutions to potential assembly issues. By doing so, the user aims to ensure that the design can be successfully realized and applied in real-world scenarios.

 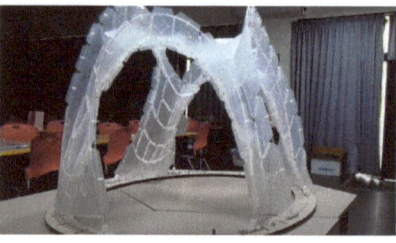

(a) (b)

Figure 10. Users discussing the prototype and the final result. (**a**) Assemblers use HoloLens; (**b**) 300 × 300 × 300 prototype.

Through the analysis of the actual operation process, the scenario is divided into two parts. The first part is the user app interaction, and the other scenario is the MR device. In the user app part, the user will use the agent to describe the interaction process. A modular individual reasoning architecture is represented by each agent. The benefits of that depiction are the simplicity with which operational reasoning or tactics for various interaction scenarios within the building may be put up, modified, and expanded. Users (Human Agents) must make a reservation on the machine they intend to use in a time slot with the appropriate dataset before they visit the area. In the co-fabrication scenario, a set of machine options must be selected, and the system must be given the power, speed, duration, and consumption for each machine. In the interaction scenario of SENS+ (Figure 11), users enter each space of the co-fabrication space with an app via the QR code, accessed via the screen on the door.

1. After scanning the QR code, the energy consumption, the generation of the location, and the reserve item are displayed in the system with basic location information (Figure 11a).
2. When the user approaches the electronic devices that are in the room, there are two ways to see the current consumption load of the device. First, through the system, it is displayed as energy consumption information on the user's mobile device. Second, the demand and use are displayed on an MR with a notification sent to the user. Users can update their design to the system. The system will automatically analysis and disassemble the component and will provide a construction methods suggestion (Figure 11b).
3. After that, the system will generate three types of document for the user. First, an E-design document is generated by the designer 3D model. This document can help designers immediately examine their design (Figure 11c).
4. The e-design document is provided by the designer before fabrication, and this document can help the designer check their design directly. The second type of archive is the e-fabrication document (Figure 11d).
5. This file is provided for the fabricator to see. This stage is the file produced by solving the communication and imagination problems of the traditional digital fabrication process. These files will provide a detailed fabrication process and advice on how to break down each component. The last e-assembly document mainly provide the basis and step understanding for assembler assembly (Figure 11e).
6. Finally, the designer checks that their file is complete and can use their preferences and habits to choose their methods and fabrication machine (Figure 11f).

The energy suggestion is the system through which the 3D model file's complex range provides the available machines and evaluates the consumption energy of different machines so that users can choose. When the machine completes the task, users can evaluate whether they are satisfied with the results by themselves. If the user is not satisfied with the results, MR will provide other options for construction method suggestions to the user. After the user evaluates and decides to switch to other construction method suggestions, they can make a new round of appointments through the app again. Through the app, users have the option to evaluate how much energy their chosen construction method can save, and the MR and app will tell users which devices are accessible in the area. In the scenario where the user evaluates the equipment usage and selects energy production, the system will display the energy usage as well as the estimated monthly savings from the energy source.

To create and test a behavior-based co-fabrication space environment, the system integrates smart sensing, a sensor network design, and user experiences. Within shared co-making spaces, the prototype integrates both energy generation and consumption. Each user's unique usage patterns, habits, and energy use in the co-fabrication spaces have an impact on the everyday environment and energy usage. An option will also affect the consumption result because higher energy consumption occurs over longer periods of time, but the machine result will be more detailed, as desired by the maker. The optimization and standard options for the machine path recorded can provide energy consumption patterns as well as potential user awareness of the patterns. As a result, each agent will interact with one another, have a distinct mission and goal, and offer a chance for improvement to the user. The optimized alternatives consume less energy, even if they frequently do not produce optimal results.

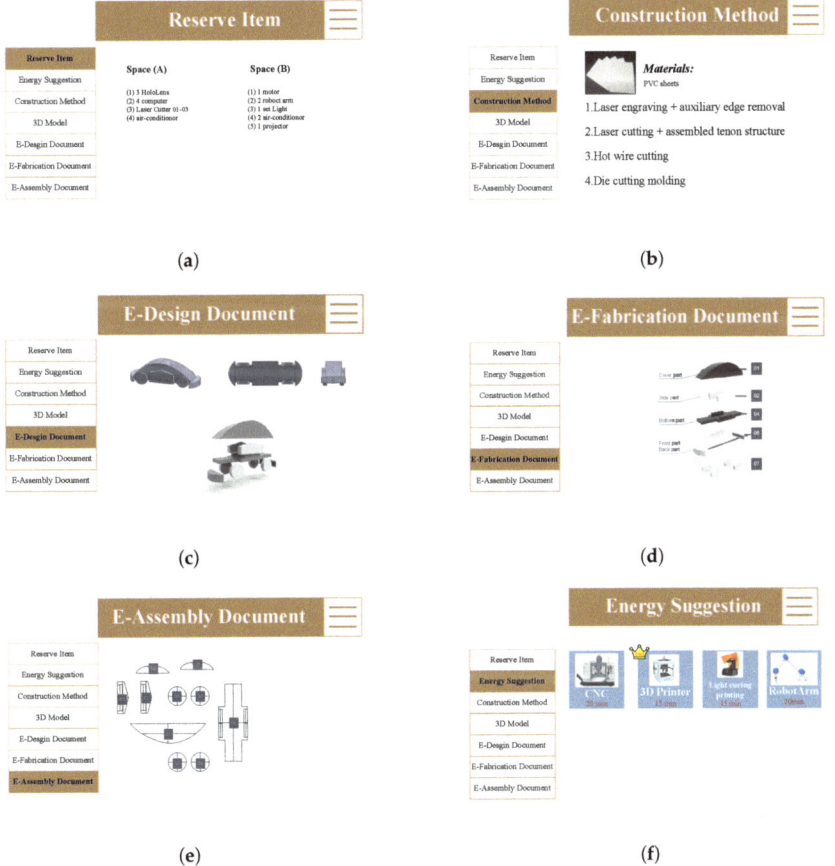

Figure 11. User scenario and UI. (**a**) Reserve Item; (**b**) Construction Method; (**c**) Design Document; (**d**) Fabrication Document; (**e**) Assembly Document; (**f**) Energy Suggestion.

5. Conclusions and Future Work

This paper presents an integration of the Co-Existing Interactive Fabrication Tool and proposes a Dynamic Interaction Process for Fabrication Design, a tool that supports designers, manufacturers, and assemblers in digital fabrication behaviors through virtual–physical integration technology. Linking the three phases of the DFA design framework and sharing information through the concept of digital twins allows the three phases to be continuously retraced and tested, enabling a real-time preview of the finished product in advance as well as a testing of multiple material applications. The tool framework illustrates the possibilities and flexibility of an immersive interactive environment and provides users with the ability to instantly view and confirm the production details based on the information provided. Through human–computer collaboration, the difference between the virtual and the physical in the cyber-physical coexistence space can be reduced, and the fabrication of complex spatial structures can be realized while reducing cost, risk, and complexity. The incorporation of method recommendations into the fabrication process can enhance user knowledge and ability for preparing fabrication. Users can discuss and adjust the graph and 3D model with fabricators immediately. The system provides a systematic perspective on disassembly, allowing fabricators to receive accurate fabrication documentation and complete accurate component fabrication in a multimodal environment.

In a future study, biosignal processing and activity modeling for multimodal human activity recognition [67] will provide further affective information needed for the SENS system. Additionally, the information-sharing function can also improve, such that other design members, participants from different fields, or remote designers can participate in the design or share that design together. The following three research directions are proposed.

1. **Fabrication process automation:** In the interactive fabrication process, fabricators are currently provided with the ability to design, manufacture, and assemble the operating components required for some of the fabrication processes one by one. After the fabrication is completed, the maker may then work out the path of the movement of the fabrication machine. To make digital fabrication easier to operate, design, and manufacture, in future development, the step-by-step design process can be automated through digital computing so that designers can control and prolong the design development process.
2. **Customized fabrication process integration:** In the current study, the fabrication process provided workflow recommendations that allowed individual machines to process and distribute, but it lacked a more precise workflow. On this basis, it is possible to extend the processing module through precise system calculations so that designers, fabricators, and assemblers can be involved in the same process by the integration of the tool, so as to achieve the aim of customizing the fabrication process integration.
3. **Digital twins for integrated method applications:** In the future, this study can focus on the fusion of real and virtual coexistence under the concept of digital twins. At present, spatial sensing as a mechanism feeds back physical products to the virtual environment, but only intercepts the current state for analysis, which is not in line with real-time physical signal transmission. The ability to receive signals in real time could be investigated in the future and combine sensors with interactive fabrication tools to allow physical feedback to occur in real time, making the finished product more compatible with virtual models.

Author Contributions: Conceptualization, T.-W.C. and S.D.; methodology, T.-W.C. and S.D.; software, H.-Y.H. and T.-W.C.; validation, H.-Y.H., C.-C.H. and T.-W.C.; formal analysis, T.-W.C., S.D. and W.N.; investigation, T.-W.C., W.N. and S.D.; resources, T.-W.C., S.D. and W.N.; data curation, H.-Y.H. and C.-C.H.; writing—original draft preparation, T.-W.C., C.-C.H. and H.-Y.H.; writing—review and editing, S.D., T.-W.C., C.-C.H. and W.N.; visualization, H.-Y.H. and C.-C.H.; supervision, T.-W.C. and S.D.; project administration, T.-W.C. and H.-Y.H. All authors have read and agreed to the published version of the manuscript.

Funding: This research received no external funding.

Conflicts of Interest: The authors declare no conflict of interest.

References

1. Marques, G. Ambient assisted living and internet of things. In *Harnessing the Internet of Everything (IoE) for Accelerated Innovation Opportunities*; IGI Global: Hershey, PA, USA, 2019; pp. 100–115.
2. Latif, S.; Driss, M.; Boulila, W.; Huma, Z.E.; Jamal, S.S.; Idrees, Z.; Ahmad, J. Deep Learning for the Industrial Internet of Things (IIoT): A Comprehensive Survey of Techniques, Implementation Frameworks, Potential Applications, and Future Directions. *Sensors* **2021**, *21*, 7518. [CrossRef] [PubMed]
3. Hakimi, S.M.; Hasankhani, A. Intelligent energy management in off-grid smart buildings with energy interaction. *J. Clean. Prod.* **2020**, *244*, 118906. [CrossRef]
4. Figueiredo, J.; Costa, J.L.S.d. A SCADA system for energy management in intelligent buildings. *Energy Build.* **2012**, *49*, 85–98. [CrossRef]
5. Yu, G.J.; Chang, T.W. Reacting with Care: The Hybrid Interaction Types in a Sensible Space. In Proceedings of the Human-Computer Interaction. Towards Mobile and Intelligent Interaction Environments, Orlando, FL, USA, 9–14 July 2011; Jacko, J.A., Ed.; Springer: Berlin/Heidelberg, Germany, 2011; pp. 250–258.
6. Zanella, A.; Bui, N.; Castellani, A.; Vangelista, L.; Zorzi, M. Internet of things for smart cities. *IEEE Internet Things J.* **2014**, *1*, 22–32. [CrossRef]
7. Liu, H.; Gamboa, H.; Schultz, T. Sensor-Based Human Activity and Behavior Research: Where Advanced Sensing and Recognition Technologies Meet. *Sensors* **2023**, *23*, 125. [CrossRef] [PubMed]
8. Leng, J.; Zhou, M.; Xiao, Y.; Zhang, H.; Liu, Q.; Shen, W.; Su, Q.; Li, L. Digital twins-based remote semi-physical commissioning of flow-type smart manufacturing systems. *J. Clean. Prod.* **2021**, *306*, 127278. [CrossRef]
9. Nikolakis, N.; Maratos, V.; Makris, S. A cyber physical system (CPS) approach for safe human-robot collaboration in a shared workplace. *Robot.-Comput.-Integr. Manuf.* **2019**, *56*, 233–243. [CrossRef]
10. Kotsiopoulos, T.; Sarigiannidis, P.; Ioannidis, D.; Tzovaras, D. Machine Learning and Deep Learning in smart manufacturing: The Smart Grid paradigm. *Comput. Sci. Rev.* **2021**, *40*, 100341. [CrossRef]
11. Sun, S.; Zheng, X.; Gong, B.; García Paredes, J.; Ordieres-Meré, J. Healthy Operator 4.0: A Human Cyber–Physical System Architecture for Smart Workplaces. *Sensors* **2020**, *20*, 2011. [CrossRef]
12. Wang, D.; Zhong, D.; Souri, A. Energy management solutions in the Internet of Things applications: Technical analysis and new research directions. *Cogn. Syst. Res.* **2021**, *67*, 33–49. [CrossRef]
13. Henderson, T.C.; Sobh, T.M.; Zana, F.; Brüderlin, B.; Hsu, C.Y. Sensing strategies based on manufacturing knowledge. In Proceedings of the ARPA Image Understanding Workshop, Monterey, CA, USA, 13–16 November 1994; pp. 1109–1113.
14. Fan, Y.; Yang, J.; Chen, J.; Hu, P.; Wang, X.; Xu, J.; Zhou, B. A digital-twin visualized architecture for Flexible Manufacturing System. *J. Manuf. Syst.* **2021**, *60*, 176–201. [CrossRef]
15. Quaid, M.A.K.; Jalal, A. Wearable sensors based human behavioral pattern recognition using statistical features and reweighted genetic algorithm. *Multimed. Tools Appl.* **2020**, *79*, 6061–6083. [CrossRef]
16. Frazzon, E.M.; Agostino, Í.R.S.; Broda, E.; Freitag, M. Manufacturing networks in the era of digital production and operations: A socio-cyber-physical perspective. *Annu. Rev. Control* **2020**, *49*, 288–294. [CrossRef]
17. Baroroh, D.K.; Chu, C.H.; Wang, L. Systematic literature review on augmented reality in smart manufacturing: Collaboration between human and computational intelligence. *J. Manuf. Syst.* **2021**, *61*, 696–711. [CrossRef]
18. Wang, L.; Liu, S.; Liu, H.; Wang, X.V. *Overview of Human-Robot Collaboration in Manufacturing*; Springer: Berlin/Heidelberg, Germany, 2020.
19. Weiser, M.; Gold, R.; Brown, J.S. The origins of ubiquitous computing research at PARC in the late 1980s. *IBM Syst. J.* **1999**, *38*, 693–696. [CrossRef]
20. Forbes, G.; Massie, S.; Craw, S. Fall prediction using behavioural modelling from sensor data in smart homes. *Artif. Intell. Rev.* **2020**, *53*, 1071–1091. [CrossRef]
21. Chua, S.L.; Foo, L.K.; Guesgen, H.W.; Marsland, S. Incremental Learning of Human Activities in Smart Homes. *Sensors* **2022**, *22*, 8458. [CrossRef] [PubMed]
22. Gupta, N.; Gupta, S.K.; Pathak, R.K.; Jain, V.; Rashidi, P.; Suri, J.S. Human activity recognition in artificial intelligence framework: a narrative review. *Artif. Intell. Rev.* **2022**, *55*, 4755–4808. [CrossRef] [PubMed]
23. Zhang, R.; V E, S.; Jackson Samuel, R.D. Fuzzy Efficient Energy Smart Home Management System for Renewable Energy Resources. *Sustainability* **2020**, *12*, 3115. [CrossRef]
24. Mekruksavanich, S.; Jitpattanakul, A. LSTM Networks Using Smartphone Data for Sensor-Based Human Activity Recognition in Smart Homes. *Sensors* **2021**, *21*, 1636. [CrossRef] [PubMed]
25. Kabalci, Y.; Kabalci, E.; Padmanaban, S.; Holm-Nielsen, J.B.; Blaabjerg, F. Internet of things applications as energy internet in smart grids and smart environments. *Electronics* **2019**, *8*, 972. [CrossRef]
26. Hřebíček, J.; Schimak, G.; Kubásek, M.; Rizzoli, A.E. (Eds.) *E-SMART: Environmental Sensing for Monitoring and Advising in Real-Time*; Springer: Berlin/Heidelberg, Germany, 2013.

27. Chang, T.W. Modeling generative interplay using actingrole model. From distributed collaboration to generative interplay. *CoDesign* **2006**, *2*, 35–48. [CrossRef]
28. Chang, T.W.; Datta, S.; Lai, I.C. Modelling Distributed Interaction with Dynamic Agent Role Interplay System. *Int. J. Digit. Media Des.* **2016**, *8*, 1–14.
29. Dutta, S.; Chukkapalli, S.S.L.; Sulgekar, M.; Krithivasan, S.; Das, P.K.; Joshi, A. Context Sensitive Access Control in Smart Home Environments. In Proceedings of the 2020 IEEE 6th Intl Conference on Big Data Security on Cloud (BigDataSecurity), IEEE Intl Conference on High Performance and Smart Computing, (HPSC) and IEEE Intl Conference on Intelligent Data and Security (IDS), Baltimore, MD, USA, 25–27 May 2020; pp. 35–41. [CrossRef]
30. Chang, T.W.; Huang, H.Y.; Hung, C.W.; Datta, S.; McMinn, T. A Network Sensor Fusion Approach for a Behaviour-Based Smart Energy Environment for Co-Making Spaces. *Sensors* **2020**, *20*, 5507. [CrossRef] [PubMed]
31. Grube, D.; Malik, A.A.; Bilberg, A. SMEs can touch Industry 4.0 in the Smart Learning Factory. *Procedia Manuf.* **2019**, *31*, 219–224. [CrossRef]
32. Fernández-Caballero, A.; Martínez-Rodrigo, A.; Pastor, J.; Castillo, J.; Lozano-Monasor, E.; López, M.T.; Zangróniz, R.; Latorre, J.; Fernández-Sotos, A. Smart environment architecture for emotion detection and regulation. *J. Biomed. Inform.* **2016**, *64*, 55–73. [CrossRef] [PubMed]
33. Hsieh, T.L.; Chang, T.W. How to collective design-and-fabricating a weaving structure interaction design—Six experiments using a design-fabrication-assembly (DFA) approach. In Proceedings of the 4th RSU National and International Research Conference on Science and Technology, Social Sciences, and Humanities 2019 (RSUSSH 2019), Pathum Thani, Thailand, 26 April 2019.
34. Piper, W.; Sun, H.; Jiang, J. Digital Twins for Smart Cities: Case Study and Visualisation via Mixed Reality. In Proceedings of the 2022 IEEE 96th Vehicular Technology Conference (VTC2022-Fall), Beijing, China, 26–29 September 2022.
35. Ríos, A.P.; Callaghan, V.; Gardner, M.; Alhaddad, M.J. Using Mixed-Reality to Develop Smart Environments. In Proceedings of the 2014 International Conference on Intelligent Environments, Shanghai, China, 30 June–4 July 2014; pp. 182–189. [CrossRef]
36. Croatti, A. Augmented Worlds: A Proposal for Modelling and Engineering Pervasive Mixed Reality Smart Environments. Ph.D. Thesis, University of Bologna, Bologna, Italy, 2019.
37. Oliff, H.; Liu, Y.; Kumar, M.; Williams, M.; Ryan, M. Reinforcement learning for facilitating human-robot-interaction in manufacturing. *J. Manuf. Syst.* **2020**, *56*, 326–340. [CrossRef]
38. Hsieh, T.L.; Chang, T.W. ViDA: A visual system of DFA process for interactive surface. In Proceedings of the 2019 23rd International Conference in Information Visualization–Part II, Adelaide, Australia, 16–19 July 2019; pp. 68–73.
39. Chang, T.W.; Hsiao, C.F.; Chen, C.Y.; Huang, H.Y., CoFabs: An Interactive Fabrication Process Framework. In *Architectural Intelligence*; Springer: Berlin/Heidelberg, Germany, 2020; pp. 271–292.
40. Croatti, A.; Ricci, A. A model and platform for building agent-based pervasive mixed reality systems. In Proceedings of the International Conference on Practical Applications of Agents and Multi-Agent Systems, Toledo, Spain, 20–22 June 2018; pp. 127–139.
41. Weichel, C.; Lau, M.; Kim, D.; Villar, N.; Gellersen, H.W. MixFab: A mixed-reality environment for personal fabrication. In Proceedings of the SIGCHI Conference on Human Factors in Computing Systems, Toronto, ON, Canada, 26 April–1 May 2014; pp. 3855–3864.
42. Taylor, A.G., What Is the Microsoft HoloLens? In *Develop Microsoft HoloLens Apps Now*; Springer: Berlin/Heidelberg, Germany, 2016; pp. 3–7.
43. Cengiz, A.B.; Birant, K.U.; Cengiz, M.; Birant, D.; Baysari, K. Improving the Performance and Explainability of Indoor Human Activity Recognition in the Internet of Things Environment. *Symmetry* **2022**, *14*, 2022. [CrossRef]
44. Ramos, R.G.; Domingo, J.D.; Zalama, E.; Gómez-García-Bermejo, J.; López, J. SDHAR-HOME: A Sensor Dataset for Human Activity Recognition at Home. *Sensors* **2022**, *22*, 8109. [CrossRef] [PubMed]
45. Kim, E.; Helal, S.; Cook, D. Human Activity Recognition and Pattern Discovery. *IEEE Pervasive Comput.* **2010**, *9*, 48–53. [CrossRef]
46. Schwartz, T. HAL. In *Proceedings of the Rob | Arch 2012*; Brell-Çokcan, S., Braumann, J., Eds.; Springer: Vienna, Austria, 2013; pp. 92–101.
47. Saeed, N.; Alouini, M.S.; Al-Naffouri, T.Y. Toward the Internet of Underground Things: A Systematic Survey. *IEEE Commun. Surv. Tutor.* **2019**, *21*, 3443–3466. [CrossRef]
48. Rafferty, L.; Iqbal, F.; Aleem, S.; Lu, Z.; Huang, S.C.; Hung, P.C. Intelligent multi-agent collaboration model for smart home IoT security. In Proceedings of the 2018 IEEE International Congress on Internet of Things (ICIOT), San Francisco, CA, USA, 2–7 July 2018; pp. 65–71.
49. Chu, G.; Lisitsa, A. Penetration testing for internet of things and its automation. In Proceedings of the 2018 IEEE 20th International Conference on High Performance Computing and Communications; IEEE 16th International Conference on Smart City; IEEE 4th International Conference on Data Science and Systems (HPCC/SmartCity/DSS), Exeter, UK, 28–30 June 2018; pp. 1479–1484.
50. Stringer, P.; Cardoso, R.C.; Huang, X.; Dennis, L.A. Adaptable and verifiable BDI reasoning. *arXiv* **2020**, arXiv:2007.11743.
51. Boulila, W.; Sellami, M.; Driss, M.; Al-Sarem, M.; Safaei, M.; Ghaleb, F.A. RS-DCNN: A novel distributed convolutional-neural-networks based-approach for big remote-sensing image classification. *Comput. Electron. Agric.* **2021**, *182*, 106014. [CrossRef]
52. Yu, G.J.; Chang, T.W.; Wang, Y.C. SAM: a spatial interactive platform for studying family communication problem. In Proceedings of the Symposium on Human Interface, San Diego, CA, USA, 19–24 July 2011; pp. 207–216.

53. Chang, T.; Jiang, H.; Chen, S.; Datta, S. Dynamic skin: interacting with space. In Proceedings of the 17th International Conference on Computer Aided Architectural Design Research in Asia, Chennai, India, 25–28 April 2012; pp. 89–98.
54. Cardoso, R.C.; Dennis, L.A.; Fisher, M. Plan library reconfigurability in BDI agents. In Proceedings of the International Workshop on Engineering Multi-Agent Systems, Online, 3–4 May 2019; pp. 195–212.
55. Sierra, P. BDI logic applied to a dialogical interpretation of human–machine cooperative dialogues. *Log. J. IGPL* **2021**, *29*, 536–548. [CrossRef]
56. Bordini, R.H.; Dennis, L.A.; Farwer, B.; Fisher, M. Automated verification of multi-agent programs. In Proceedings of the 2008 23rd IEEE/ACM International Conference on Automated Software Engineering, L'Aquila, Italy, 15–19 September 2008; pp. 69–78.
57. Castanedo, F.; Garcia, J.; Patricio, M.A.; Molina, J. A multi-agent architecture based on the BDI model for data fusion in visual sensor networks. *J. Intell. Robot. Syst.* **2011**, *62*, 299–328. [CrossRef]
58. Leask, S.; Alechina, N.; Logan, B. A Computationally Grounded Model for Goal Processing in BDI Agents. In Proceedings of the Proceedings of the 6th Workshop on Goal Reasoning (GR'2018), Stockholm, Sweden, 13 July 2018.
59. Lee, E.A. Cyber physical systems: Design challenges. In Proceedings of the 2008 11th IEEE International Symposium on Object and Component-Oriented Real-Time Distributed Computing (ISORC), Orlando, FL, USA, 5–7 May 2008; pp. 363–369.
60. Hallgrimsson, B. *Prototyping and Modelmaking for Product Design*; Laurence King Publishing: Hachette, UK, 2012.
61. Yan, Y.; Li, S.; Zhang, R.; Lin, F.; Wu, R.; Lu, Q.; Xiong, Z.; Wang, X. Rapid prototyping and manufacturing technology: Principle, representative technics, applications, and development trends. *Tsinghua Sci. Technol.* **2009**, *14*, 1–12. [CrossRef]
62. Lechuga-Jiménez, C. Las metáforas del mapa y el espejo en el arte contemporáneo. El Genio Maligno. *Rev. Humanidades Cienc. Soc.* **2016**.
63. Bechthold, M. The return of the future: a second go at robotic construction. *Archit. Des.* **2010**, *80*, 116–121. [CrossRef]
64. Gramazio, F.; Kohler, M. *Made by Robots: Challenging Architecture at a Larger Scale*; John Wiley and Sons: Hoboken, NJ, USA, 2014.
65. Lavallee, J.; Vroman, R.; Keshet, Y. Automated folding of sheet metal components with a six-axis industrial robot. In Proceedings of the 31st Annual Conference of the Association for Computer Aided Design in Architecture (ACADIA) 2011, Calgary/Banff, AL, Canada, 11–16 October 2011.
66. Schon, D.A.; Wiggins, G. Kinds of seeing and their functions in designing. *Des. Stud.* **1992**, *13*, 135–156. [CrossRef]
67. Liu, H. Biosignal processing and activity modeling for multimodal human activity recognition. Ph.D. Thesis, Universitat Bremen, Bremen, Germany, 5 November 2021. [CrossRef]

Disclaimer/Publisher's Note: The statements, opinions and data contained in all publications are solely those of the individual author(s) and contributor(s) and not of MDPI and/or the editor(s). MDPI and/or the editor(s) disclaim responsibility for any injury to people or property resulting from any ideas, methods, instructions or products referred to in the content.

Review

Understanding Naturalistic Facial Expressions with Deep Learning and Multimodal Large Language Models

Yifan Bian [1], Dennis Küster [2], Hui Liu [2] and Eva G. Krumhuber [1,*]

1. Department of Experimental Psychology, University College London, London WC1H 0AP, UK; yifan.bian.23@ucl.ac.uk
2. Department of Mathematics and Computer Science, University of Bremen, 28359 Bremen, Germany; kuester@uni-bremen.de (D.K.); hui.liu@uni-bremen.de (H.L.)
* Correspondence: e.krumhuber@ucl.ac.uk

Abstract: This paper provides a comprehensive overview of affective computing systems for facial expression recognition (FER) research in naturalistic contexts. The first section presents an updated account of user-friendly FER toolboxes incorporating state-of-the-art deep learning models and elaborates on their neural architectures, datasets, and performances across domains. These sophisticated FER toolboxes can robustly address a variety of challenges encountered in the wild such as variations in illumination and head pose, which may otherwise impact recognition accuracy. The second section of this paper discusses multimodal large language models (MLLMs) and their potential applications in affective science. MLLMs exhibit human-level capabilities for FER and enable the quantification of various contextual variables to provide context-aware emotion inferences. These advancements have the potential to revolutionize current methodological approaches for studying the contextual influences on emotions, leading to the development of contextualized emotion models.

Keywords: automatic facial expression recognition; naturalistic context; deep learning; multimodal large language model

1. Introduction

Recent advances in computer vision have greatly facilitated the development of sophisticated affective computing systems for facial expression recognition (FER) research [1]. Researchers across domains have applied various computational techniques to analyze diverse and complex mental states, including emotions [2], pain [3], physiological correlates [4], personality traits [5], and clinical disorders [6]. Nevertheless, our understanding of facial expressions remains mostly limited to inferences drawn from laboratory studies. Facial expressions produced in controlled laboratory settings may suffer from a lack of ecological validity and fail to represent the full spectrum of facial behaviors observed in real-life scenarios [7,8]. There is a growing emphasis on investigating naturalistic facial behaviors coupled with advanced computational techniques to spark theoretical advancements in affective science [9].

Naturalistic facial expressions can be observed in connection with a wide range of psychologically significant contexts encountered in everyday situations. Naturalistic facial expressions may more accurately reflect the complex and dynamic nature of emotional experiences in the real world than expressions elicited by experimental manipulations that are often artificial or short-lived (e.g., receiving an electrical shock). One approach toward studying more ecologically valid facial expressions relies on materials sourced from third-party media such as reality shows, vlogs, movies, and documentaries. Such sources often comprise millions of facial expressions accompanied by perceptually rich contexts, which are being made available in several newly developed datasets for FER research [10]. For surveys of existing naturalistic facial expression databases, readers are referred to [11–13].

Examining naturalistic expressions presents vast conceptual and empirical opportunities to yield further insights into how emotions emerge and reconfigure in naturalistic situations [14], discover new emotion categories that are rarely observed in laboratory environments [15], and document a comprehensive taxonomy of facial behaviors [16].

However, there are two main challenges in the studies of naturalistic facial expressions. The first challenge relates to the difficulties in detecting and tracking facial behaviors in unconstrained environments. Naturalistic expressions are often captured in uncontrolled settings with unexpected variations in head orientation, illumination, complex background, and facial occlusions, which may result in errors in detecting facial behaviors. More robust affective computing models are required to effectively analyze facial behaviors captured in wild conditions, as traditional models often exhibit drastic drops in performance in uncontrolled environments [17]. The second challenge concerns the interpretation of naturalistic facial expressions. Unlike laboratory studies that can validate the underlying emotional experiences of expressions through self-report measurements, it is not feasible to collect retrospective data to validate naturalistic expressions sourced from the Internet. Hence, accurately inferring the emotion states of naturalistic expressions can be more challenging than working with laboratory data. Nevertheless, this issue can be addressed through multimodal annotation and a comprehensive analysis of situational contexts to specify the underlying psychological states [18].

This review is divided into two sections to discuss the applications of deep learning-based FER toolboxes and multimodal large language models (MLLMs) for tackling these challenges. The first section evaluates several newly developed and easy-to-use FER toolboxes for facial expression analysis in unconstrained environments. To support researchers in making informed decisions regarding the selection of appropriate toolboxes, we critically review the performance of five FER toolboxes, namely OpenFace 2.0 [19], Affdex 2.0 [20], Py-Feat [21], LibreFace [22], and PyAFAR [23], along with their underlying neural architectures and databases used for model training. The second section discusses the potential utilization of MLLMs for analyzing and interpreting naturalistic expressions in association with contextual cues. Naturalistic expressions can be rendered meaningful by referencing the specific contexts in which they occur and interact [18]. MLLMs such as GPT-4V [24] and LLaVA [25] exhibit promising capabilities for quantifying contextual variables, which can serve as contextualized FER models for robust, explainable emotion inferences of naturalistic expressions.

2. Analyzing Naturalistic Facial Expressions with Deep Learning

The general processes of facial expression analysis consist of three components: face detection, feature extraction, and the prediction of facial action units (AUs) [26] and/or emotions. In particular, feature extraction has been regarded as the most crucial component in the FER process, which can be categorized into handcrafted and learned features [15]. Handcrafted features can be extracted with methods such as histograms of oriented gradients (HOGs) and local binary patterns (LBPs) for capturing facial textures and shapes. FER models trained based on handcrafted features using shallow learning approaches (e.g., support vector machines (SVMs)) achieve good performance in facial expressions produced in laboratory settings but have recently been outperformed by models trained on learned features [27]. Learned features are directly extracted from raw data through deep neural networks (DNNs) comprising multiple layers that hierarchically learn more complex, representative spatial–temporal features than the previous layers. Accumulating research evidence suggests that DNN models consistently surpass most of the shallow learning models based on handcrafted features by a large margin [28]. Compared to shallow learning models, cross-domain experiments reveal that DNN models such as convolutional neural networks (CNNs) and vision transformers (ViTs) achieve superior generalizability and accuracy for emotion recognition [29–32] and AU detection [33–35] on unseen datasets with different demographics, camera views, and emotion-eliciting contexts. More importantly, deep FER models are fairly robust to variations in brightness, head poses, and occlusions [27], whereas shallow learning models trained on handcrafted features (e.g., LBPs) may be substantially impacted by variations in luminance [21] and head rotation [36].

Despite their impressive performance, accessing state-of-the-art (SOTA) deep learning models for FER research has been hindered by several obstacles. SOTA models are rarely released for public use in repositories such as GitHub or are rarely open source for end users to fine-tune the models with their own datasets for new tasks. Open source models often lack graphical user interfaces or documentation for easy implementation for users who may lack specific programming knowledge. These SOTA models might be too heavy to run for real-time analysis, restricting their potential application in practical settings. Additionally, early FER models have been primarily trained on datasets captured in controlled recording conditions with invariant illumination and fixed camera position [37]. These models may learn features that rarely align with real-world situations, which limits their effectiveness in analyzing naturalistic expressions. To bridge this gap between cutting-edge FER techniques and their implementation, we discuss a selection of the most prominent, publicly accessible, lightweight, and user-friendly toolboxes that incorporate SOTA models suitable for analyzing facial expressions in the wild.

In the remainder of this section, we will first provide an overview of two FER toolboxes that incorporate both shallow and deep learning models for facial behavior analysis, namely OpenFace 2.0 and Py-Feat. Next, we will discuss three FER toolboxes that are primarily trained on deep learning methods: Affdex 2.0, LibreFace, and PyAFAR. Table 1 summarizes the main characteristics and access information of each FER toolbox.

Table 1. FER toolbox comparison on functionality, neural architecture, and type of dataset used for training emotion or AU models. The deep learning models are shown in bold. The datasets are categorized as W (wild), S (spontaneous), or P (posed), representing facial expressions sourced from the Internet or nonlaboratory environments, induced by experimental procedures, or deliberately mimicked by actors in laboratory settings, respectively. * Please consult the website for complete documentation of all face detection models incorporated in Py-Feat. ** The emotion model of Affdex 2.0 is not specified as it is based on the activation of AUs. *** The dataset used to train AU models in Affdex 2.0 is considered spontaneous despite being captured in nonlaboratory settings.

	OpenFace 2.0	Py-Feat	Affdex 2.0	LibreFace	PyAFAR
Face Detection	CE-CLM	MTCNN, RetinaFace *	R-CNNs	MediaPipe	MediaPipe, Facenet
Emotion Recognition		ResMasNet, SVM	**	ViT, ResNet-18	
Action Unit	SVM, SVR	SVM, XGB	CNNs	ViT, ResNet-18	ResNet-50
Datasets	P, S	P, S, W	S ***	S, W	S
Open Source/Free	Yes	Yes		Yes	Yes
Graphical User Interface	Yes		Yes	Yes	Yes
Website	github.com/TadasBaltrusaitis/OpenFace (accessed on 29 November 2023)	py-feat.org (accessed on 29 November 2023)	www.affectiva.com (accessed on 29 November 2023)	github.com/ihp-lab/LibreFace (accessed on 29 November 2023)	affectanalysisgroup.github.io/PyAFAR2023 (accessed on 29 November 2023)

2.1. FER Toolboxes Based on Mixed Learning Models

OpenFace 2.0 [19]: OpenFace 2.0 is a representative tool of the SOTA shallow learning models based on handcrafted features (i.e., HOGs) for AU recognition. OpenFace 2.0 is capable of a variety of facial analysis tasks, including facial landmark detection, gaze and pose estimation, and AU detection. OpenFace 2.0 uses a deep convolutional expert-constrained local model (CE-CLM) for facial landmark detection and tracking, which is trained on wild datasets with nonfrontal faces and varying illumination. It can better detect profiles or severely occluded faces than its previous version [37]. For AU detection, OpenFace 2.0 relies on shallow learning algorithms, including linear SVM for binary detection (presence or absence) and support vector regression (SVR) for intensity probability estimation. The AU models are trained in seven laboratory databases with AU annotations, which contain posed (Bosphorus [38]; FERA 2011 [39]), spontaneous (CK+, also containing posed expressions [40]; UNBC-McMaster [3]; DISFA [41]; and BP4D/FERA 2015 [42]), and conversational (SEMAINE [43]) expressions. Research by [36] demonstrated that OpenFace 2.0 outperformed several commercial FER toolboxes such as FaceReader 7.0 [44] for AU detection on datasets that contain posed (DISFA+ [45]), conversational (GFT [46]), and wild (Aff-wild2 [47]) expressions. However, OpenFace 2.0 was found to have inferior generalizability ability in AU detection on unseen datasets compared to deep FER models [48]. This is likely due to the limited discriminative values of handcrafted features or the inability of shallow learning algorithms to capture intricate, nonlinear patterns of facial behaviors.

Py-Feat [21]: Py-Feat provides various pretrained models based on both handcrafted and learned features, allowing users to flexibly decide which combinations of models to use according to specific task requirements. Py-Feat includes several face detection models such as multitask convolutional neural networks (MTCNNs [49]) and RetinaFace [50], which demonstrate robustness on partially obscured or nonfrontal faces. For AU detection, Py-Feat uses popular shallow learning methods for computing binary results with SVM and continuous results with optimized gradient boosting (XGB [51]), which are pretrained on both laboratory (BP4D, DISFA, CK+, and UNBC-McMaster) and wild (Aff-wild2 [47]) datasets. Its AU models have been shown to be slightly less accurate than the reported results of OpenFace 2.0 benchmarking on the DISFA+ dataset. However, Py-Feat yielded more consistent and reliable AU estimation for faces with varying head orientations, whereas the performance of OpenFace 2.0 drops dramatically when head angles are larger than 45 degrees [36]. The superior performance of Py-Feat on AU detection for occluded faces might be attributed to the inclusion of the wild dataset [47], which contains nonstandardized facial images and videos for pretraining models. Py-Feat provides direct emotion inferences for six basic emotions with emotion models trained on datasets with spontaneous (CK+ [40]), posed (JAFFE [52]), and wild (ExpW [53]) expressions. Since not all facial regions contribute equally to emotion perception, and some facial regions may be occluded in unconstrained conditions, Py-Feat exploits the residual masking network (ResMasNet [54]), a deep learning method that utilizes attention mechanisms to adaptively weight and select the most emotionally salient regions of the face while ignoring irrelevant features (e.g., facial occlusion). The ResMasNet outperforms another shallow learning model (i.e., SVM) adopted in Py-Feat and a commercial FER toolbox (FACET [55]) on a wild dataset (AffectNet [56]). Py-Feat provides numerous functional packages for data preprocessing, statistical analyses (e.g., time-series correlation and regressions), and visualization, which facilitate data exploration. Py-Feat is written in the Python programming language. Although a graphical user interface is currently not available in Py-Feat, it is relatively easy to use following step-by-step tutorials.

2.2. Deep FER Toolboxes

Affdex 2.0 [20]: Affdex 2.0 is a commercial software program designed to analyze facial behaviors in the wild. For face detection, it exploits region-based convolutional neural networks (R-CNNs [57]), which perform better on challenging conditions (e.g., variations in illumination, hand occlusions, etc.) compared to Affdex 1.0 [58]. For AU detection, Affdex 2.0 utilizes large samples of a private dataset collected using a web-based approach [59] to train CNN models, which showed enhanced accuracy and less biased results on ethnic minorities (e.g., African and South Asian) than its previous version. On benchmark evaluation for AU detection, Affdex 2.0 outperforms numerous SOTA methods on the DISFA dataset, which contains spontaneous expressions induced by emotion-evoking videos. Although Affdex 2.0 performs slightly worse than OpenFace 2.0, it is important to note that the AU models of Affdex 2.0 have not been trained on the DISFA, which may bias the result. Unlike other FER toolboxes that directly predict emotions with separate models specifically trained on emotional expression datasets, Affdex 2.0 estimates seven basic emotions [60] based on the reverse inference of the activation of AUs with assigned weights postulated in EMFACS [61]. Negative weights are assigned for emotion prediction when opposite AUs occur to reduce a false-positive rate. For instance, the presence of a "lip corner lowerer" (AU15) may decrease the likelihood of predicting "happiness" based on the activation of a "lip corner raiser" (AU12). Affdex 2.0 outperforms several SOTA emotion models on the Aff-wild2 dataset [47]. In addition to basic emotions, Affdex 2.0 can predict other affective states such as confusion, sentiment, and engagement based on predefined rules for AU activation. Although it is a commercial software program, Affdex 2.0 has been trained on a large sample of wild dataset with 11 million annotated images of people of all ages, genders, and ethnicities, which might be otherwise difficult to collect and annotate without financial support. By contrast, many open source systems are trained on a comparatively limited number of publicly accessible laboratory datasets with a small number of participants and a narrow range of demographic diversity. They are thus at a disadvantage with respect to the development of robust, unbiased models. However, since Affdex 2.0 is not open source, it does not allow users to further fine-tune the models for downstream tasks.

LibreFace [22]: LibreFace is a newly developed toolkit that incorporates several SOTA deep networks for facial AU and emotion expression analysis. LibreFace utilizes MediaPipe [62] for precise face detection with 468-point 3D landmark registration and normalization with geometric transformation. Model performances including feature extraction, robustness, and generalizability are enhanced by several pretraining processes. Specifically, a ViT-base model and a ResNet-18 model [63] were pretrained on several large wild datasets sourced from the Internet, including the training set of AffectNet [56], EmotioNet [64], and FFHQ [65], which consist of millions of facial images with a wide range of variations in demographic features, illumination, and head orientation. After the pretraining phase, the models were fine-tuned on the DISFA dataset [41] for AU intensity estimation and the BP4D-Spontaneous datasets [66] for AU detection. LibreFace leverages the Swin Transformer model [67] to capture the spatial correlations and interactions between different facial features from a global perspective, which could achieve better performance than traditional CNNs that focus on local patterns and features within facial subregions [68,69]. Moreover, LibreFace bypasses the need for large, labeled datasets by exploiting the masked autoencoder (MAE) method that allows for learning representative facial features through the process of image and video reconstruction [70]. LibreFace utilizes feature-wise knowledge distillation to reduce computational costs, thereby boosting the inference efficiency for real-time facial expression analysis. LibreFace outperforms OpenFace 2.0 [19] and other SOTA deep learning models [35,71] on the DISFA and BP4D datasets for predicting the activation of AU. For emotion prediction, LibreFace achieves competitive results comparable or superior to more complicated and heavier SOTA models [72] on two wild datasets, AffectNet and RaF-DB [73]. LibreFace is currently open source in Python, and for Windows users, an easy-to-use graphical user interface is available using the OpenSense platform [74].

PyAFAR [23]: PyAFAR is developed for facial AU detection and intensity estimation in addition to head orientation and facial landmark detection. For face detection, it uses MediaPipe and Facenet [75], which can identify and track individuals even if they exit and re-enter the field. PyAFAR adopts two separate DNN models based on ResNet50 [63], with increased depth of neural architecture to perform more effective convolution operations for complex facial feature representation. The models have been pretrained on the ImageNet [76] dataset for detecting 12 Aus in adults and 9 Aus in infants. The adult model is fine-tuned on BP4D+ [77], an expansion of the BP4D-Spontaneous dataset containing spontaneous facial expressions induced by both active (e.g., singing) and passive (e.g., watching emotionally loaded videos) tasks. The infant model is trained on the MIAMI [78] and CLOCK [79] databases, which capture infants' responses induced by experimental procedures, such as the removal of attractive toys and still face paradigms [80]. Both adult and infant models achieve accurate results on within-domain validation [23]. PyAFAR shows superior cross-database performance on the GFT dataset [46] compared to OpenFace

2.0 and the previous version of AFAR [81]. An executable interface and a step-by-step visual instruction guide are available for the easy implementation of the toolbox.

These FER toolboxes empower researchers to effectively address the challenge of unexpected variation in naturalistic behaviors acquired from unconstrained environments. However, it is important to note that some FER toolboxes (e.g., Affdex 2.0, LibreFace, and Py-AFAR) have only been recently created and validated by their developers. Further empirical research conducted by independent researchers is required to compare and validate the cross-domain performance of these toolboxes [1].

What is still missing in the studies of naturalistic facial expressions is the lack of a comprehensive analysis of contextual information critical for a naturalistic understanding of emotions [82]. Prior works have focused extensively on the analysis of facial features using FER toolboxes for emotion inferences, while contextual variables have been largely ignored [83]. Emotion inferences made solely based on decontextualized faces are ecologically invalid and meaningless. For instance, a smile can be reliably recognized as expressing "happiness" by FER toolboxes, but it is difficult to evaluate the meaning of the emotion without referencing the emotional stimuli or surrounding contexts (e.g., smile as reflecting the anticipation of a music festival [84,85]). Elucidating the interaction between facial behaviors and concurrent contexts is an important research question for affective science.

A comprehensive analysis of contextual elements can provide important cues for an accurate assessment of the underlying emotional experience associated with naturalistic facial expressions [18]. For example, naturalistic facial behaviors (e.g., a smile) are often accompanied by contextual cues presented in various forms such as clothing (e.g., a gown), scenery (e.g., wedding venues), activities (e.g., marriage proposal), voices (e.g., "I love you"), body postures (e.g., holding hands), other faces, and so forth, which shape how faces are perceived (e.g., the enjoyment of interpersonal connection). Contextual variables can be measured and quantified by human annotators [86]. When facial expressions are presented with perceptually rich contextual information, human annotators show substantially greater agreement for labeling facial expressions than decontextualized faces [87]. This indicates that the current limitations of evaluating facial expressions with FER systems could be addressed by including contextual cues, as human perceivers can make more robust, reliable emotion inferences. However, annotating naturalistic expressions and their contexts can be more complicated and labor-intensive than laboratory datasets. Advanced multimodal annotation tools [88] may help provide multimodal annotation to evaluate facial expressions together with other nonverbal modalities and rich contextual information to provide accurate portrayals of the interaction between facial expressions and contexts. In the following section, we discuss the novel applications of MLLMs that could circumvent the need for extensively annotated datasets, fostering further advancement in naturalistic affective research.

3. Advancing Naturalistic Affective Science with Multimodal Large Language Models

Recent advancements in MLLMs have demonstrated remarkable versatility and capability in various domains and tasks. Although MLLMs are not specifically programmed for emotion recognition tasks, such capabilities emerge as the result of data scaling [89]. The main idea of using MLLMs for emotion recognition is to use powerful large language models as an intelligent brain to process and align textual, visual [25,90], and/or auditory [91,92] information to perform emotion inferences [93]. There is an increasing number of open source MLLMs, including LLaVA [25] and MiniGPT-4 [90], available on platforms such as HuggingFace. Many MLLMs also provide a user-friendly web interface that enables more flexible interactions with the user. This section provides an in-depth discussion on the emergent novel applications of MLLMs for context-aware emotion recognition, generalizable facial expression analysis, and adaptability to other related tasks such as the classification of nuanced emotion categories.

3.1. MLLMs as a Contextualized Emotion Model

MLLMs can serve as a contextualized model quantifying contextual variables to provide robust emotional reasoning for naturalistic expressions (see Figure 1 for example). Several studies have demonstrated the exceptional capabilities of MLLMs in identifying emotionally evoking context and comprehending how these contextual cues may influence the emotional state of a person [94]. To quantify contextual variability, MLLMs can be used to perform a wide range of visual reasoning tasks such as spatial relationship analysis and object recognition to understand the visual world with a simple prompt, i.e., "describe the image". This can provide a detailed description of the situational contexts, which can then be used to infer relevant emotional states and related antecedents or consequences. Such contextualized inferences are more in line with how humans naturally perceive emotions in real-life situations by synthesizing concurrent multisensory information from the face and contextual cues [95]. This potential application is exemplified in a recent technical report of GPT-4V in the sense that it can make accurate emotion inferences based on integrated contextual cues such as "protest crowd" and "presence of policies" for inferring "injustice" and "anger" [24]. Moreover, research by Etesam et al. [94] has revealed that MLLMs outperformed vision language models in context-aware emotion recognition tasks using the EMOTIC dataset [96], which contains contextually rich images annotated with 26 emotion categories. While vision language models like CLIP [97] are effective at detecting immediately visible characteristics such as facial expressions, body postures, and activities, they fail to reason about the underlying causal relationships of these contextual data for emotion inferences. In contrast, MLLMs like LLaVA [25] not only identify these visible characteristics but also integrate and capture the complex relationships among these contextual cues for emotion inferences. For instance, while CLIP may perceive "raising arms" as signifying "surprise" and "fear", MLLMs may reason that this body posture actually reflects "happiness" and "excitement" given the context of skiing. This empirical evidence supports the practical utility of MLLMs for affective research, which can further improve our understanding of how naturalistic emotions manifest in real-life scenarios by considering contextual variability.

Furthermore, the contextual reasoning generated from MLLMs can complement the results obtained from FER toolboxes to produce context-aware emotion inferences, which are more robust and insightful than simply analyzing the face alone. This is particularly important for accurately identifying and interpreting complex or vague expressions that convey mixed emotional signals such as sarcasm and Schadenfreude indicated by the incongruence between facial behaviors and contexts (e.g., a polite smile accompanied by sarcastic statements [98]). More importantly, future studies can utilize MLLMs and FER toolboxes for examining the relationship between different types of contexts and facial expressions. Past research by Cowen et al. [14] has utilized sophisticated DNNs to automatically classify facial expressions and contexts from over millions of videos sourced from the Internet. They found that sixteen types of facial expressions consistently occur in specific contexts, suggesting substantial contextual dependence on facial expressions. Nevertheless, in this study, the DNN models for context classification could only predict the topics of videos. For instance, a context like "the bride dressed in a gown dancing with the groom dressed in a suit in a banquet hall filled with tables and wine glasses" may be simplified into a label like "wedding". This may lead to the oversimplification of contextual variability that could significantly alter emotion perception. A small object shown in the wedding context, such as a gun, may completely shift the emotion perception from "joy" to "fear". Therefore, future studies are encouraged to utilize MLLMs to generate detailed context descriptions, as illustrated in Figure 1, to fully examine the complex relationships between facial expressions and contextual variability. In this way, researchers can not only leverage contextual information for inferring the emotional states of naturalistic expressions but also utilize facial information to guide the prediction of situational contexts (e.g., appeasement smile indicating submissive contexts [99]).

Figure 1. Contextualized emotion inferences generated from LLaVA-7B [25] (image from OASIS [100]). Temperature is set to zero for reproducible response; note that the model can identify the association between the emotion of "excitement" and the activity of "swing through the air", but it mistakenly describes the person as wearing a black helmet. Prompt: "Describe the image? What emotions can be inferred based on the context?". LLaVA-7B: "The image features a woman wearing an orange vest and a black helmet, hanging from a rope or wire. She appears to be enjoying herself as she swings through the air, possibly on a rope swing or a zip line. The woman's smile and the fact that she is wearing a helmet suggest that she is having a fun and adventurous time. The scene conveys a sense of excitement and joy, as the woman embraces the thrill of the activity."

3.2. MLLMs as Generalizable FER Toolboxes

Several technical reports have documented the impressive human-level capabilities of MLLMs for analyzing facial expressions [24]. Researchers can conduct FER tasks using MLLMs with a simple query "What is the facial expression?" MLLMs can detect emotions from facial expressions and can generalize well to unseen data because MLLMs are developed to address a diverse range of tasks. MLLMs can also analyze a wide spectrum of emotion categories based on natural language processing, which allows researchers to capture more nuanced emotions (e.g., awe, shame, emotional suppression, etc. [24]). By contrast, prior emotion models have typically been programmed to predict a limited set of basic emotions, which restrict their potential applications in more generalized domains (e.g., monitoring student engagement in the classroom or pain in the ICU). MLLMs can also provide multiple emotion labels for comprehensive descriptions of complex emotional states. Such responses are more aligned with the human perception of facial expressions that have been conceptualized as a manifold representation of varying degrees of emotions (e.g., a face may be perceived as 51% anger and 49% disgust [101]). In comparison, most existing emotion models rely on majority voting to determine the single most representative label [102]. Although such approaches may enhance the overall reliability of emotion prediction, they may exclude other relevant but less dominant emotions and fail to detect compound emotions [103]. Moreover, unlike prior works that mainly focus on emotion prediction, MLLMs can take a further step to provide reasonable explanations of the facial features being used for emotion inferences (see Figure 2, for example). For instance, after inferring a facial image as showing "fear" or "anxiety", GPT-4V can offer detailed descriptions of the visual cues to interpret these emotional states (e.g., "wide-open eyes" revealing "a sense of shock or surprise" and "hands covering the mouth" indicating "suppressing scream" [24]). It can also modify its emotional inferences based on hand gestures, which are often ignored or treated as facial occlusion by existing FER toolboxes. Furthermore, Lian et al. [93] examined the performance of several MLLMs, including Video-LLaMA [91], PandaGPT [92], and Valley [104], for generating explainable emotional inferences from a subset of the MER2023 dataset [105] comprising dynamic facial expressions sourced from the Internet. Specifically, they evaluated the abilities of MLLMs to identify emotion-related cues (e.g., "lifted eyebrows", "smiling face", etc.) and predict emotional states (e.g., "happiness") based on the identified cues. In addition, the plausibility of the emotion reasoning process was validated by human observers. Valley consistently achieved the best performance, with 72.1% accuracy in identifying emotion-related cues and 57.8% accuracy in predicting emotional states. It should be noted that the paper did not report the base rates for emotion prediction, and the authors used ChatGPT to relabel more subtle emotions, going beyond the original labels introduced in MER2023 as a baseline. The emotion reasoning process demonstrated by Valley also aligned most closely with human perception, receiving a plausibility score of 65.0%. This study provides empirical evidence demonstrating the capabilities of MLLMs in identifying explainable cues for emotion inferences, which supports their potential utility in FER research. However, further efforts are required to improve the model's performance. With the enhanced interpretability of emotion recognition processes, FER researchers can be more confident in asserting the psychological states of naturalistic expressions, thereby achieving greater reliability and specificity.

Figure 2. Explainable facial expression recognition generated from LLaVA-7B [25] (image from OASIS [100]). Temperature is set to zero for reproducible response. Prompt: "What emotions are shown on the baby's face? What specific facial cues are used to infer the emotions?" LLaVA-7B: "The baby's face shows a mix of emotions, including sadness, fear, and possibly discomfort. The baby's facial cues include a frowning expression, tears in the eyes, and a slightly puckered mouth. These cues suggest that the baby is experiencing negative emotions or discomfort, possibly due to the situation or the environment around them".

3.3. Adaptability of MLLMs for Different Emotion Recognition Tasks

MLLMs have strong adaptation capabilities for more challenging emotion recognition tasks through few-shot in-context learning (ICL). Few-shot ICL refers to the ability to quickly adapt to novel tasks given a short instruction and few examples without fine-tuning the models and abundant labeled data. After few-shot ICL, MLLMs such as Flamingo [106] can outperform some of the SOTA contrastive models [97] despite using only around 30 task-specific examples, which are about 1000 times fewer data inputs required for fine-tuning the models. There are numerous potential applications of few-shot ICL for naturalistic affective research such as the classification of nuanced emotion categories. For instance, several large-scale datasets contain a broad spectrum of facial behaviors captured in the wild, which present new opportunities for investigating the complexity and variability of emotional experiences and their underlying psychosocial processes in real-life scenarios. However, many of these datasets are only annotated with a few emotion classes, which limits their potential utilization in affective research. With the advanced ICL functionality, researchers can further exploit these datasets by applying MLLMs to identify new emotion categories for a more fine-grained analysis of human emotions [93]. Specifically, FER researchers can provide a few demonstration examples of facial expressions categorized by specific emotions in the format of image–test or video–text pairs to extrapolate to new emotion recognition tasks with a visual query such as "identify the images that display the same facial expression illustrated in the above examples". This approach may enable FER researchers to evaluate more specific and contextualized expressions in the wild that may not be accounted for by the limited set of expressions detected by existing FER toolboxes. This could also pave the way for examining the assumptions of various emotion theories, e.g., appraisal theories [107], the theory of constructed emotions [82], and the behavioral ecology view [108]. Coupled with ICL techniques, researchers can also examine if facial expressions produced in laboratory settings generalize to naturalistic contexts, thereby gaining deeper insights into the ecological validity of these facial displays and their potential implications in real-world situations. Unfortunately, to the best of our knowledge, in-context learning tailored specifically for FER tasks has not received much empirical attention. Future studies should leverage the intriguing emergent ability to further advance the field of naturalistic affective science.

3.4. Limitations of MLLMs

While MLLMs have the potential to revolutionize the domain of FER research, it is essential to acknowledge their limitations and work toward addressing them. For facial behavior analysis, it is unclear whether MLLMs can provide FACS-like inferences [26] that are precise enough for accurate facial behavior analysis. It is important to test if MLLMs can capture the variations in facial parameters [109] and distinguish the subtle differences between various types of facial behaviors [110]. For instance, smiles can be characterized by different facial configurations such as Duchenne smiles (e.g., AU6 + 12), "selfie smiles" (e.g., AU13), or "miserable" smiles (e.g., AU12 + 14 or 12 + 15), which are associated with distinct psychological states [111]. Further empirical investigation is required to systematically examine the similarity between the facial behaviors described by MLLMs and the facial AUs detected by FER toolboxes, as well as explore methods to fine-tune the models to achieve comparable or superior results. Before such empirical testing is carried out, it is recommended to incorporate both MLLMs and FER toolboxes for fine-grained facial behavior analyses. In addition, it remains unclear to what extent the contextual perception of MLLMs aligns with human perception in terms of emotion inferences [93,112]. It is possible that contextualized emotion inferences may be biased by random noise in the context. Variability also exists in the susceptibility to contextual influences among individuals [113], as evidenced in the case of depressed individuals who often exhibit facial responses insensitive to contextual cues [114]. Therefore, it is critical to formulate theories and models to elucidate the mechanisms underlying the integration of contextual information during the process of facial expression recognition [115]. Finally, it is crucial to acknowledge that despite an extensive search for empirical evidence across various domains, some arguments concerning MLLMs remain hypothetical, particularly regarding their adaptability, as discussed in Section 3.3. Although MLLMs have demonstrated promising capabilities in addressing numerous important research questions in affective science, as illustrated in Sections 3.1 and 3.2, they are still in the early phases of development, awaiting further improvement. Further empirical research is required to explore the versatility of MLLMs for emotion recognition tasks, thereby bridging the gap between their potential applications and practical implementation.

4. Conclusions

In this review, we have provided an overview of publicly accessible and user-friendly FER toolboxes for robust facial analyses under unconstrained conditions and an introduction to the potential applications of MLLMs to further advance the field of affective science. However, it is important to emphasize that most of these techniques have only been recently developed. Additional empirical investigations are needed to validate their practical utility across various domains and further improve their performance. Future studies are encouraged to utilize these sophisticated techniques to expand our knowledge of naturalistic facial expressions and develop contextualized emotion models to achieve a comprehensive understanding of emotional experiences in the real world [116].

Author Contributions: Conceptualization, Y.B. and E.G.K.; writing—original draft preparation, Y.B., D.K., H.L. and E.G.K.; writing—review and editing, Y.B. and E.G.K. All authors have read and agreed to the published version of the manuscript.

Funding: This research received no external funding.

Institutional Review Board Statement: Not applicable.

Informed Consent Statement: Not applicable.

Data Availability Statement: No new data were created or analyzed in this study. Data sharing is not applicable to this article.

Conflicts of Interest: The authors declare no conflicts of interest.

References

1. Dupré, D.; Krumhuber, E.G.; Küster, D.; McKeown, G.J. A performance comparison of eight commercially available automatic classifiers for facial affect recognition. *PLoS ONE* **2020**, *15*, e0231968. [CrossRef] [PubMed]
2. Krumhuber, E.G.; Küster, D.; Namba, S.; Skora, L. Human and machine validation of 14 databases of dynamic facial expressions. *Behav. Res. Methods* **2021**, *53*, 686–701. [CrossRef] [PubMed]
3. Lucey, P.; Cohn, J.F.; Prkachin, K.M.; Solomon, P.E.; Matthews, I. Painful data: The UNBC-McMaster shoulder pain expression archive database. In Proceedings of the 2011 IEEE International Conference on Automatic Face & Gesture Recognition (FG), Santa Barbara, CA, USA, 21–23 March 2011; pp. 57–64.
4. Chang, C.Y.; Tsai, J.S.; Wang, C.J.; Chung, P.C. Emotion recognition with consideration of facial expression and physiological signals. In Proceedings of the 2009 IEEE Symposium on Computational Intelligence in Bioinformatics and Computational Biology, Nashville, TN, USA, 30 March–2 April 2009; pp. 278–283.
5. Biel, J.I.; Teijeiro-Mosquera, L.; Gatica-Perez, D. Facetube: Predicting personality from facial expressions of emotion in online conversational video. In Proceedings of the 14th ACM International Conference on Multimodal Interaction 2012, Santa Monica, CA, USA, 22–26 October 2012; pp. 53–56.
6. Fisher, H.; Reiss, P.T.; Atias, D.; Malka, M.; Shahar, B.; Shamay-Tsoory, S.; Zilcha-Mano, S. Facing Emotions: Between- and Within-Sessions Changes in Facial Expression During Psychological Treatment for Depression. *Clin. Psychol. Sci.* **2023**, 21677026231195793. [CrossRef]
7. Küster, D.; Steinert, L.; Baker, M.; Bhardwaj, N.; Krumhuber, E.G. Teardrops on my face: Automatic weeping detection from nonverbal behavior. *IEEE Trans. Affect. Comput.* **2022**, in press. [CrossRef]
8. Krumhuber, E.G.; Skora, L.I.; Hill, H.C.H.; Lander, K. The role of facial movements in emotion recognition. *Nat. Rev. Psychol.* **2023**, *2*, 283–296. [CrossRef]
9. Lin, C.; Bulls, L.S.; Tepfer, L.J.; Vyas, A.D.; Thornton, M.A. Advancing naturalistic affective science with deep learning. *Affect. Sci.* **2023**, *4*, 550–562. [CrossRef] [PubMed]
10. Ren, Z.; Ortega, J.; Wang, Y.; Chen, Z.; Whitney, D.; Guo, Y.; Yu, S.X. VEATIC: Video-based Emotion and Affect Tracking in Context Dataset. *arXiv* **2023**, arXiv:2309.06745.
11. Siddiqui, M.F.H.; Dhakal, P.; Yang, X.; Javaid, A.Y. A survey on databases for multimodal emotion recognition and an introduction to the VIRI (visible and InfraRed image) database. *Multimodal Technol. Interact.* **2022**, *6*, 47. [CrossRef]
12. Guerdelli, H.; Ferrari, C.; Barhoumi, W.; Ghazouani, H.; Berretti, S. Macro-and micro-expressions facial datasets: A survey. *Sensors* **2022**, *22*, 1524. [CrossRef]
13. Weber, R.; Soladié, C.; Séguier, R. A Survey on Databases for Facial Expression Analysis. In Proceedings of the 13th International Joint Conference on Computer Vision, Imaging and Computer Graphics Theory and Applications (VISIGRAPP 2018), Madeira, Portugal, 27–29 January 2018; pp. 73–84.
14. Cowen, A.S.; Keltner, D.; Schroff, F.; Jou, B.; Adam, H.; Prasad, G. Sixteen facial expressions occur in similar contexts worldwide. *Nature* **2021**, *589*, 251–257. [CrossRef]
15. Zhu, Q.; Mao, Q.; Jia, H.; Noi, O.E.N.; Tu, J. Convolutional relation network for facial expression recognition in the wild with few-shot learning. *Expert Syst. Appl.* **2022**, *189*, 116046. [CrossRef]
16. Srinivasan, R.; Martinez, A.M. Cross-cultural and cultural-specific production and perception of facial expressions of emotion in the wild. *IEEE Trans. Affect. Comput.* **2018**, *12*, 707–721. [CrossRef]

17. Dhall, A.; Goecke, R.; Joshi, J.; Wagner, M.; Gedeon, T. Emotion recognition in the wild challenge 2013. In Proceedings of the 15th ACM on International Conference on Multimodal Interaction 2013, Sydney, Australia, 9–13 December 2013; pp. 509–516.
18. Barrett, L.F. Context reconsidered: Complex signal ensembles, relational meaning, and population thinking in psychological science. *Am. Psychol.* **2022**, *77*, 894. [CrossRef] [PubMed]
19. Baltrusaitis, T.; Zadeh, A.; Lim, Y.C.; Morency, L.P. Openface 2.0: Facial behavior analysis toolkit. In Proceedings of the 2018 13th IEEE International Conference on Automatic Face & Gesture Recognition (FG 2018), Xi'an, China, 15–19 May 2018; pp. 59–66.
20. Bishay, M.; Preston, K.; Strafuss, M.; Page, G.; Turcot, J.; Mavadati, M. Affdex 2.0: A real-time facial expression analysis toolkit. In Proceedings of the 2023 IEEE 17th International Conference on Automatic Face and Gesture Recognition (FG), Waikoloa Beach, HI, USA, 5–8 January 2023; pp. 1–8.
21. Cheong, J.H.; Jolly, E.; Xie, T.; Byrne, S.; Kenney, M.; Chang, L.J. Py-feat: Python facial expression analysis toolbox. In *Affective Science*; Springer: Berlin/Heidelberg, Germany, 2023; pp. 1–16.
22. Chang, D.; Yin, Y.; Li, Z.; Tran, M.; Soleymani, M. LibreFace: An Open-Source Toolkit for Deep Facial Expression Analysis. *arXiv* **2023**, arXiv:2308.10713.
23. Hinduja, S.; Ertugrul, I.O.; Cohn, J.F. PyAFAR: Python-Based Automated Facial Action Recognition for Use in Infants and Adults. 2023. Available online: https://www.jeffcohn.net/wp-content/uploads/2023/08/ACII_2023_paper_242-2.pdf (accessed on 29 November 2023).
24. Yang, Z.; Li, L.; Lin, K.; Wang, J.; Lin, C.C.; Liu, Z.; Wang, L. The dawn of lmms: Preliminary explorations with gpt-4v (ision). *arXiv* **2023**, arXiv:2309.17421.
25. Liu, H.; Li, C.; Li, Y.; Lee, Y.J. Improved Baselines with Visual Instruction Tuning. *arXiv* **2023**, arXiv:2310.03744.
26. Ekman, P.; Friesen, W.V. Facial Action Coding System. Environmental Psychology & Nonverbal Behavior. 1978. Available online: https://www.paulekman.com/facial-action-coding-system/ (accessed on 29 November 2023).
27. Karnati, M.; Seal, A.; Bhattacharjee, D.; Yazidi, A.; Krejcar, O. Understanding deep learning techniques for recognition of human emotions using facial expressions: A comprehensive survey. *IEEE Trans. Instrum. Meas.* **2023**, *72*, 5006631.
28. Sajjad, M.; Ullah, F.U.M.; Ullah, M.; Christodoulou, G.; Cheikh, F.A.; Hijji, M.; Muhammad, K.; Rodrigues, J.J. A comprehensive survey on deep facial expression recognition: Challenges, applications, and future guidelines. *Alex. Eng. J.* **2023**, *68*, 817–840. [CrossRef]
29. Li, S.; Deng, W. A deeper look at facial expression dataset bias. *IEEE Trans. Affect. Comput.* **2020**, *13*, 881–893. [CrossRef]
30. Georgescu, M.I.; Ionescu, R.T.; Popescu, M. Local learning with deep and handcrafted features for facial expression recognition. *IEEE Access* **2019**, *7*, 64827–64836. [CrossRef]
31. Hasani, B.; Mahoor, M.H. Spatio-temporal facial expression recognition using convolutional neural networks and conditional random fields. In Proceedings of the 2017 12th IEEE International Conference on Automatic Face & Gesture Recognition (FG 2017), Washington, DC, USA, 30 May–3 June 2017; pp. 790–795.
32. Mollahosseini, A.; Chan, D.; Mahoor, M.H. Going deeper in facial expression recognition using deep neural networks. In Proceedings of the 2016 IEEE Winter Conference on Applications of Computer Vision (WACV), Lake Placid, NY, USA, 7–10 March 2016; pp. 1–10.
33. Büdenbender, B.; Höfling, T.T.; Gerdes, A.B.; Alpers, G.W. Training machine learning algorithms for automatic facial coding: The role of emotional facial expressions' prototypicality. *PLoS ONE* **2023**, *18*, e0281309. [CrossRef]
34. Cohn, J.F.; Ertugrul, I.O.; Chu, W.S.; Girard, J.M.; Jeni, L.A.; Hammal, Z. Affective facial computing: Generalizability across domains. In *Multimodal Behavior Analysis in the Wild*; Academic Press: Cambridge, MA, USA, 2019; pp. 407–441.
35. Zhao, K.; Chu, W.S.; Zhang, H. Deep region and multi-label learning for facial action unit detection. In Proceedings of the IEEE Conference on Computer Vision and Pattern Recognition, Las Vegas, NV, USA, 27–30 June 2016; pp. 3391–3399.
36. Namba, S.; Sato, W.; Osumi, M.; Shimokawa, K. Assessing automated facial action unit detection systems for analyzing cross-domain facial expression databases. *Sensors* **2021**, *21*, 4222. [CrossRef] [PubMed]
37. Baltrušaitis, T.; Robinson, P.; Morency, L.P. Openface: An open source facial behavior analysis toolkit. In Proceedings of the 2016 IEEE Winter Conference on Applications of Computer Vision (WACV), Lake Placid, NY, USA, 10 March 2016; pp. 1–10.
38. Savran, A.; Alyüz, N.; Dibeklioğlu, H.; Çeliktutan, O.; Gökberk, B.; Sankur, B.; Akarun, L. Bosphorus database for 3D face analysis. In *Biometrics and Identity Management: First European Workshop, BIOID 2008, Roskilde, Denmark, 7–9 May 2008*; Revised Selected Papers 1; Springer: Berlin/Heidelberg, Germany, 2008; pp. 47–56.
39. Valstar, M.F.; Jiang, B.; Mehu, M.; Pantic, M.; Scherer, K. The first facial expression recognition and analysis challenge. In Proceedings of the 2011 IEEE International Conference on Automatic Face & Gesture Recognition (FG), Santa Barbara, CA, USA, 21–23 March 2011; pp. 921–926.
40. Lucey, P.; Cohn, J.F.; Kanade, T.; Saragih, J.; Ambadar, Z.; Matthews, I. The extended cohn-kanade dataset (ck+): A complete dataset for action unit and emotion-specified expression. In Proceedings of the 2010 IEEE Computer Society Conference on Computer Vision and Pattern Recognition-Workshops, San Francisco, CA, USA, 13–18 June 2010; pp. 94–101.
41. Mavadati, S.M.; Mahoor, M.H.; Bartlett, K.; Trinh, P.; Cohn, J.F. Disfa: A spontaneous facial action intensity database. *IEEE Trans. Affect. Comput.* **2013**, *4*, 151–160. [CrossRef]
42. Valstar, M.F.; Almaev, T.; Girard, J.M.; McKeown, G.; Mehu, M.; Yin, L.; Pantic, M.; Cohn, J.F. Fera 2015-second facial expression recognition and analysis challenge. In Proceedings of the 2015 11th IEEE International Conference and Workshops on Automatic Face and Gesture Recognition (FG), Ljubljana, Slovenia, 4–8 May 2015; Volume 6, pp. 1–8.

43. McKeown, G.; Valstar, M.; Cowie, R.; Pantic, M.; Schroder, M. The semaine database: Annotated multimodal records of emotionally colored conversations between a person and a limited agent. *IEEE Trans. Affect. Comput.* **2011**, *3*, 5–17. [CrossRef]
44. Skiendziel, T.; Rösch, A.G.; Schultheiss, O.C. Assessing the convergent validity between the automated emotion recognition software Noldus FaceReader 7 and Facial Action Coding System Scoring. *PLoS ONE* **2019**, *14*, e0223905. [CrossRef] [PubMed]
45. Mavadati, M.; Sanger, P.; Mahoor, M.H. Extended disfa dataset: Investigating posed and spontaneous facial expressions. In Proceedings of the IEEE Conference on Computer Vision and Pattern Recognition Workshops, Las Vegas, NV, USA, 27–30 June 2016; pp. 1–8.
46. Girard, J.M.; Chu, W.S.; Jeni, L.A.; Cohn, J.F. Sayette group formation task (gft) spontaneous facial expression database. In Proceedings of the 2017 12th IEEE International Conference on Automatic Face & Gesture Recognition (FG 2017), Washington, DC, USA, 30 May–3 June 2017; pp. 581–588.
47. Kollias, D.; Zafeiriou, S. Aff-wild2: Extending the aff-wild database for affect recognition. *arXiv* **2018**, arXiv:1811.07770.
48. Ertugrul, I.O.; Cohn, J.F.; Jeni, L.A.; Zhang, Z.; Yin, L.; Ji, Q. Crossing domains for au coding: Perspectives, approaches, and measures. *IEEE Trans. Biom. Behav. Identity Sci.* **2020**, *2*, 158–171. [CrossRef]
49. Zhang, N.; Luo, J.; Gao, W. Research on face detection technology based on MTCNN. In Proceedings of the 2020 International Conference on Computer Network, Electronic and Automation (ICCNEA), Xi'an, China, 25–27 September 2020; pp. 154–158.
50. Deng, J.; Guo, J.; Zhou, Y.; Yu, J.; Kotsia, I.; Zafeiriou, S. Retinaface: Single-stage dense face localisation in the wild. *arXiv* **2019**, arXiv:1905.00641.
51. Chen, T.; Guestrin, C. Xgboost: A scalable tree boosting system. In Proceedings of the 22nd ACM Sigkdd International Conference on Knowledge Discovery and Data Mining 2016, San Francisco, CA, USA, 13–17 August 2016; pp. 785–794.
52. Lyons, M.; Kamachi, M.; Gyoba, J. The Japanese Female Facial Expression (JAFFE) Dataset. 1998. Available online: https://zenodo.org/records/3451524 (accessed on 29 November 2023).
53. Zhang, Z.; Luo, P.; Loy, C.C.; Tang, X. From facial expression recognition to interpersonal relation prediction. *Int. J. Comput. Vis.* **2018**, *126*, 550–569. [CrossRef]
54. Pham, L.; Vu, T.H.; Tran, T.A. Facial expression recognition using residual masking network. In Proceedings of the 2020 25th International Conference on Pattern Recognition (ICPR), Milan, Italy, 10–15 January 2021; pp. 4513–4519.
55. iMotions. Facial Expression Analysis: The Definitive Guide. 2016. Available online: https://imotions.com/facialexpression-guide-ebook/ (accessed on 29 November 2023).
56. Mollahosseini, A.; Hasani, B.; Mahoor, M.H. AffectNet: A Database for Facial Expression, Valence, and Arousal Computing in the Wild. *IEEE Trans. Affect. Comput.* **2019**, *10*, 18–31. [CrossRef]
57. Ren, S.; He, K.; Girshick, R.B.; Sun, J. Faster R-CNN: Towards Real-Time Object Detection with Region Proposal Networks. *IEEE Trans. Pattern Anal. Mach. Intell.* **2015**, *39*, 1137–1149. [CrossRef]
58. McDuff, D.; Mahmoud, A.; Mavadati, M.; Amr, M.; Turcot, J.; Kaliouby, R.E. AFFDEX SDK: A cross-platform real-time multi-face expression recognition toolkit. In Proceedings of the 2016 CHI Conference Extended Abstracts on Human Factors in Computing Systems 2016, New York, NY, USA, 7–12 May 2016; pp. 3723–3726.
59. McDuff, D.; Kaliouby, R.; Senechal, T.; Amr, M.; Cohn, J.; Picard, R. Affectiva-mit facial expression dataset (am-fed): Naturalistic and spontaneous facial expressions collected. In Proceedings of the IEEE Conference on Computer Vision and Pattern Recognition Workshops 2013, Portland, OR, USA, 23–28 June 2013; pp. 881–888.
60. Ekman, P. An argument for basic emotions. *Cogn. Emot.* **1992**, *6*, 169–200. [CrossRef]
61. Friesen, W.V.; Ekman, P. *EMFACS-7: Emotional Facial Action Coding System*, University of California at San Francisco: San Francisco, CA, USA, 1983; *unpublished work*.
62. Lugaresi, C.; Tang, J.; Nash, H.; McClanahan, C.; Uboweja, E.; Hays, M.; Zhang, F.; Chang, C.L.; Yong, M.; Lee, J.; et al. Mediapipe: A framework for perceiving and processing reality. In Proceedings of the Third Workshop on Computer Vision for AR/VR at IEEE Computer Vision and Pattern Recognition (CVPR) 2019, Long Beach, CA, USA, 17 June 2019; Volume 2019.
63. He, K.; Zhang, X.; Ren, S.; Sun, J. Deep residual learning for image recognition. In Proceedings of the IEEE Conference on Computer Vision and Pattern Recognition (2016), Las Vegas, NV, USA, 27–30 June 2016; pp. 770–778.
64. Fabian Benitez-Quiroz, C.; Srinivasan, R.; Martinez, A.M. Emotionet: An accurate, real-time algorithm for the automatic annotation of a million facial expressions in the wild. In Proceedings of the IEEE Conference on Computer Vision And Pattern Recognition (2016), Las Vegas, NV, USA, 27–30 June 2016; pp. 5562–5570.
65. Karras, T.; Laine, S.; Aila, T. A style-based generator architecture for generative adversarial networks. In Proceedings of the IEEE/CVF Conference on Computer Vision and Pattern Recognition (2019), Long Beach, CA, USA, 15–19 June 2019; pp. 4401–4410.
66. Hang, X.; Yin, L.; Cohn, J.F.; Canavan, S.; Reale, M.; Horowitz, A.; Liu, P.; Girard, J.M. Bp4d-spontaneous: A high-resolution spontaneous 3d dynamic facial expression database. *Image Vis. Comput.* **2014**, *32*, 692–706.
67. Liu, Z.; Lin, Y.; Cao, Y.; Hu, H.; Wei, Y.; Zhang, Z.; Lin, S.; Guo, B. Swin transformer: Hierarchical vision transformer using shifted windows. In Proceedings of the IEEE/CVF International Conference on Computer Vision (2021), Montreal, BC, Canada, 17 October 2021; pp. 10012–10022.
68. Xue, F.; Wang, Q.; Guo, G. Transfer: Learning relation-aware facial expression representations with transformers. In Proceedings of the IEEE/CVF International Conference on Computer Vision (2021), Montreal, BC, Canada, 17 October 2021; pp. 3601–3610.
69. Gao, J.; Zhao, Y. Tfe: A transformer architecture for occlusion aware facial expression recognition. *Front. Neurorobot.* **2021**, *15*, 763100. [CrossRef] [PubMed]

70. He, K.; Chen, X.; Xie, S.; Li, Y.; Dollár, P.; Girshick, R. Masked autoencoders are scalable vision learners. In Proceedings of the IEEE/CVF Conference on Computer Vision and Pattern Recognition, New Orleans, LA, USA, 24 June 2022; pp. 16000–16009.
71. Gudi, A.; Tasli, H.E.; Den Uyl, T.M.; Maroulis, A. Deep learning based facs action unit occurrence and intensity estimation. In Proceedings of the 2015 11th IEEE International Conference and Workshops on Automatic Face and Gesture Recognition (FG), Ljubljana, Slovenia, 4–8 May 2015; Volume 6, pp. 1–5.
72. Yu, J.; Lin, Z.; Yang, J.; Shen, X.; Lu, X.; Huang, T.S. Generative image inpainting with contextual attention. In Proceedings of the IEEE Conference on Computer Vision and Pattern Recognition, Salt Lake City, UT, USA, 18–22 June 2018; pp. 5505–5514.
73. Li, S.; Deng, W.; Du, J. Reliable crowdsourcing and deep locality-preserving learning for expression recognition in the wild. In Proceedings of the IEEE Conference on Computer Vision and Pattern Recognition, Honolulu, HI, USA, 21–26 July 2017; pp. 2852–2861.
74. Stefanov, K.; Huang, B.; Li, Z.; Soleymani, M. Opensense: A platform for multimodal data acquisition and behavior perception. In Proceedings of the 2020 International Conference on Multimodal Interaction, Virtual Event, The Netherlands, 25–29 October 2020; pp. 660–664.
75. Schroff, F.; Kalenichenko, D.; Philbin, J. Facenet: A unified embedding for face recognition and clustering. In Proceedings of the IEEE Conference on Computer Vision and Pattern Recognition, Boston, MA, USA, 12 June 2015; pp. 815–823.
76. Deng, J.; Dong, W.; Socher, R.; Li, L.J.; Li, K.; Fei-Fei, L. Imagenet: A large-scale hierarchical image database. In Proceedings of the 2009 IEEE Conference on Computer Vision and Pattern Recognition, Miami, FL, USA, 20–25 June 2009; pp. 248–255.
77. Zhang, Z.; Girard, J.M.; Wu, Y.; Zhang, X.; Liu, P.; Ciftci, U.; Canavan, S.; Reale, M.; Horowitz, A.; Yang, H.; et al. Multimodal spontaneous emotion corpus for human behavior analysis. In Proceedings of the IEEE Conference on Computer Vision and Pattern Recognition, Las Vegas, NV, USA, 27–30 June 2016; pp. 3438–3446.
78. Hammal, Z.; Cohn, J.F.; Messinger, D.S. Head movement dynamics during play and perturbed mother-infant interaction. *IEEE Trans. Affect. Comput.* **2015**, *6*, 361–370. [CrossRef] [PubMed]
79. Luquetti, D.V.; Speltz, M.L.; Wallace, E.R.; Siebold, B.; Collett, B.R.; Drake, A.F.; Johns, A.L.; Kapp-Simon, K.A.; Kinter, S.L.; Leroux, B.G.; et al. Methods and challenges in a cohort study of infants and toddlers with craniofacial microsomia: The CLOCK study. *Cleft Palate-Craniofacial J.* **2019**, *56*, 877–889. [CrossRef] [PubMed]
80. Adamson, L.B.; Frick, J.E. The still face: A history of a shared experimental paradigm. *Infancy* **2003**, *4*, 451–473. [CrossRef]
81. Ertugrul, I.O.; Jeni, L.A.; Ding, W.; Cohn, J.F. Afar: A deep learning based tool for automated facial affect recognition. In Proceedings of the 2019 14th IEEE International Conference on Automatic Face & Gesture Recognition (FG 2019), Lille, France, 14–18 May 2019.
82. Barrett, L.F.; Adolphs, R.; Marsella, S.; Martinez, A.M.; Pollak, S.D. Emotional expressions reconsidered: Challenges to inferring emotion from human facial movements. *Psychol. Sci. Public Interest* **2019**, *20*, 1–68. [CrossRef] [PubMed]
83. Lange, J.; Heerdink, M.W.; Van Kleef, G.A. Reading emotions, reading people: Emotion perception and inferences drawn from perceived emotions. *Curr. Opin. Psychol.* **2022**, *43*, 85–90. [CrossRef]
84. Krumhuber, E.G.; Hyniewska, S.; Orlowska, A. Contextual effects on smile perception and recognition memory. *Curr. Psychol.* **2023**, *42*, 6077–6085. [CrossRef]
85. Day, S.E.; Krumhuber, E.G.; Shore, D.M. The bidirectional relationship between smiles and situational contexts. *Cogn. Emot.* **2023**; in press.
86. Lee, J.; Kim, S.; Kim, S.; Park, J.; Sohn, K. Context-aware emotion recognition networks. In Proceedings of the IEEE/CVF International Conference on Computer Vision, Seoul, Republic of Korea, 27 October–2 November 2019; pp. 10143–10152.
87. Cabitza, F.; Campagner, A.; Mattioli, M. The unbearable (technical) unreliability of automated facial emotion recognition. *Big Data Soc.* **2022**, *9*, 20539517221129549. [CrossRef]
88. Mason, C.; Gadzicki, K.; Meier, M.; Ahrens, F.; Kluss, T.; Maldonado, J.; Putze, F.; Fehr, T.; Zetzsche, C.; Herrmann, M.; et al. From human to robot everyday activity. In Proceedings of the 2020 IEEE/RSJ International Conference on Intelligent Robots and Systems (IROS), Las Vegas, NV, USA, 25–29 October 2020; pp. 8997–9004.
89. Yin, S.; Fu, C.; Zhao, S.; Li, K.; Sun, X.; Xu, T.; Chen, E. A Survey on Multimodal Large Language Models. *arXiv* **2023**, arXiv:2306.13549.
90. Zhu, D.; Chen, J.; Shen, X.; Li, X.; Elhoseiny, M. Minigpt-4: Enhancing vision-language understanding with advanced large language models. *arXiv* **2023**, arXiv:2304.10592.
91. Zhang, H.; Li, X.; Bing, L. Video-llama: An instruction-tuned audio-visual language model for video understanding. *arXiv* **2023**, arXiv:2306.02858.
92. Su, Y.; Lan, T.; Li, H.; Xu, J.; Wang, Y.; Cai, D. Pandagpt: One model to instruction-follow them all. *arXiv* **2023**, arXiv:2305.16355.
93. Lian, Z.; Sun, L.; Xu, M.; Sun, H.; Xu, K.; Wen, Z.; Chen, S.; Liu, B.; Tao, J. Explainable multimodal emotion reasoning. *arXiv* **2023**, arXiv:2306.15401.
94. Etesam, Y.; Yalcin, O.N.; Zhang, C.; Lim, A. Emotional Theory of Mind: Bridging Fast Visual Processing with Slow Linguistic Reasoning. *arXiv* **2023**, arXiv:2310.19995.
95. Wieser, M.J.; Brosch, T. Faces in context: A review and systematization of contextual influences on affective face processing. *Front. Psychol.* **2012**, *3*, 471. [CrossRef]
96. Kosti, R.; Alvarez, J.M.; Recasens, A.; Lapedriza, A. Context based emotion recognition using emotic dataset. *IEEE Trans. Pattern Anal. Mach. Intell.* **2019**, *42*, 2755–2766. [CrossRef]

MDPI AG
Grosspeteranlage 5
4052 Basel
Switzerland
Tel.: +41 61 683 77 34

Sensors Editorial Office
E-mail: sensors@mdpi.com
www.mdpi.com/journal/sensors

Disclaimer/Publisher's Note: The title and front matter of this reprint are at the discretion of the Guest Editors. The publisher is not responsible for their content or any associated concerns. The statements, opinions and data contained in all individual articles are solely those of the individual Editors and contributors and not of MDPI. MDPI disclaims responsibility for any injury to people or property resulting from any ideas, methods, instructions or products referred to in the content.